ENERGY AND ENVIRONMENT

ENERGY AND ENVIRONMENT

GARY O. ROBINETTE

Executive Director
A.S.L.A. Foundation

KH

KENDALL/HUNT PUBLISHING COMPANY
DUBUQUE, IOWA

Library of Congress Catalog Card Number: 73-76344

ISBN 0—8403—0689—X

Printed in the United States of America

Dedicated to
Margaret Ann Roth Robinette

my sculpture in the landscape
for the past quarter century

Contents

Preface, ix

Introduction, 1

Power Generation Facilities, 15

Power Transmission Facilities, 111

Power Transformation Facilities, 219

Power Distribution Facilities, 267

Conclusion, 275

Appendix A, 281

Appendix B, 289

Appendix C, 297

Selected References, 301

Preface

This is not a negative book of fault-finding, but a positive book of hope, of guidance and direction. It is largely the chronicling of one profession dealing with some of the current problems of one industry. It is a sometime perfunctory delineation of the way that the profession of landscape architecture has worked with the electric utility industry in dealing with some of the problems associated with the generation and distribution of the necessary power in an age of increasing concern with the preservation of the environment.

This book represents a positive approach to delineating these two basic problems in our society at the present time and indicates how they have been solved in some instances. The purpose of the book is to identify positive processes, approaches and solutions. It has been written with no one's money and under no one's direction. Hopefully, it will assist in cutting through much of the obfuscation, accusation and eco-pornography. It is also the intent of this book to cut out a seemingly finite slice of a seemingly infinite problem. It is a link in a chain in which some other publications are contributory and precedent, and from which some other publications will develop.

The entire problem is so overwhelmingly large and complex, and related to so many technical areas that it is not possible in any one volume or even a library, to gather the desired significant data on the entire subject. Due to the intense interest in the subject a great number of studies are underway simultaneously, involving the economic, scientific, political, legal and planning communities. This creates a pletheora of reports, studies, documents and other publications often with conflicting, more often overlapping, information which is seldom communicated, indexed, cataloged or disseminated fully to all the others with an interest in the field or the subject. It is, therefore, the purpose of this book to take one small segment of the problem, both in breadth and depth, and to illustrate it in a manner that will present positive endeavors to a wider audience. Some of the answers, optimum processes or directions have not been found as yet. It is not the purpose of this book to present more than a cataloging and a report of progress in certain aspects in energy generation, transformation and transmission and how these aspects have been dealt with in order to preserve and enhance the environment.

This book answers basically the three following questions:

1. What is happening in regard to problem identification or definition?
2. What is developing relative to process development or clarification?
3. What is being presented relative to positive forms or design solutions to the problems identified.

The two problems of energy and environment are not new nor unrecognized by the larger community. In fact, in the decade of the 1950's there was increasing concern in regard to the energy needs of the burgeoning American population. In the 1960's the concern intensified in the areas of the quality of the environment and the identification of the various agencies, industries and instruments causing the despoilation and destruction of the environment for future generations. Over the years, since that time, the voices have become more shrill and the demands more harsh, the lines on either side of the question more clearly drawn. The approaches to the problem have ranged from hand-wringing to accusation, to self-defensive statements, to eco-pornography, to ill-founded judgments and placing of blame. Information on this subject and examples of positive solutions to the problem are scattered and not easily collectable. This is so because the opposing sides on some of the current environmental questions do not communicate in a positive way. This is also because many of the various disciplines and their members have limited interchange and interaction. They know little of what is happening or what has been done in fields which are not their own or closely related to it. This is because much of the publication in regard to the problem and its solutions is done largely within limited interest groups. In order to share data among the disciplines it is often necessary to publish commercially in order to reach a larger and a wider audience.

Because of the large number of public utilities in the United States, each feels a natural constraint to conduct studies in environmental problems caused by energy generation, transformation, transmission or distribution facilities. Many times studies in regard to these problems are done for similar purposes, though often without the knowledge of what other clients or consultants are doing. Therefore, each consultant faced with these problems either "re-invents the wheel" or else makes an independent study without the knowledge that the "wheel" exists. There is therefore, much overlap, duplication and little sharing or communication. There have been a number of seminars outlining guidelines and long-range planning. On the other hand a number of closely analytical studies have been done. There has been very little communication of what has been done in a positive way by the various parts of the industry and by some of the professionals most capable of making an optimum contribution in meshing the demands of energy generation and delivery, with the urgent need for environmental preservation and enhancement.

The scope of this book deals with the physical design, the process, the studies and the guidelines, primarily of the physical accoutrements of the electric utilities. The purpose of this book, as mentioned previously, is a positive one to give credit, guidance, direction and encouragement, to define the condition of the art, and to pull together all of the information available on the subject at the present time. There are certain limitations in an attempt of this type. Some of these are determined by the fact that none of the intricacies of the science and the chemistry of water and air pollution can be dealt with here. The positive contributions of many of the "hard" scientists are neglected in a study of this type. There is no overview or feedback in this publication regarding the effectiveness of these proposed environmental solutions, since the studies are so recent that it is not possible at this early date to determine their effectiveness. The studies illustrated in this book, however, go far beyond the normal concept of landscape development and illustrate the potential contribution of the landscape architect to some of the current environmental problems. Hopefully, it will be of assistance both to environmental designers and persons in the electric utility industry as well as those with a larger desire to know more of the positive solutions to the problem identified by many responsible persons and groups.

Reston, Virginia

Introduction

Problem Statement

The words "Energy" and "Environment" denote, at least in the past decade, arenas of conflict. Though the words to some writers are mutually exclusive the most concise statement of the twin problems and their relationship to each other is made in a very succinct manner in a report entitled *Environmental Criteria for Electrical Transmission Systems,* jointly sponsored in 1970 by the U.S. Departments of Agriculture and Interior. That publication stated the following: "The electric utility industry is faced with a complex challenge in the 1970's. It must provide the generation and transmission facilities that are and will be needed to meet the ever-growing demand for reliable electrical power, and it must do this in a decade dedicated to the restoration and protection of our environment."[1] These, then are the optimum goals; the generation of the necessary energy for the sustenance and growth of contemporary technology and at the same time the preservation of the essential environment. Many times these have been conflicting and contradictory demands. Obviously, these two needs do not have to be in conflict, but are because of lack of clear identification of the diametrically opposed needs of both energy and environment, and the lack of clear guidelines and precedents as to how these two needs may be accommodated. There has arisen a number of points of contention between the demands of energy and environment. Basically the demands for the in-

1. U.S. Department of the Interior, U.S. Department of Agriculture, *Environmental Criteria for Electrical Transmission Systems,* (Washington, D.C., U.S. Government Printing Office, Superintendent of Documents, 1970), p. iii.

creased generation and utilization of energy, and the needs and demands for preservation of environmental values have always been in conflict. As the destruction of the environment becomes more widespread and the need for energy becomes more critical, the two opposing demands come into sharper conflict.

The growth and demand for energy have long been in one direction. That is for reasonably priced energy production in a hurry. This has caused disastrous environmental consequences. The need for energy has been expressed and demanded through contemporary life styles. The need and desire to preserve the environment has been expressed through literature, protests, legislation and administrative guidelines. There has been a vast variety of responses to the recognition and apprehension of these twin problems which are on collision course. Amongst these responses has been resignation, frustration, name-calling, threats, intimidation, analysis and investigation and eco-pornography (This is the chauvinistic overinflation of minor environmental efforts through minimal expenditure on correction of environmental problems and maximum expenditure on public relations programs).

There is extensive literature identifying the problem, but too few attempts have been made to collect, catalog and communicate the positive directions, attempts and successes in finding solutions to the problems of developing the necessary energy while preserving the essential environment and the quality of life. This book deals with the problems and the solutions of one industry and one profession at one level.

An overview in some depth of the twin problems as seen from the eyes of a number of observers seems imperative at the outset of this book. What are the needs for energy and most particularly electrical energy at the present time and in the future? This need and demand has been stated eloquently by a number of spokesmen for the electric utility industry over the past few years. Probably one of the most cogent statements in this regard was by Frank B. Burggraf, Jr., Chief of Transmission Facilities of the New York State Public Service Commission, in an article called "Power to the People" in the July-August, 1971, issue of *Planning News*. Mr. Burggraf stated at that time:

> Almost nothing has contributed more to the quality of life in rural and urban America than electric energy. And almost nothing is taken more for granted than that the lights will go on when the switch is turned. Public dissatisfaction with power brownouts and marginal telephone service is seemingly aggravated by the fact that there has always been quality service to enjoy in the past.
>
> The demand for energy has about doubled every 10 years since 1930. The Federal Power Commission's projected staff estimates of needed generating capacity indicated 665,000 megawatts in 1980, and 1,260,000 megawatts in 1990, as compared to 340,000 megawatts in 1970.
>
> Demand for power increases because population increases. But demand also increases because per capita consumption increases. If environmental tradeoffs such as electric cars become a reality, power consumption could grow out of all expectations. Imagine the impact of thousands of commuters coming home and plugging in their electric car in the garage, going in to their electric heated and air-conditioned home to watch television while supper is taken from the electric freezer and refrigerator and cooked on the electric stove.[2]

The Federal Power Commission has stated the past, present and future needs for electric power in the following words:

> Historically, the use of electric power has been doubling about every ten years. For the past two years total use has grown even faster—nine percent per year. To meet this demand, new and larger plants have been built. If current trends persist in the next two decades, the present demands for electric power will triple or quadruple. While population growth is responsible for part of this expanding need, the per capita consumption of electric power has been increasing roughly five times as fast as population growth.[3]

2. Frank B. Burgraff, Jr., "Power to the People," *Planning News* (Albany, New York, New York State Planning Federation, Vol. 35, No. 4, July-August 1971), p. 1.
3. Federal Power Commission, *The 1970 National Power Survey: Guidelines for Growth of the Electric Power Industry,* (Washington, D.C., U.S. Government Printing Office, Superintendent of Documents, 1971), p. 0.

One of the strongest statements concerning the present energy crisis was in a statement by the National Rural Electric Cooperative Association. This rather extensive statement of the problem is as follows:

Man's continuing progress in improving the world's standard of living depends on a continuously growing availability and use of energy. In fact, the average amount of energy consumed by each citizen is a good indicator of a nation's economic strength. The United States with one-seventeenth of the earth's population consumes one-third of its energy to support the world's highest standard of living. Per person use of energy in the United States is six times the world average while our standard of living is five times the world average.

However, our goal of an ever-improving standard of living—for all nations, not only the U.S.—may be jeopardized by a conflict between our desire for both more energy and a high quality environment. Yet, these twin aims need not be incompatible. We need to strive to achieve a workable balance between increased productivity and improved environmental quality. We need to devote our technological know-how to protecting our environment as well as providing our material needs. In this way, our standard of living will not be unnecessarily sacrificed for lack of sufficient energy and power nor will our environment be needlessly damaged in pursuit of more energy.

Energy (and the materials and fuels required to produce it) is the foundation for an active, growing economy. It is in essence, 'the life blood' of the economy on which our national security depends.

Ever-increasing amounts of energy are used in manufacturing consumer and industrial products, in farming, in mining and extracting raw materials, in transporting and distributing goods in providing services and selling merchandise, in processing and disposing of wastes, in supplying warmth or coolness, in powering a multitude of household appliances, and in a variety of other ways. In fact, the U.S. has become so dependent on various forms of power that human labor now provides energy for less than 1% of the work done in the nation's factories.

Among the different types of energy, there is a growing dependence on electricity as a source of power. The share of total energy consumed in this country provided by electric utilities rose from about 19% to 24% in 1970 according to the Federal Power Commission (FPC). By the end of the next 10 years, the FPC projects that electrical power will comprise 34% of the total energy utilized in the U.S. Historically, over-all demand for electricity has doubled every ten years and this trend is expected to continue. Thus, the need for electric power in 1980 will be twice the requirements for 1970, and by 1990 the demand will be quadrupled.

Furthermore, the demand for electrical power per capita has been growing five times faster than the population. Therefore, a slowdown in the rate of population growth will have no material effect in diminishing the rate of increase in the demand for electrical energy. In addition, although the greatest thrust in the demand for electricity has been generated by the burgeoning number of new and improved household appliances, there has been and will continue to be a strong industrial demand for more and more electrical power. In the five-year period between 1965 and 1970 FPC statistics reveal that the total amount of electricity used by industry rose by 40% and that industry consumption equaled about two-fifths of the total in both years. FPC projections for the next twenty years show an annual average gain of 7% for industrial consumption of electric power.[4]

They further stated the problem in another way in regard to the needs and demands for electrical energy:

The proportion of energy required in the form of electricity is currently about 25%; present projections indicate an increase to 43% in 1990.

Electric energy consumption is expected to increase by 284% during the period 1970-1990. In the next two decades the electric utility industry must quadruple existing

4. Stanley H. Ruttenberg and Associates, Inc., *The Electric Power Crisis: Its Impact on Workers and Consumers,* (Washington, D.C., National Rural Electric Cooperative Association, 1971), p. 5.

capacity to meet this demand—an addition of nearly 1,000,000,000 kilowatts involving a major national investment of financial and natural resources.[5]

The Federal Power Commission, once again, in the 1970 survey of energy demands, under Conclusions and Recommendations, state at the outset of their basic findings the following:

> Electric power use in the United States has been doubling about every ten years for many decades. The new concern over the environment or other factors may alter this historical rate of growth and some suggest that growth rates should be reduced. However, if prevailing growth patterns and pricing policies continue, electric power capacity may need to triple or quadruple in the next two decades.[6]

In an apt summarization in the Annual Report of the Citizen's Advisory Committee on Recreation and Natural Beauty, the following statement was made:

> Energy, from whatever source and in whatever form, is essential to the quality of our life. Coal, oil, natural gas, uranium, and falling water provide directly, or through conversion to electricity, and energy without which we would not have the heat, light transport, and industrial production we take for granted. Yet, the production of energy—at the mineral extraction stage at the point of conversion, and in the process of distribution or delivery—causes significant damage to our environment in such forms as strip mining, air pollution, overhead transmission lines, and oil spills. The public objects to the damage, and often it objects even to the construction of essential new facilities. The public also objects, however, when the energy is not delivered and when a shortage of fuel or power adversely affects public necessities or conveniences.
>
> The problems are compounding. There is a limited supply of resources to produce energy, but a rapidly accelerating demand. Not only is the population growing, there is a growing per-capita use of energy—more appliances, more automobiles, and more industrial and commercial consumption. If present trends do not change, a recent study suggests, by 1980 there is likely to be an increase of about 50 percent in the United States energy requirements; by 2000, 300 percent.
>
> The objective should be to assure that necessary energy will be produced with the best technology available and with minimum demand to the environment, but without unnecessary delay. There must be a proper balance between our needs for energy and our concern for environmental protection.
>
> Increased concern in recent years for environmental quality was much needed and has been welcomed by all responsible citizens. It has made people aware of many serious problems and has led to the start of numerous corrective programs and the acceleration of others. In some instances, however, there has been a tendency to exaggerate this concern at the expense of urgent social and economic needs. This has resulted in prolonged delay or complete stoppage of projects that had long been planned to meet urgent needs. We believe this is short-sighted.[7]

This latter part of the quote makes mention of the other facet of the basic problem—the preservation of the environment and how this may be accomplished.

The growth of energy needs and supplies, since the dawn of man, have been shown in a rather interesting graphic manner in two different, recent publications. These are by the Oak Ridge National Laboratory.

The second indication of the growth of energy indicates the tie between energy, resources and supply in the growth of man's culture and technology.

The methods of energy generation have changed dramatically in the century since the invention of the electric light by Thomas Edison. The production of plentiful and inexpensive energy forms

5. Ibid., p. 12.
6. Federal Power Commission, *The 1970 National Power Survey: Guidelines for Growth of the Electric Power Industry,* (Washington, D.C., U.S. Government Printing Office, Superintendent of Documents, 1971), p. I-1-12.
7. Citizens Advisory Committee on Environmental Quality, *Report to the President and the Council on Environmental Quality,* (Washington, D.C., Citizens Advisory Committee on Environmental Quality, 1971), p. 17.

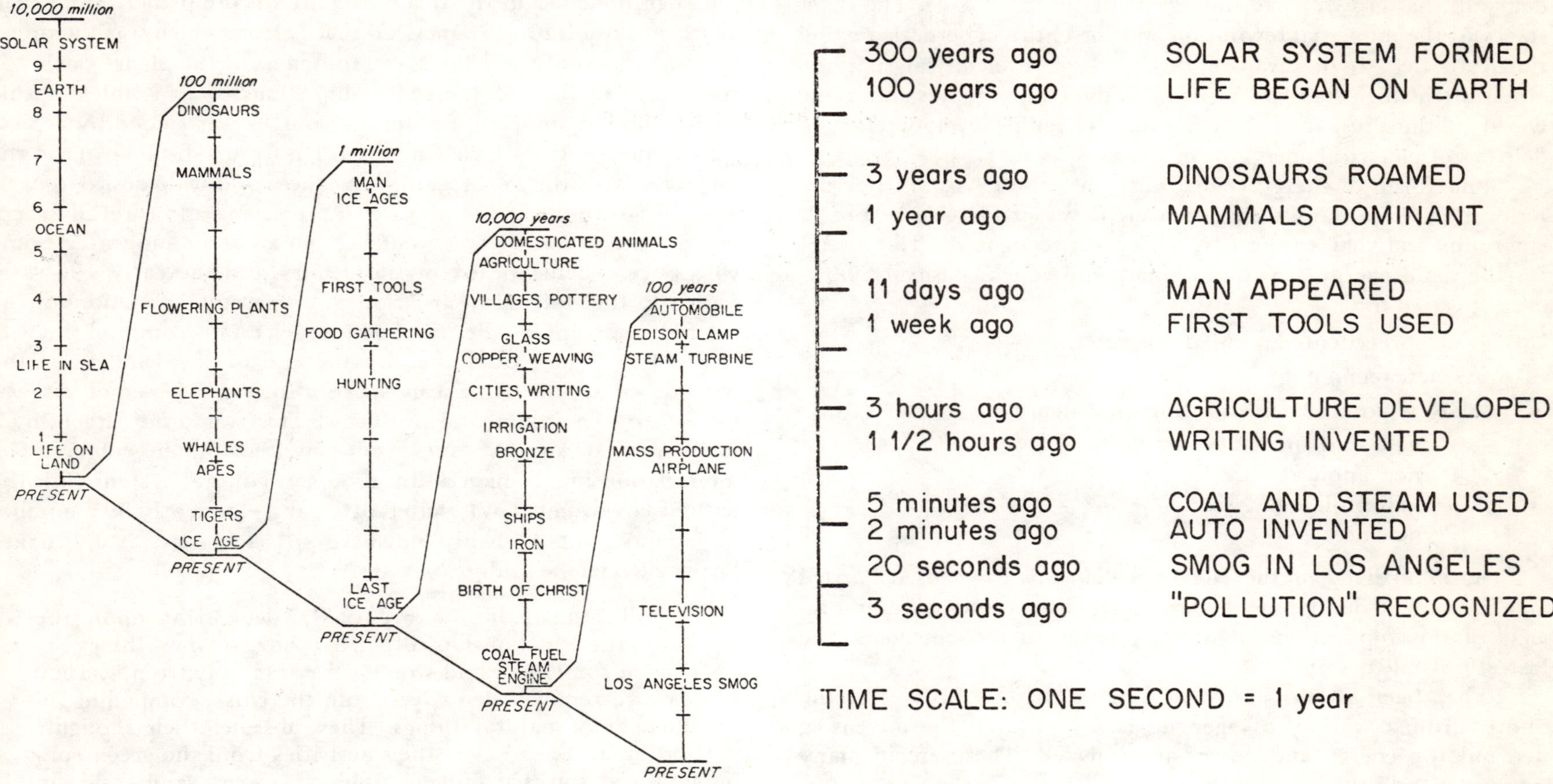

An Historical Perspective of Man and Energy[8]

History on a Compressed Time Scale[9]

8. Subcommittee on Communications and Power of the Committee on Interstate and Foreign Commerce, House of Representatives, Ninety-Second Congress, First Session, *Powerplant Siting and Environmental Protection,* Hearings before the Subcommittee on Bills Relating to Powerplant Siting and Environmental Protection from May 4-27, 1971, (Washington, D.C., U.S. Superintendent of Documents, Government Printing Office, Printed for use of the Committee on Interstate and Foreign Commerce, Part 1 [Serial No. 92-31, Part 2 (Serial No. 92-32, Part 3,) (Serial No. 92-33, 1972)], p. 1122.

9. Ibid., p. 1123.

is inextricably intertwined with the culture and technology of Western society. Electrical energy, both in generation, needs and demands, has grown at an unprecedentedly rapid pace. The latter stages of the industrial revolution and the entire cybernetic revolutions are tied to the availability of adequate electrical supplies. Much of the high standard of living in the United States and in the world at the present time is due to the rapid development and delivery of electrical energy from a variety of sources.

Some forms of energy transformation cause environmental deterioration. Even those which do, were developed largely before all environmental values were firmly fixed or recognized. The reason for the rapid energy growth are many and varied. Basically, however, it is tied to:

- Unprecedented population growth
- New technology
- New popular demands for convenience
  - new appliances
  - new applications
  - new building
  - new uses

The immediacy of the electrical demands has caused hurried and ill studied development and inadequate consultation and review of the implications of meeting many of those needs in the fastest possible manner.

It has been shown above in a number of sources and the quotes from a variety of agencies have indicated the increasing demand for energy and power as projected. These are in many cases, just and justified, and show few chances of a slowing.

The logical question is asked, in view of the massive needs for electrical energy and its resultant damage to the environment, as to what are the various alternatives in the growth patterns of the past. The obvious answers seem to deal with the reduction of energy usage and rechanneling of the energy, the more efficient use of the existing energy and the development of techniques, standards and guidelines for the increased development of energy with the lessening of the deleterious environmental impact.

Probably no other subject in the history of mankind has been so well documented in such a short time as the entire question of the environment and the quality of life on the planet. More than any other result of the American space effort which was the major national thrust of the 1960's, the finiteness of the planet earth was realized. The view of the space ship "Earth" as a finite capsule with a limited amount of natural resources was gained by achieving the perspective of viewing the earth from a distance out in the universe. With this realization came the world-wide desire to exercise the greater amount of care in preserving and carefully shepherding the limited nonrenewable resources of air and water. Some voices were being heard in the 1950's and the early 1960's in regard to the isolated destruction of either certain portions of the natural landscape or the despoilation of certain commodities necessary for human existence. By the end of the 1960's and the wind-down of the American space effort, greater emphasis was focused on environmental problems. This was done largely by a number of perspective and erudite writers on the subject. The interrelationship of man and the human cultural system with the natural environment was stated often and eloquently by a number of writers, but probably nowhere so succinctly as by Baker Brownell when he said:

> All human life is eventually dependent upon the symbiotic relationship of man and growing things . . . though some men and women live remotely from it. They learn to remove themselves from the close, compelling influences of natural things. They abstract their thoughts and interests and even their activities from the green context and spiritual millieu of life in nature, and reside in massive aggregations or in other ways remote from rural interests. But they have not taken with them into those aggregations the stable community or the naturally integrated life. They have renounced Nature at a price. That price would seem to be something close to spiritual impoverishment and phyletic discontinuity and decadence.[10]

10. Baker Brownell as quoted in a speech by Mr. Fred Smith entitled, "A Practical Program for Protecting the Environment," before the Institute of Electric and Electronic Engineers, in May 1969.

Oftentimes to the environmental interests the industrialists who were meeting the needs of culture as they saw them were the enemies and despoilers of the environmental system. Dr. Rene Dubos made the statement:

> Throughout ages, and in all climates, man has expressed a reverence for nature, acknowledging his dependence on air, water and earth by personalizing them as dieties. We still worship Nature, but with a sense of guilt. One of the most painful dilemmas of our times is that we still regard Nature as the ultimate source of beauty and other fundamental blessings, yet we exploit and despoil it for the sake of wealth and power. . . . We accept the belief that economic profits justify the creation of ugliness.[11]

By early 1972 the problem of the environment had become so acute that the United Nations called a conference in Stockholm, Sweden, on the human environment. The world-wide problems were summarized as they applied to the United States.

> Despite man's efforts to reach to the stars and distant worlds in outer space, his environment today and in the forseeable future—and perhaps forever—must continue to be defined by the natural limitations of one finite planet.
>
> So richly endowed eons ago with the resources to provide abundant life, this planet—within the memory of living man—has been forced to endure stresses and depletions that have strained the quality of its life-support systems.
>
> Thus, man suddenly finds himself confronted with the perplexing problems of determining how best to stem the tide of environmental decline and to preserve as much as possible the quality of life to which he has become accustomed. In a larger sense, 20th-century man has come to the realization that the dimensions of human environment include not only all ecological systems, whether natural or manmade, but also a multitude of social and cultural values—all of which react and interact between and among one another.
>
> Man's status is measured, therefore, by the quality of environment in which he exists. Whereas, in primitive societies, man's activities had little impact on the pristine quality of the environment, the advent of technology and larger masses of people have brought undesirable changes in that quality. Moreover, man's exploitation of his environment has also brought a degree of affluence that has drawn dangerously near to embracing a philosophy of the invincible sovereignty of man without regard for the inevitable reactions that govern all other life systems.[12]

The 1970 National Power Survey by the Federal Power Commission clearly states the changes and the relationships between power and the environment in the following way:

> *The 1964 outlook.* In 1964, apart from traditional conservationist interest in land use and the beginnings of concern by ecologists about other fundamental environmental questions, concern over the environmental effects of electric power operations was largely confined to the electric utilities themselves and to those agencies of government involved directly or indirectly with the power industry. The 1964 Survey Report contained an extensive discussion of the effects of power operations on air and water quality. It is anticipated that these effects might present difficulties as the scale of power operations increased, and called for increased industry emphasis on pollution controls and for research in this general field of power technology, especially in the area of ecological studies. While it cited the possibility that added costs of environmental protection might prevent full realization of the targeted reduction in the average cost of electricity, it concluded that the nation's capacity to produce needed electrical energy will not be impaired because of these environmental considerations.
>
> *The 1970 Survey findings.* As was brought out in Part A (of the survey) public interest in environmental questions was awakened since 1964 and there is today what

11. Rene Jules Dubos, Man and His Environment: Adaptations and Interactions in *The Fitness of Man's Environment,* Smithsonian annual II, Smithsonian Institution Press, Washington, D.C., 1967, p. 233.

12. U.S. Department of State, U.S. National Report on the Human Environment, Prepared for: United Nations Conference on Human Environment, June, 1972, Stockholm, Sweden, p. 1.

amounts to a national crusade to improve environmental quality. As stated in the Introduction of the report to the Commission by the National Power Survey Task Force on Environment:

> 'In our time, we have seen commanding new social values arise, and among the most important of these is a new respect for the conservation of the environment, and the need to adapt our energy sources and supply to the restrictions this imposes upon us.'

As already observed, the problem our society now faces is how to protect and, where possible, to upgrade the quality of the outdoor environment without cancelling out the gains that have been made in the quality of our indoor environment and, equally important, without denying these same gains to new generations or to those among us who have not yet shared them but seek to do so.

While electric power is almost ideal insofar as the indoor environment is concerned, its production and supply may, and usually does, have undesirable effects on the outdoor environment.[13]

The extent of environmental pollution and just one aspect—that of the estimated nationwide discharges of airborne pollutants in 1968 is shown in the following chart.

The chart below from the Federal Power Survey was accompanied by the following quote:

> As these figures reflect, apart from sulfur dioxide emission to which electric power operations are the major contributions, the electric power industry is far from being the major source of air pollutants. It is nonetheless a significant source and, moreover, it is a concentrated source. And in view of the over-all seriousness of the nation's air pollution problems, especially in or near major urban and industry centers, there is obvious need to do all that is

13. Federal Power Commission, *The 1970 National Power Survey: Guidelines for Growth of the Electric Power Industry,* (Washington, D.C., U.S. Government Printing Office, Superintendent of Documents, 1971), p. I-1-22.

Estimated Nationwide Discharges of Airborne Pollutants—1968[14]

[Million tons]

Source	Carbon Monoxide	Particulate Matter	Sulfur Oxides *	Hydro-Carbons	Nitrogen Oxides *	Total
Transportation	63.8	1.2	0.8	16.6	8.1	90.5
Power plants	0.1	5.6	16.8	Neg.	4.0	26.5
Other fuel combustion in stationary sources	1.8	3.3	7.6	0.7	6.0	19.4
Industrial processes	9.7	7.5	7.3	4.6	0.2	29.3
Solid waste disposal	7.8	1.1	0.1	1.6	0.6	11.2
Miscellaneous	16.9	9.6	0.6	8.5	1.7	37.3
Total	100.1	28.3	33.2	32.0	20.6	214.2

Source: National Air Pollution Control Administration. (now Air Pollution Control Office, Environmental Protection Agency)

*Sulfur oxides expressed as tons of sulfur dioxide and nitrogen oxides as tons of nitrogen dioxide.

14. Ibid., p. I-11-2.

within technology's power to reduce the emission of any and all significant air pollutants, from whatever source, to the lowest levels possible.[15]

One of the institutions most readily attacked by active environmentalists was, of course, the electric utility industry. The reasons for this will be discussed later. Suffice it now to summarize the basic criticisms by the environmentalists of the electric utilities and the environmental problems which they create. The following quotation from a paper by Mr. Michael McCloskey, entitled "The Energy Crisis: The Issues and a Proposed Response" states most clearly and most emphatically the environmentalists view of the utilities and their activities:

> Today we consume 15 times the energy we did 100 years ago, though our population has only tripled in that time. Over the past decade the average growth rate in the consumption of energy in all its forms has been more than four percent annually, climbing to about five percent annually over the last five years. Growth has been particularly phenomenal in the electrical energy sector, at about seven percent annually in recent years. Projections based on that rate of growth call for a doubling of electric power production about every ten years.
>
> With these growth rates we may soon find that we are reaching absolute limits in physical space for power plants. It has been calculated that even with large 1,000 megawatt power plants, each of which requires an area of only 1,000 feet on a side, in less than 20 doublings (less than 200 years) all the available land space in the United States should be occupied by such plants. In California, where power production is expected to double every eight years, if power were to be supplied by 110 megawatt plants on 80 acre sites, the entire land area would be covered in only 122 years. Similar startling projections could doubtless be made with respect to other forms of energy use, such as the amount of space that will need to be paved over to accommodate our automobile oriented transportation system.[16]

Other quotes from the same paper by Mr. McCloskey continue to emphasize the other side of the coin in regard to the environmental crisis. In summary, a statement by Sylvia Crowe, in her earlier book on energy generation in the British Isles and the impact on that environment is appropriate at this time:

> Our ancestors were convinced that all they did was justified by the economic results; their creed was to produce at a minimum cost. The fallacy lay in the fact that the real cost was hidden, and only revealed itself after generations in a spoilt country and unhealthy people. If the tragedy is not to be repeated on a far greater scale, the welfare of the landscape must this time be taken into consideration as a basic factor in the planning of these new erections of the machine and atomic age.[17]

Because of the rapidity of communications, no issue in mankind has moved from the problem identification stage to the development of administrative guidelines and legislation faster than has the environmental issue. State legislatures, the Federal Congress and all administrative agencies at all levels of government have been inundated by rules, regulations and procedures to protect and enhance the environment. Certainly this is true in regard to those environmental factors controlled and affected by the electric utility industry. Probably in the remainder of the decade additional other new procedures will be laid down whereby greater control will be exercised over the utilities and their operations. This then is the essence of the problem. It is the classic case of the irresistible force meeting an immovable object. The environmental advocates and their concerns meeting the persons charged with the responsibility for the generation and distribution of adequate en-

15. Federal Power Commission, *The 1970 National Power Survey: Guidelines for Growth of the Electric Power Industry*, (Washington, D.C., U.S. Government Printing Office, Superintendent of Documents, 1971), p. I-1-23.
16. Michael McCloskey, "The Energy Crisis: Issues and a Proposed Response," in *Environmental Affairs*, Vol. 1, No. 3, (Brighton, Massachusetts, The Environmental Center, Boston College Law School, 1971), pp. 587-588.
17. Sylvia Crowe, *The Landscape of Power* (London, The Architectural Press, 1958), p. 10.

ergy to assist in the growth of technology, culture and the society. This is one argument in which nearly everyone is right. The environmental critics are correct—there have been excesses and oversights. The defenders in the electric utility industry are also correct. The environmental values had not previously been defined, delineated or quantified.

There are over 3,600 electric utilities in the United States. These are divided into three basic types—the investor owned, cooperatives, and the publicly owned. All of these have the concern for meeting the needs of their users in the fastest, least expensive possible way. The critics concerned with the environment have performed a service. They have prodded the industry to positive action. They have created within the users a willingness to pay for environmental considerations, study, research and work by the electric utility industry. They have prompted concerned designers and environmentalists to make themselves available to the doers rather than to side with the critics. As mentioned at the outset of this discussion, the problem in essence is, as quoted previously from the Environmental Criteria for Electric Transmission Systems: "The electric utility industry is faced with a complex challenge in the 1970's. It must provide the generation and transmission facilities that are and will be needed to meet the ever-growing demand for reliable electrical power, and it must do this in a decade dedicated to the restoration and protection of our environment."

Extent

The problem of energy and environment is certainly worldwide, and within the United States it is nationwide. It permeates all areas of the country and it is all pervasive throughout wilderness areas, rural areas, suburban areas, urban areas and transportation corridors. Power generation is usually in relatively isolated or in wilderness areas. This is because it is quite often near the source of power. Hydroelectric plants are on or near bodies of water. Fossil fuel plants are either near the source of the fuel or near the ultimate users of the power. One of the advantages of nuclear power plants is their relative freedom from finite geographical locations. They are usually located somewhat near the users of the power. Of course in highly populated areas citizen rejection of local nuclear power plants has happened with some regularity.

Electric utility transmission lines are the arteries of the system and various types and sizes traverse all of the landscape. Transformation facilities intersperse residential, commercial, industrial and transportation areas. Distribution facilities, which are the capillaries of the system carry the developed power directly to the users. Therefore, probably more than any other environmental problem, those created by the electric utility industry pervade a greater portion and to a greater extent the environment and landscape.

Causes

The basic causes for the over-all problem are extremely complex and varied. In the most elemental form they exist because of the intense need felt and expressed and the resultant rapid growth of the electric utilities and their accouterments and the lack of environmental guidelines to assist in their proper growth and development. The problems created by the electric utility industry are one of the most visible of the environmental problems. The industry is one most susceptible to control and also the one getting the lion's share of study and analysis. Little has been written on a comprehensive scale documenting the progress, the solutions and the positive approaches. The conflicting demands on the environment have caused problems of air pollution, water pollution and visual pollution. Most of the reasons and causes for these problems have to do with:

- The speed of growth and development
- The lack of consideration for the environment in this growth
- The lack of knowledge as to ways in which accommodations could be made
- The lack of guidelines to assist growth while preserving the environment
- The lack of criteria

- The lack of examples
- The lack of regulation

Part of the problem is the fault of the designers. They have often aligned themselves on the side of the critics and not on the side of those attempting to solve the problems. As a physical environmental design problem the objects utilized in electric utility industry are larger in size, have little esthetic design input, and have largely been unrecognized or unaccepted as the sculptural objects they are. Industry did little to consider or ask for assistance, and the designers were derelict in their duties for not volunteering the competence. Once it was brought to the attention of designers, legislators and the industry things began to change. This is a relatively new area of work for designers of all types whether they be industrial designers, architects, landscape architects or environmental planners. Because of the extensiveness of the industry and its components, it can cause really the greatest despoilation of the environment on a large scale. The objects it uses are greater in scale and in magnitude, and the extent and pervasiveness of the various parts of the overall circulation systems infringe on the environment at all scales and in all locations.

These functional landscape elements have not had the study and care that architecture and traditional landscape architectural elements have had. Once it was brought to the attention of the industry, legislators, administrators and designers this whole problem opened an entire new area of work for designers of all types. The destruction of the environment is more pervasive, it exists at more scales and is more obvious to the general public in the utility than in many others. This industry is also more easily harassed and controlled. From the wilderness where the hydroelectric plants are found, to the farm land right-of-ways for utility lines, to the suburban transforming facilities and the neighborhood distribution lines, certain parts of the natural environmental processes are destroyed. There are a variety of ways in which these electric utility accouterments destroy environmental values.

Generating plants cause water pollution, large smoke stacks pollute the air and cause visual pollution. Unscreened transforming facilities cause visual ugliness and discordance, and local distribution facilities clutter the skyscape. The obviousness of the environmental destruction occurring in a relatively controlled industry which has experienced quantum growth compounds the entire environmental problem of the electric utility industry. This, coupled with the fact that it is a controlled and controllable industry, an industry with a conscience or an imposed conscience, has emphasized the problem even more.

There was a great inadvertent destruction of various parts of the environment when no one had put a value or a price on the environment. The early warning environmentalists have preceded the scientific fact producers by a far enough margin to cause pressure without the necessary facts to assist in correcting the problems identified. Even though this has been a problem building for a long time it needed to reach certain limits before it was recognized. The space program and the far view of the planet as a closed and finite system, assisted materially in bringing the problems to a head. One of the concomitant problems had to do with the lack of valid rational communication between the opponents. To the utility people the environmentalists are obstructionists, to the environmentalists the utility people are philistines.

History

How long have the two crises been building up? Probably in their most basic form since Edison found the filament which worked to produce an electric light. The need for power and its development and delivery has always caused some destruction of either land, water, air or vision. When the land, water or air needs of the industry were small, when the population was low and the resources seemingly limitless, the problem was not critical. In the early years of the 20th century the demands for electrical facilities and services were quite small because of the lower population in the United States and the lower level of technology. As technology grew it demanded ever-increasing quantities of power, and as the population grew the demands were multiplied. But as these problems were being caused, other environmental destruction was taking place. In the mid-1960's the concern for the environment began to increase with the number of vigorous voices recognizing and explaining the problems to a larger audience. By the late 1960's the problem had been clearly and loudly stated, and legislative and administrative guidelines were begun. By the beginning of the 1970's the decade of environment was launched. The history is a problem of priorities. Energy production, which was essential, had spoiled the environment which had little or no value at that particular time. While solving problems of technology in society energy production had created problems in the environment and hence it has caused a disruption of the natural processes and a marring of the beauty of the natural environment. Little concern existed in the 1940's, '50's and early '60's for environmental problems, little knowledge or hard data was available concerning natural processes. The environmental disruption on a small scale by itself was not as serious as was larger scale activity coupled with other pollution and environmental deterioration. The rapid imperative and immediate need for electric power has caused much structural utility development to take place before adequate environmental data was available. There are basically at least four critical points in the energy environment interface. These are:

- Generating plants
- Transforming facilities
- Transmission lines
- Distribution facilities

In essence, the history of the problem is as follows: The need for energy of all types is great. To meet this has been a staggering problem and an impressive achievement. The demand has been unprecedented, the electric utility industry has been meeting the need for increased power and an admirable way in most cases. The power demands of increased population and accelerated technology has heightened the environmental damage. This growth of the energy production and delivery system has taken an infrastructure which in turn has been placed in a natural environment with a great deal of insensitivity.

The Electric Power Industry 1970[18]

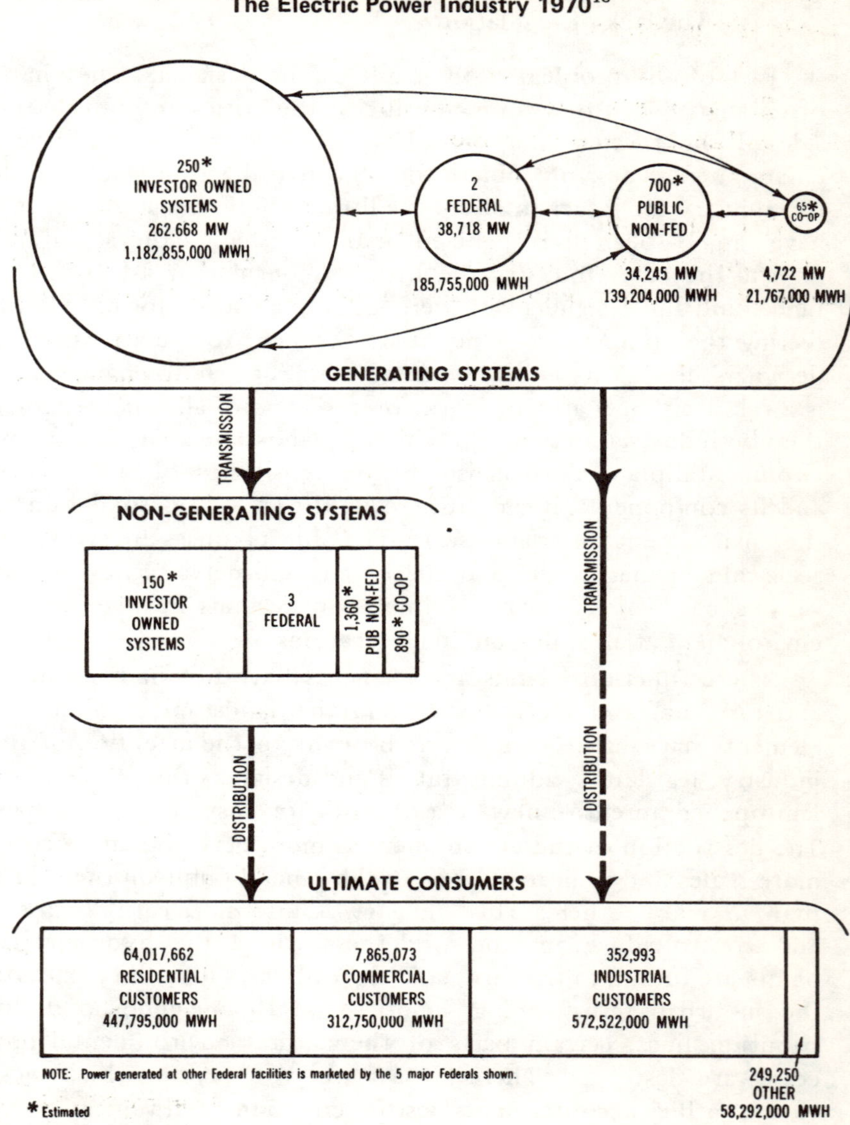

18. Federal Power Commission, *The 1970 National Power Survey: Guidelines for Growth of the Electric Power Industry,* (Washington, D.C., U.S. Government Printing Office, Superintendent of Documents, 1971), p. I-1-11.

Points of Conflict

As mentioned a number of times previously in this introduction the sore points, or the points of greatest environmental damage or disruption are:

1. The points of generation
2. The points of transformation
3. The lines of transmission
4. The means of distribution

This is basically a gross oversimplification but in the hierarchical pyramid of environmental considerations and priorities these are some of the major points of impact or disruption. Each of these problem areas and the innovative ways in which they have been approached make up the next four chapters of this book. Energy needs can be and have been realized while the natural landscape has been preserved and at times enhanced. This has been done at each of these points of conflict by a variety of abilities and a number of consultants. The ways in which this has been done hopefully will make up an instructive series of case studies for the edification and guidance of others faced with problems of the same type or magnitude.

U.S. Energy Consumption 1920-1990[19]

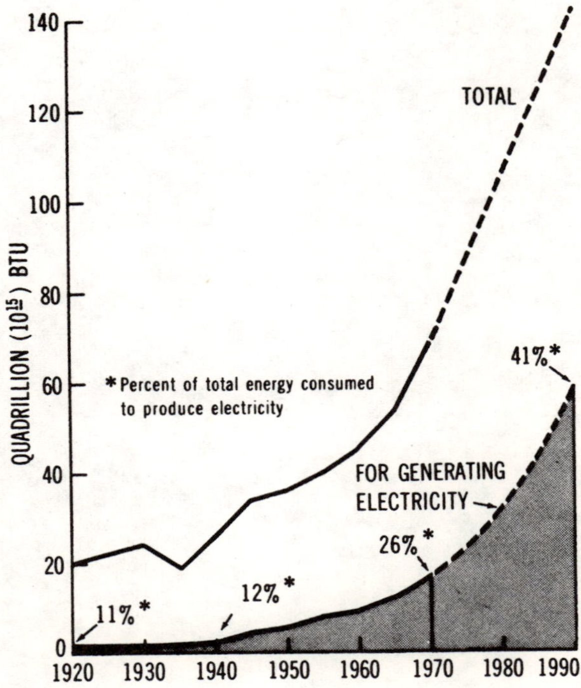

19. Ibid., p. I-3-4.

Power Generation Facilities

Introduction

The start, the heart, and the "generator" (if the double use of the word will be allowed) of the problem of energy and environment is in the power generating facilities. The beginning of solving any need for new power is in the development of new generating plants. The increase in technology has, while solving certain problems of transforming natural forms of energy to electrical power, assisted in the proliferation of ways by which this may be done. Where in times gone by the means for converting energy were simple, they have not only multiplied in type but increased in complexity, size and potential environmental damage. The most succinct and clearly stated outline of present and future power needs is stated in the document "Considerations Affecting Steam Power Plant Selection" sponsored by the Energy Policy Staff of the Office of Science and Technology of the White House, which was prepared in cooperation with a number of federal agencies with an interest in both energy and the environment.

> The total installed electric generating capacity in the United States is now slightly less than 300,000 megawatts (a megawatt is 1,000 kws). Estimates of future loads indicate the installed generating capacity of the United States in the year 1990 will be a little over 1 million megawatts. Changes in electric power system technology anticipated increases in sizes of individual generating units and total capacity at single locations will affect the requirements for future generating plant sites. The use of higher pressures

and temperatures and the trend toward much larger unit sizes and total plant capacity may reduce the relative amount of required space per unit of power production capacity. It is obvious, however, that the total electric power requirements of the nation during the next 20 years will necessitate many new plant sites.[1]

This statement is accompanied by the following chart which tells the same story on a regional basis throughout the United States.

Region	Ratio of 1980 to 1970		Ratio of 1990 to 1970	
	Peak demand	Kw.-hr.	Peak demand	Kw.-hr.
Northeast	1.8	1.8	3.2	3.2
East Central	1.9	1.8	3.4	3.4
Southeast	2.1	2.2	4.1	4.1
West Central	2.0	2.0	3.8	3.8
South Central	2.3	2.3	4.5	4.7
West	2.0	2.0	4.0	4.0

Basically, a power generating facility is a place of conversion of various forms of harnessed natural energy into electrical energy for transformation, transmission and distribution to the ultimate users. Traditionally, these places of energy generation have been tied to the various natural sources of energy. In more recent years with the advent of nuclear power, where possible, the places of energy generation are transferred nearer to the locations of the ultimate users. There has been an ever-growing continuum in the historical development of means of electrical conversion or generation. The most elemental forms consisted of the use of natural hydroelectric power. In time, more sophisticated uses of pumped storage hydroelectric, of fossil fuels such as oil, coal and natural gas begin to be utilized to an even greater degree. In more recent years nuclear-fueled power generating facilities have been developed, sophisticated and perfected to an ever-greater degree. At the same time this is taking place there has been a growing trend to increase the sizes and to sophisticate the types of equipment used in generation. This is best illustrated in a quote from the previously mentioned publication:

> The trend toward larger and larger nuclear and fossil-fueled generating units to capture the economies of scale is to a large degree shaping the plans for expansion of electric power systems throughout the Nation. In the late 1950's a 300-mw generating unit was considered maximum, but one short decade later orders have been placed for units as large as 1,300 mw. Equipment manufacturers indicate that significant extrapolation of present basic designs to still larger sizes is feasible and will produce further economies of scale. Generating units with capacities as large as 2,000-mw were assumed in the preparation of 1990 generating plant site estimates.[3]

The following chart shows the projected increases in maximum generating unit size. (Page 17)

The relative numbers of the various types of generating facilities in relation to the forms of energy generation are shown in another chart from that same publication. (See page 18)

In order to understand some of the problems created in the generation of energy and the preservation of environmental quality it seems advisable to review briefly the basic elements concerning the various types of electrical generating facilities, and to review the environmental problems caused by distinctive generating facilities associated with each of these types of energy production.

In the most basic form the types of electric generation systems may be categorized as follows:

A. Conventional hydroelectric
B. Pumped storage hydroelectric
C. Fossil fuel-oil, coal and gas
D. Geothermal
E. Gas turbine, diesel and total energy systems
F. Nuclear fuel

1. The Energy Policy Staff, Office of Science and Technology, *Considerations Affecting Steam Power Plant Selection,* (Washington, D.C., U.S. Government Printing Office, Superintendent of Documents, 1968), p. 1.
2. Ibid., p. 1.
3. Ibid., p. 3.

Maximum Generating Unit Size[4]

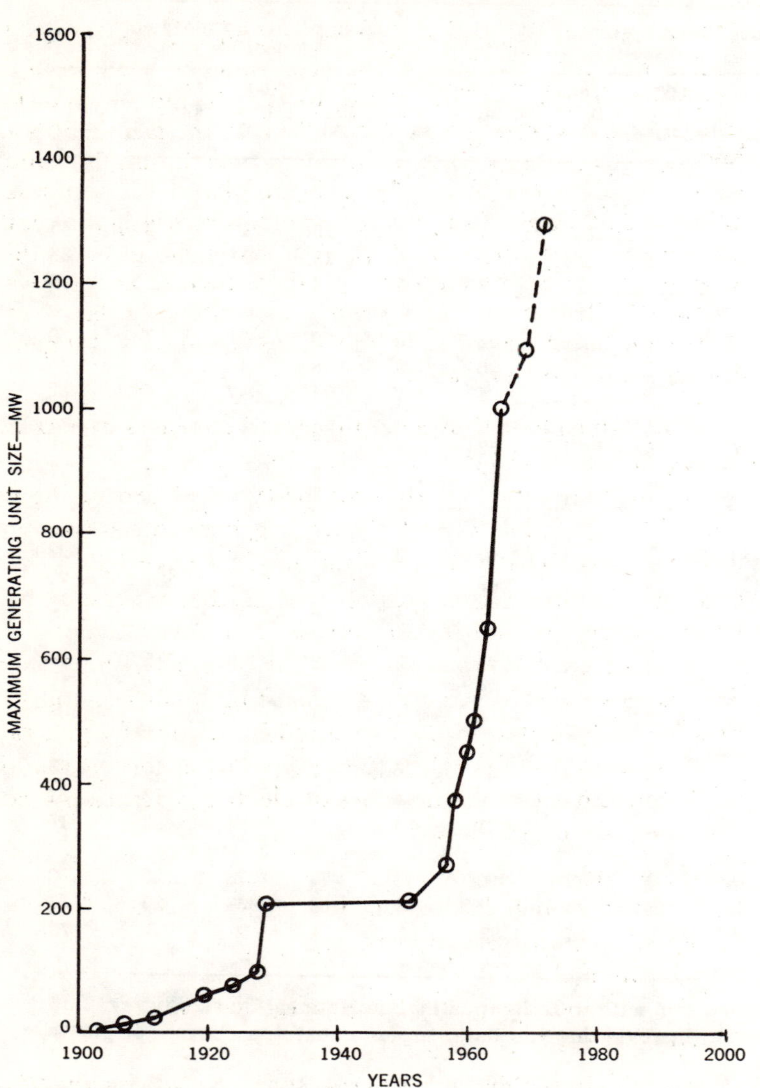

4. The Energy Policy Staff, Office of Science and Technology, *Considerations Affecting Steam Power Plant Selection,* (Washington, D.C., U.S. Government Printing Office, Superintendent of Documents, 1968), p. 3.

Environmental problems caused by energy generating facilities in essence may be outlined as follows:

A. General land use and land requirements
B. Water pollution—either chemical or thermal
C. Air pollution, particulate, gaseous, odoriferous
D. Visual pollution embodying either the plant itself, its supply facilities or the switch-yard for generated energy.
E. "Fish, wildlife and esthetic"
F. Radiological effects

Little concern was exhibited for the environmental problems caused by some of the earlier and more rudimentary forms of electrical energy generation. Even though they have caused environmental problems, little has been or can be done to redesign these facilities to make them less environmentally objectionable since they represent substantially concrete and relatively permanent infrastructure. On the other hand, since some of the newer forms of electrical power generation have been developed in era of great environmental concern, they are better designed and documented in regard to solving some of the currently recognized problems with regard to the quality of human life on the planet.

Hydroelectric Power Generation

Hydroelectric power, or the use of the latent energy in falling or moving water was one of the first harnessed sources of power in the developmental continuum of energy generation. Hydroelectric power generation, by its very nature, is tied to water bodies. Oftentimes there is a finite geographical location available for such purposes. These may be far from the sources of the ultimate users of the electric power. There is also a limited number of potential locations in the future. Therefore, even though hydroelectric power is one of the earliest forms of electrical generation, it will not be able to be expanded to meet future generation needs. This fact was stated in a chapter on "Conventional and Pumped Storage Hydroelectric Power," from the 1970 National Power Survey by the Federal Power Commission, in the following words:

Estimated* Number of Thermal Generating Plant Sites 500-Megawatt Capacity and Above for Year 1990[5]

NPS Region	Fossil-fuel plants by megawatt capacities					Nuclear plants by megawatt capacities				
	500 to 1,000	1,000 to 2,000	2,000 to 4,000	Over 4,000	Total	500 to 1,000	1,000 to 2,000	2,000 to 4,000	Over 4,000	Total
Northeast:										
Total sites	22	15	4		41	7	19	17	2	45
New sites		1	4		5	6	18	14		38
Cooling towers	3	5			8	2	4	3		9
Southeast:										
Total sites	15	12	7		34	10	22	21	7	60
New sites	3	3			6	8	18	14	5	45
Cooling towers		1	3		4	7	14	7	4	32
East Central:										
Total sites	30	26	5	1	62	1	10	8	2	21
New sites	5	9	1	1	16		8	7	2	17
Cooling towers	7	8	1	1	17		4	1	1	6
South Central:										
Total sites	37	31	23	1	92	3	9	9	1	22
New sites	17	12	19	1	49	3	9	9	1	22
Cooling towers	17	7	3		27	1	2	1	1	5
West Central:										
Total sites	11	12	2		25	3	6	9	1	19
New sites	1	5			6	2	4	5		11
Cooling towers	1	4	1		6	1	4	3		8
West:										
Total sites	14	19	4	1	38	4	7	9	13	33
New sites	4	4	1		9	4	6	9	12	31
Cooling towers	10	10	1		21	4	5	4	2	15
Total U.S.:										
Total sites	129	115	45	3	292	28	73	73	26	200
New sites	30	34	25	2	91	23	63	58	20	164
Cooling towers	38	35	9	1	83	15	33	19	8	75

*** Estimates are based on preliminary information assembled by FPC staff in connection with work in updating the National Power Survey. The staff of the Water Pollution Control Administration in the Department of the Interior, in reviewing the data, suggests that the number of plants requiring cooling towers may be greater than the FPC staff estimate.**

5. The Energy Policy Staff, Office of Science and Technology, *Considerations Affecting Steam Power Plant Selection,* (Washington, D.C., U.S. Government Printing Office, Superintendent of Documents, 1968), p. 5.

Distribution of Generating Capacity[7]

Type	Percent of Total Capacity *				
		1970–1980		1980–1990	
	In Service End 1970	Additions During Period	In Service End 1980	Additions During Period	In Service 1990
Hydraulic					
Hydroelectric	15	5	10	2	7
Pumped storage	1	7	4	7	6
Sub-totals (rounded)	16%	11%	14%	9%	12%
Thermal					
Steam-electric					
Fossil-fuel-fired	77	45	59	34	44
Nuclear	2	40	22	53	40
Gas turbine & diesel	6	3	5	3	4
Sub-totals (rounded)	84%	89%	86%	91%	88%
Total	100%	100%	100%	100%	100%

*Since different types of plant are operated at different capacity factors, this capacity breakdown is not directly representative of share of kilowatt-hour production. For example, since nuclear plants are customarily used in base-load service and therefore operate at comparatively high capacity factors, nuclear power's contribution to total electricity production would be higher than its capacity share (see table 1.4).

Conventional hydroelectric developments use dams and waterways to harness the energy of falling water in streams to produce electric power. Pumped storage developments utilize the same principle for the generating phase, but all or part of the water is made available for repeated use by pumping it from a lower to an upper pool. At the end of 1970, conventional hydroelectric capacity totaled 52,323 megawatts compared to only 3,689 megawatts of pumped storage capacity. During the next 20 years, however, the installation of pumped storage capacity is expected to exceed greatly the installation of new conventional capacity. By 1990, conventional hydroelectric capacity is expected to total approximately 82,000 megawatts and pumped storage capacity about 70,000 megawatts.

Over the years, hydroelectric plants have provided a substantial but declining proportion of the nation's electric power supply. This trend is expected to continue despite the construction of many large pumped storage plants. Hydroelectric plants, which now account for 16 percent of total generating capacity, are expected to provide about 12 percent of the total capacity in 1990.[6]

The generating capacity of the various methods utilizing the sources of natural energy are shown on the above chart.

6. Federal Power Commission, *The 1970 National Power Survey: Guidelines for Growth of the Electric Power Industry,* (Washington, D.C., U.S. Government Printing Office, Superintendent of Documents, 1971), p. I-7-1.
7. Ibid.

At the same time the projected growth of utility generating capacity for each of these sources is shown on the following chart.

Projected* Growth of Utility Generating Capacity[8]

(all figures in thousands of megawatts)

Type of Plant	Class of Service (see text)			Installed Capacity End 1970	1970–1980			1980–1990		
	Base Load	*Intermediate*	*Peaking*		Projected Additions	Projected Retirements	Projected Installed Capacity End 1980	Projected Additions	Projected Retirements	Projected Installed Capacity End 1990
Hydroelectric										
Conventional	seasonal	seasonal	seasonal	52	16	....	68	14	....	82
Pumped storage			√	4	23	....	27	44	....	71
Thermal										
Steam-electric										
Fossil-fuel-fired	√ (newer units)	√ (older units)	....	260	157	(24) **	393	225	(61) **	557
Nuclear	√			6	141	....	147	353	....	500
Gas-turbine & diesel			√	19	12	....	31	20	....	51
Totals (rounded)				340	349	(24)	665	656	(61)	1260

*Keyed to electrical demand projections made by Regional Advisory Committee studies carried out in the 1966–1969 period. Premised on average gross reserve margin of 20%.

** Units placed in service prior to 1955.

8. Ibid., P. I-1-17.

The three following charts indicate the projected generation mix to 1990, the estimated electric utility generation by the primary energy sources, and the projected nuclear, hydro and fossil-fueled capacity.

ESTIMATED ANNUAL ELECTRIC

Estimated Annual Electric Utility Generation by Primary Energy Sources[9]

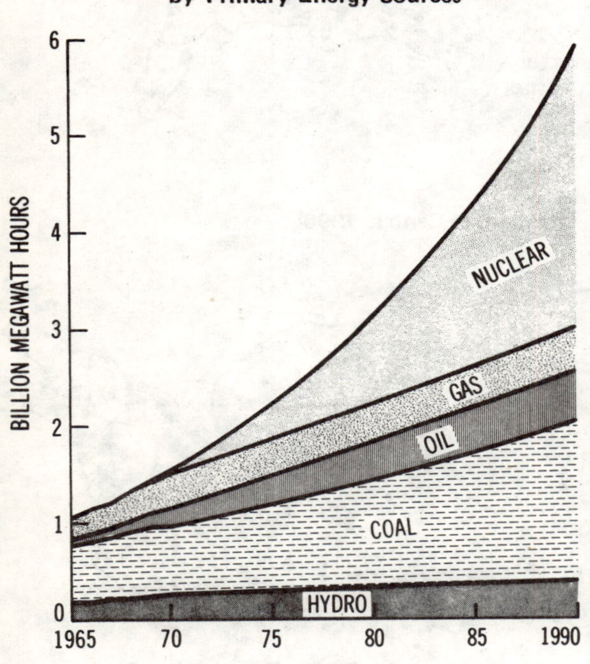

Projected Generation Mix[10]

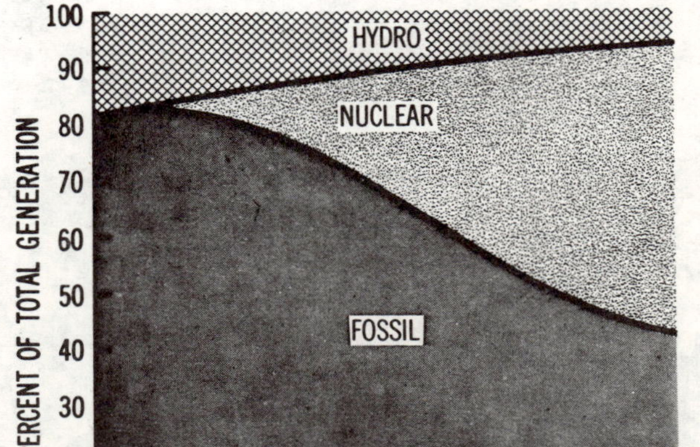

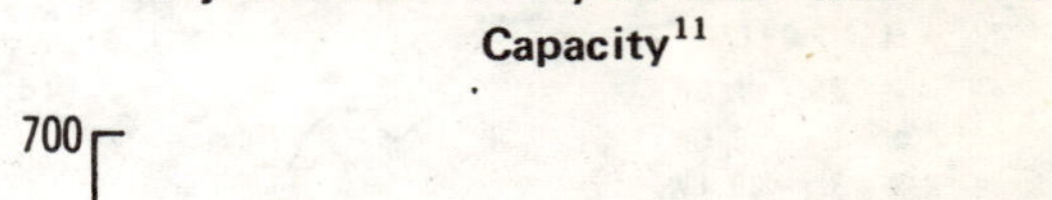

Projected Nuclear Hydro and Fossil-Fueled Capacity[11]

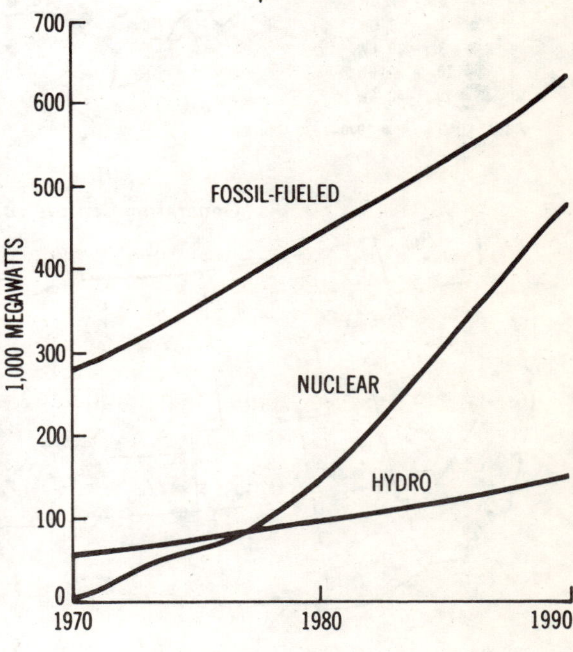

9. Ibid., p. I-4-1.

10. Ibid., p. I-4-1.

11. Ibid., p. I-6-7.

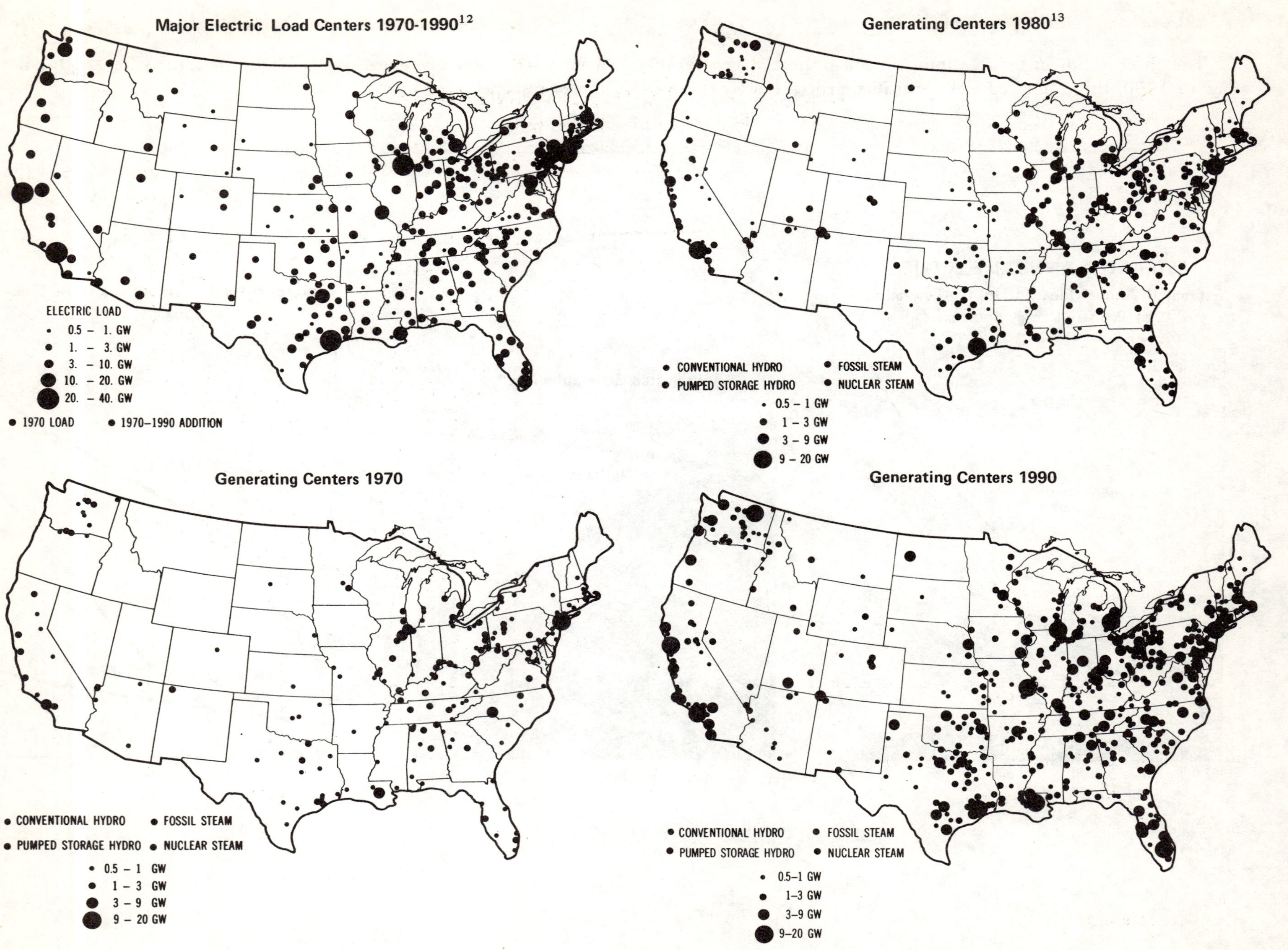

12. Ibid., p. I-18-4.

13. Ibid., p. I-18-5.

NUCLEAR ELECTRIC GENERATING UNITS[14]

Existing and Scheduled

June 30, 1971

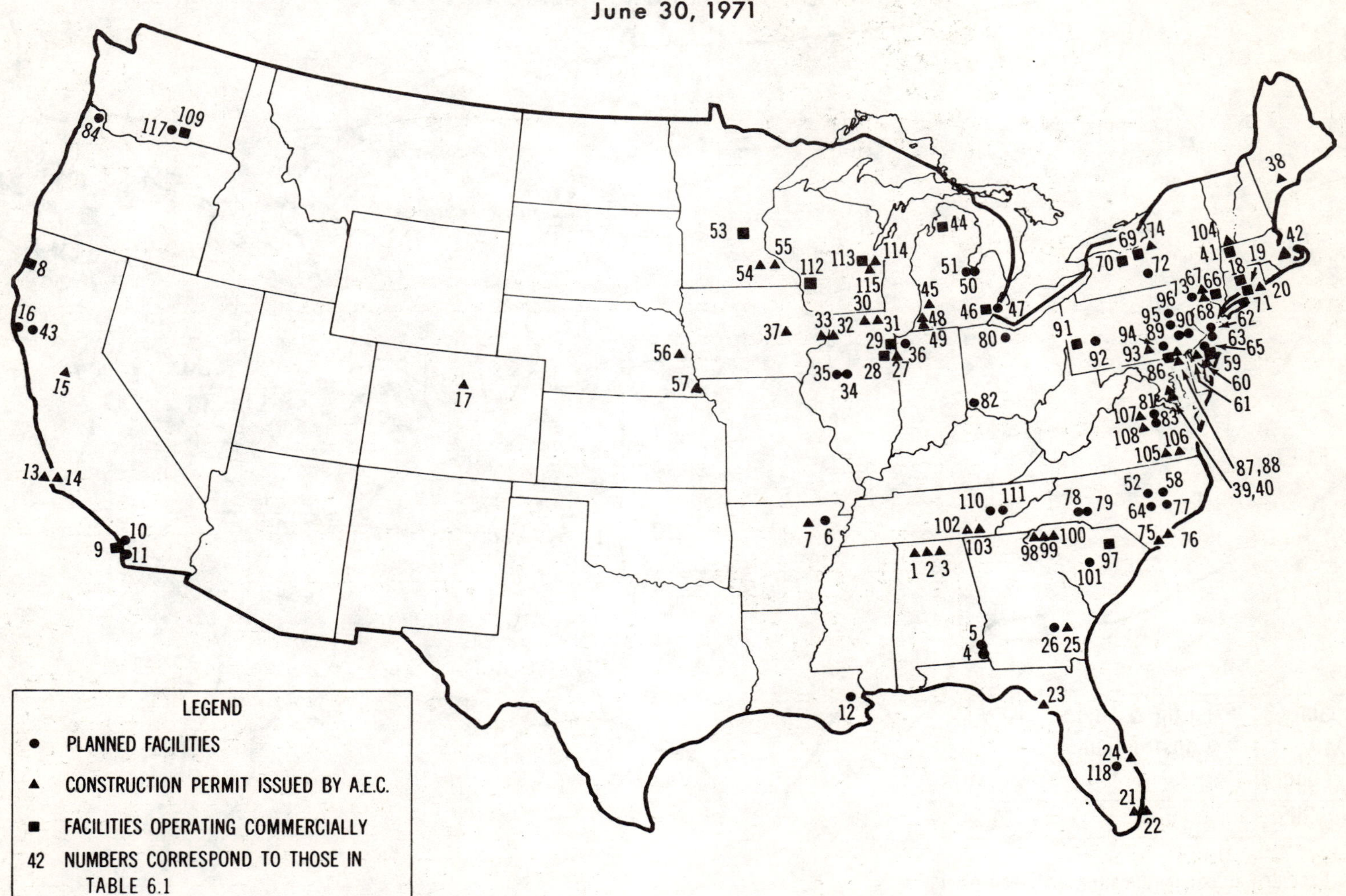

14. Ibid., I-6-6.

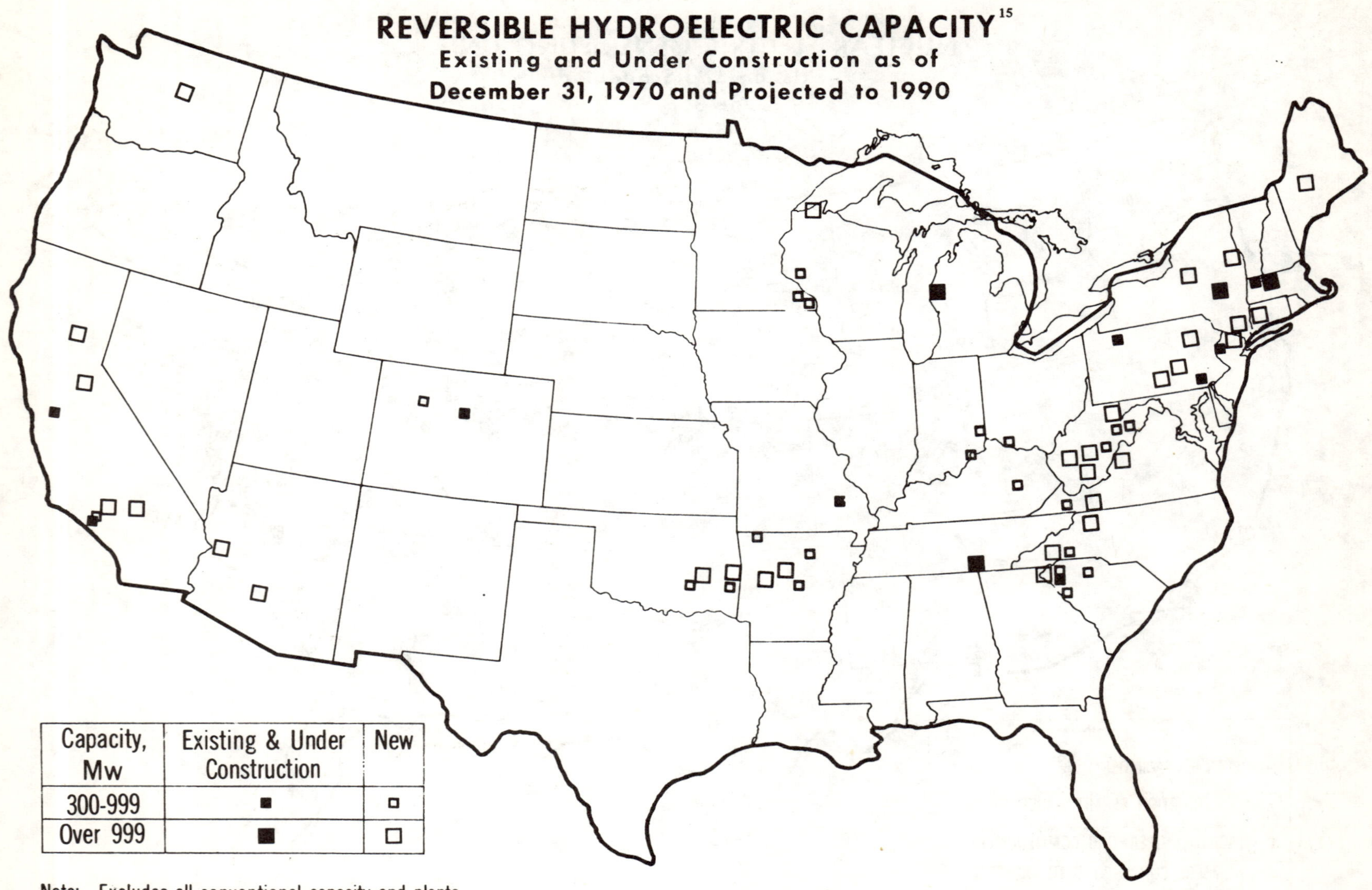

Note: Excludes all conventional capacity and plants with reversible capacity less than 300 mw.

15. Ibid., p. I-7-29.

This map indicates the conventional hydroelectric capacity (new and expanded) projected to 1990.

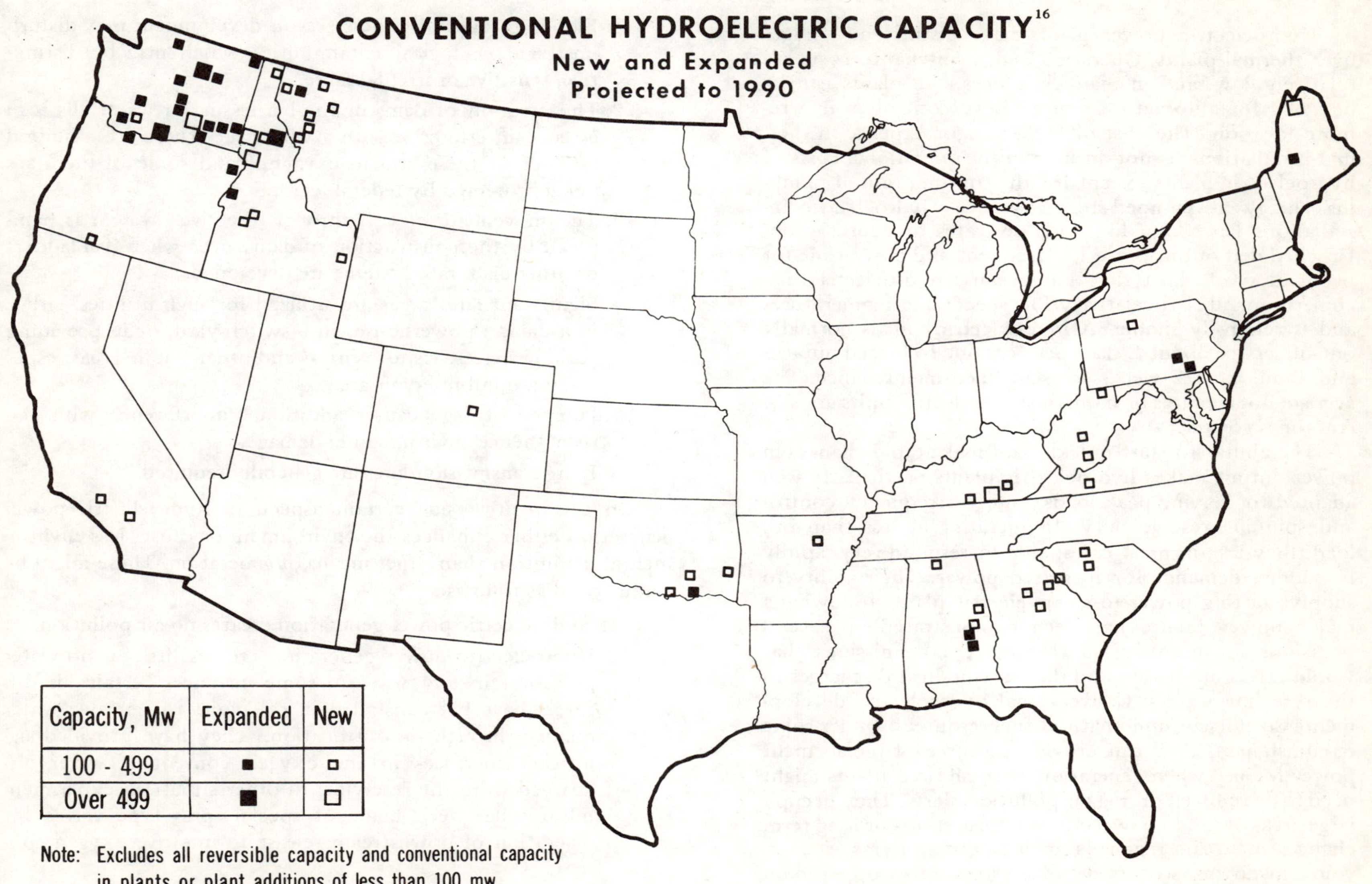

Note: Excludes all reversible capacity and conventional capacity in plants or plant additions of less than 100 mw.

16. Ibid., p. I-7-28.

The following summarizes most aptly the operating characteristics of hydroelectric power plants:

> Hydroelectric power plants have distinct advantages over thermal plants. Operation and maintenance costs are relatively low, and in many instances, the plants can be designed for automatic or supervisory control from a remote location. The cost of fuel, a major expense in thermal installations, is not an item in the operational costs of hydroelectric plants except for the consumption of pumping energy at pumped-storage plants. Hydroelectric installations have long life and low rates of depreciation. Unscheduled outrages are less frequent and down-time for overhaul is of short duration because hydroelectric machinery operates at relatively low speeds and temperatures and is relatively simple. A hydroelectric unit is normally out of service about 2 days per year due to forced outages and about 1 to 2 weeks for scheduled maintenance. The average outage rates of modern steam-electric units are several times greater.
>
> The ability to start quickly and make rapid changes in power output makes hydroelectric plants particularly well adapted for serving peak loads, and for frequency control and spinning reserve duty. If operating at less than full load, they are, in most cases, able to respond very rapidly to sudden demands for increased power. Their ability to supply starting power to steam-electric plants following a major power failure has been demonstrated on several occasions in recent years. There are no emissions that would affect air quality and there are no heat discharges to the receiving waters. Conventional hydroelectric developments do not consume natural fuel resources. Under some circumstances, they can provide a source of replacement power for use when generation at fossil-fired plants might need to be reduced during air pollution alerts. They occupy large areas of land, however, and cause short- or long-term changes in stream regimens, including such items as reservoir drawdowns, so they are often strenuously opposed on esthetic or ecological grounds.[17]

The environmental impact of hydroelectric generating facilities are able to be evaluated both negatively and positively. On the minus side of the following are negative environmental impacts on such facilities:

1. Flooding caused by the reservoir development may disturb upstream ecological communities which may be distinctive, sensitive or irreplaceable.
2. The creation of dams or the harnessing of waterfalls is in essence interference with a wild river. There are a limited number of these still in existence and some of them are being protected by federal legislation.
3. The movement of fish through the river system is hampered by the construction of dams, even when fish ladders or other elaborate systems are developed.
4. Significant land areas are utilized for such utilities, either in a dam, a powerhouse, in a switch-yard, or in providing parking spaces, visitor centers and other public areas, especially around reservoir areas.
5. Pumped storage causes additional interference with the river—hence environmental damage.
6. Long transmission lines are generally required.

On the positive side certain aspects of hydroelectric power generation either enhances the environment or causes less environmental disruption than other means of generation. These might be summarized as follows:

1. Hydroelectric power generation creates no air pollution.
2. Hydroelectric power generation creates little or no water pollution in any way. In some instances certain air diffusers have been installed in reservoirs to improve water quality through destratification. They have provided significant increases in the oxygen content of water discharged from the reservior. In other situations experimental use has been made of special spray-type valves for reaeration of downstream release to improve oxygen content.
3. Reservoir development often has multiple usage as recreation areas for surrounding residents.

17. Ibid., p. I-17-1.

4. The development of dams or other hydroelectric generating facilities acts to reduce flooding on a river by curtailing seasonal variation of waterfall.

In summary, hydroelectric generation probably creates a minimum environmental impact of any means of energy generation. Because of the distant location of the hydroelectric facilities from the ultimate users, the environmental cost is greatest in the transportation of generated electricity. Unfortunately, there are not enough sources of hydroelectric plants nor is there enough power available from that source for future energy needs.

The positive environmental aspects of hydroelectric power generation facilities is summarized in the following quote:

> Many associated benefits are provided by conventional hydroelectric developments. Reservoirs are used extensively for recreation, and many provide water for domestic and industrial water supply and condenser cooling water for thermal-electric plants. Some projects provide flood control, and some are operated to supplement natural flows during low-flow periods to provide benefits related to such non-power functions as water quality control, navigation, municipal water, irrigation water, and fish and wildlife.
>
> Streamflow regulation by reservoirs may have adverse as well as beneficial effects, but special measures can usually be taken to minimize adverse effects. For example, a reregulating reservoir may provide uniform flows downstream from a peaking hydroelectric plant. Facilities may be constructed to provide for passage and protection of anadromous fish runs. However, even with fish passage facilities, the cumulative effect of a series of dams may be substantial reductions of such runs, as has been the case on the Columbia River. At some reservoirs fish passage has not been successful but the runs are being maintained with fish hatcheries financed by the owner of the power facility.[18]

One last citation summarizes effectively the esthetic and land use characteristics of hydroelectric power generating facilities:

> The topography, geology, and other site conditions of the region generally determine not only the location of a hydroelectric project but also the type of major structures such as the dam, spillway, and outlet facilities, that are required. Once the general plan is established, however, there are a number of opportunities for esthetic treatment.
>
> In the case of massive concrete structures, there are opportunities for architectural treatment of the facilities so as to present an attractive and coordinated design for the entire project. For best results all design professions must work together to achieve a unified design. The designers must also work in close cooperation with the contractor to assure the preservation of natural features and provide compatibility between the structures and the surrounding landscape.
>
> For projects including large embankments, the appearance can be improved by landscaping. Ridge lines need not be uniformly straight, and they may include some undulations. Landscaping along the crest and the toe of embankments can blend them with their surroundings. Graded plantings and grasses can be utilized to conceal the scars of construction and of fluctuating water levels.
>
> When a powerhouse is located at a dam, it should be designed to harmonize with the dam design.
>
> Hydroelectric projects tend to attract large numbers of visitors. Therefore, parking space, visitors' centers, and other facilities should be provided to take care of the visiting public. The design of these facilities should conform to the over-all design of the project. Recreational use of the land and water of a project also requires facilities, which should be designed, operated, and maintained, to the extent possible, to blend with, preserve, and enhance the natural landscape.[19]

Fossil Fuels

Fossil fuels of all types have been in more recent years one of the most common sources of electrical power used in generation

18. Ibid., p. I-7-1.
19. Ibid., p. I-12-10.

facilities. At the same time they also cause some of the most severe environmental problems. The 1970 National Power Survey makes the following statement concerning fossil fuel for energy generation:

> By the year 1990, electric power generation is expected to account for over 40 percent of the Nation's annual energy requirements. During that year, fossil-fueled electric generating plants may require an estimated 25 quadrillion (25×10^{15}) Btu in the form of coal, gas and oil—an 87 percent increase over 1970. In spite of this increase, the relative role of fossil fuels in over-all thermal electric power generation will diminish from about 82 percent of total generation in 1970, to only 44 percent of total generation by 1990. The projected changes in the proportions of total generation in 1970, to only 44 percent of total generation by 1990.[20]

There are three basic fossil fuels—these are coal, oil and gas. To better understand the environmental problems caused by fossil fuel generation facilities, it is necessary to understand initially the processes by which fossil fuel is converted to electrical energy, the physical facilities and process this entails, and to understand in greater detail the inherent characteristics of the materials themselves. In the process of converting fossil fuels to electrical energy there are at least eight basic steps—all of which can and have caused physically perceivable environmental problems. These steps are:

1. The extraction of the fuel
2. The transportation of the fuel
3. The delivery of the fuel
4. The storage of the fuel
5. The combustion of the fuel
6. The conversion of the energy derived from combustion into electrical energy
7. The disposal and storage of waste products from the combustion process
8. The transmission from the generated energy

The amount of land required for the fossil generating plants, keyed according to the plant fuel type is shown on the following chart:

Plant fuel[21]	Number of acres of land	Remarks
Coal	900 to 1200	Assumes onsite coal storage and ash disposal.
Nuclear	200 to 400	(See Ch. III: Nuclear Power Reactor Plant Siting.)
Gas	100 to 200	Assumes pipeline delivery and modest onsite fuel storage tanks.
Oil	150 to 350	Assumes adequate onsite fuel storage.

The projected distribution of fuel used for thermal power generation from 1970 to 1990 is shown on the following chart:

Projected Distribution of Fuel Use for Thermal Power Generation 1970-1990[22]

[Percent of Heat Input]*

Fuel	Year		
	1970	1980	1990
Coal	54%	41%	30%
Natural gas	29%	14%	8%
Residual fuel oil	15%	14%	9%
Nuclear	2%	31%	53%
Totals	100%	100%	100%

*Staff Estimates. The projected uses of each of these fuels in 1980 and 1990 are based on assumed prices and availabilities. If the assumptions vary, then the mix of the fuels actually used may also vary.

20. Ibid., p. I-4-1.
21. The Energy Policy Staff, Office of Science and Technology, *Considerations Affecting Steam Power Plant Selection,* (Washington, D.C., U.S. Government Printing Office, Superintendent of Documents, 1968), p. 11.
22. Federal Power Commission, *The 1970 National Power Survey: Guidelines for Growth of the Electric Power Industry,* (Washington, D.C., U.S. Government Printing Office, Superintendent of Documents, 1971), p. I-1-19.

On the following graph the estimated annual fossil fuel requirements for electric power generation is indicated:

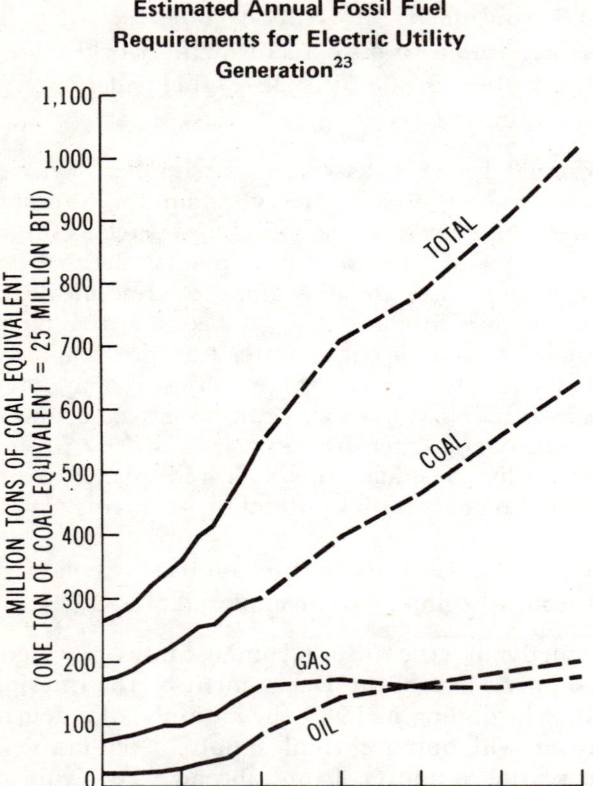

Estimated Annual Fossil Fuel Requirements for Electric Utility Generation[23]

A brief review of the essential characteristics of the basic fossil fuels as they relate to environmental problems may be assembled from a series of discontinuous quotes from the 1970 National Power Survey:

Coal is the most abundant indigenous fossil fuel and currently provides the primary energy for about 56 percent of fossil-fueled electric generation, or slightly less than one-half of total electric generation.

The United States is well-endowed with coal resources.

Because of its domestic abundance, widespread geographical distribution, and chemical versatility, coal is destined to play an important role in the energy economy of the United States for many years to come. However, the competitive position of coal in the electric utility market is highly affected by the geographical distribution of the deposits in relation to major electric load centers and by the physical and chemical characteristics of those deposits. Thus, the trends in consumption of coal for electric generation can be expected to vary from one region to another.

The need to control sulfur oxide emissions has created demands for low-sulfur coal as a replacement fuel for existing units, and for new plants.[25]

The relative use of coal, gas and oil in projected fossil fuel generation could be indicated in the following manner.

Relative Use of Coal, Gas, and Oil in Projected Fossil-Fueled Generation[24]

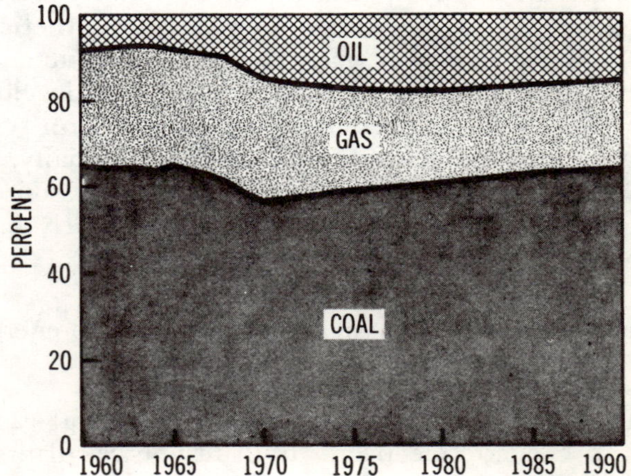

23. Ibid., p. I-4-2.
24. Ibid., p. I-4-4.
25. Ibid., p. I-4-2—I-4-4.

Therefore, coal will continue to be an important energy source, but the greatest search will be for low sulfur coal which has less disastrous air polluting characteristics. One other important development which has taken place in the use of coal in power generation is mine-mouth plants.

> The construction of generating plants at the mine mouth is not new. In recent years, however, economies of scale deriving from advances in high-voltage transmission together with the relatively higher cost of air pollution abatement and control in urban centers, have improved the relative economics of mine-mouth plants in a number of United States Locations.
>
> The number of potential sites for mine-mouth plants, where commercial coal deposits are situated near adequate water supply, is limited.[26]

Coal conversion is one other form or means of potential of obtaining fuel for power generation. The National Power Survey had this to say about coal conversion:

> Research in the area of coal conversion to synthetic gaseous and liquid fuels has been going on for a considerable time. In the past, unfavorable economics has been a major deterrent to the development of a synthetic fuels industry. Since, however, most synthetic products can be made relatively ash and sulfur free, there has developed a greater incentive for speeding up coal conversion research and for the use of synthetic products. Potentially, therefore, synthetic fuels may become significant sources of primary energy for steam generation, particularly in areas with serious air pollution control problems.[27]

Natural gas is another fossil fuel widely used in energy generation.

> In 1970, 3.9 trillion cubic feet of natural gas were burned to generate electricity in utility power plants. This was nearly 18 percent of all natural gas consumed in the country and about 26 percent of all gas used for industrial purposes. The use of natural gas for electric power generation has grown rapidly. In the 1930's power plant consumption of natural gas amounted to only about 7 percent of total consumption.[28]

The U.S. consumed 22.1 trillion cubic feet of natural gas in 1971. This amounted to 49 percent of the world's consumption. Residual fuel oil is one other fossil fuel widely used in electric power generation.

> Residual fuel oil, as the name implies, is a residual refinery product and normally competes directly with natural gas and coal for heavy-fuel uses, such as the generation of steam at electric power plants. Because residual fuel oil is quite viscous and cannot be economically moved by pipeline over long distances, its competitive position is greatest in areas with cheap water transport facilities or in areas adjacent to petroleum refineries. Some low-sulfur residual oil is less viscous and can be moved economically by pipeline over greater distances.
>
> Nationally, residual fuel oil-fired plants contribute about 15% to generation by fossil-fueled plants.[29]

Oil shale is the latent fossil fuel source for energy generation which is just now beginning to be explored.

> In analyzing the future United States petroleum demand-supply picture, the Department of the Interior forecasts that beginning in 1980 the nation's total demand for petroleum will outpace total supply. Even in the face of ever-increasing imports from abroad. To avoid future shortage it may become necessary to develop supplemental indigenous sources of oil supply such as oil shale, tar sands and synthetic liquid fuels from coal.[30]

The actual generation facilities to provide electrical energy creates environmental destruction, but at the same time the fuel

26. Ibid., p. I-4-10.
27. Ibid., p. I-4-10.
28. Ibid., p. I-4-12.
29. Ibid., p. I-4-17.
30. Ibid., p. I-4-20.

transport facilities which bring the fuel supplies from their natural locations to the generating plants also cause widespread destruction of the environment.

Except where nuclear fuel is employed to generate electricity in power plants located near load centers, the transport of energy in one or another form constitutes a significant part of the total cost of electricity. Either the fossil fuels—coal, gas and oil—must be transported from the source to the generating plant; or where the electric energy is generated at the source of fuel, as in the case of mine-mouth plants (coal), well-head plants (gas), or plants located near refineries (residual fuel oil), the generated electricity must be transported to load centers by wire. The choice of energy transport is usually one of economics with the environmental impact now receiving increasing emphasis.

Coal is the largest single item of all-railroad freight traffic.

Coal movements on inland waterways are essentially limited to the central eastern portions of the country.

The movement of natural gas from the reservoir to the ultimate user requires a continuous pipe between these points. More than 891,000 miles of pipeline have been installed by gas companies to accomplish this movement. More than 248,000 miles of these pipes are classified as transmission mains of which about 72 percent are subject to the jurisdiction of the Federal Power Commision.[31]

Fossil-Fueled Steam-Electric Generation

Fossil-fueled steam-electric plants long have been the mainstay of the electric power industry. They currently account for about 76 percent of total generating capacity and more than 80 percent of total generation. With increased reliance on nuclear power, they are expected to account for only about 44 percent of both capacity and generation by 1990. Nevertheless, the total capacity of these plants is expected to increase from 259,000 megawatts in 1970 to 390,000 megawatts in 1980, and 558,000 megawatts in 1990.

Engineering criteria concerned with supplies of cooling water, adequacy of fuel supply, fuel delivery and handling facilities, and proximity of load centers have always been important factors in the selection of power plant sites. More recently, however, environmental factors have gained in influence and now often dominate in the selection of sites.

Environmental problems involved in power plant siting include the discharge of objectionable gasses and particulates to the atmosphere, the rejection of waste heat to natural bodies of water, the discharge of chemical and sanitary wastes, and esthetic considerations.[32]

Some of the environmental problems caused in the development of facilities for electrical generation by steam-electric plants, are indicated in the following statement:

Although generally more flexible than the selection of hydroelectric plant sites, the selection of the sites for steam-electric plants involves consideration of such factors as economy, the availability of cooling water and fuel, safety, reliability of service, environmental effects, and esthetics. Obviously, sites selected for steam-electric plants must be consistent with state, regional, and local land use plans and zoning regulations. Until recently, environmental effects and esthetics were given only secondary consideration in selecting many plant sites. The industry now generally recognizes the need to improve the appearance of its facilities and reduce environmental impacts. In order to gain full public acceptance, the trend toward environmental improvement needs to be broadened and expedited.

The location and design of facilities on a selected site are important factors in minimizing the impact on the natural environment. Appearance can be enhanced by the establishment of buffer zones around the plant. These zones, which should be considered by regulatory bodies as integral parts of the site, can be used to screen the plant facilities by means of trees, vegetation, and other landscap-

31. Ibid., p. I-4-20–I-4-21.
32. Ibid., p. I-5-1.

ing. In many cases, picnic areas can be provided, and the addition of artfully designed visitor centers at power plants can improve public acceptance of the facilities. It is important also that transmission lines extending from the plant be visually acceptable.

Both nuclear and fossil-fueled steam-electric plants include large buildings and other structures, and fossil plants often include very tall stacks. The esthetics of power plants are enhanced by good architectural design and landscaping treatment. Strategic profiling and positioning of buildings and structures, use of appropriate materials and colors, and use of screening and blending of landscaping are among the techniques available to moderate visual impact. Even the stacks, which are difficult to treat esthetically, can be designed, located, and colored to complement other structures. Outside lighting may also be used to improve plant appearance.

Substantial improvements can be made in the appearance of fossil-fueled plants. Coal-burning plants in particular can be upgraded by improved designs for coal storage facilities. Use of retaining walls and additional conveyors would permit trimmer and smaller coal yards and thus reduce the random appearance of such areas at many existing plants. Landscaping of coal yards and ash disposal areas would also improve the appearance, and fuel transportation facilities can be given appropriate esthetic treatment.

The type of cooling system used greatly influences the esthetic qualities of steam-electric plants. Once-through systems usually cause the least noticeable change in the natural environment. The required structures include an intake with screens, a conduit or canal leading to the condensers, a discharge conduit or canal, and possibly a dispersion outlet. These rather modest structures are normally located at the edge of a water body with a major part of the installation being placed underground or underwater. However, care must be taken to assure that the warm water discharges will not adversely affect the ecology or appearance of the water source.

Cooling ponds are more expensive than the normal once-through systems but the structural requirements are essentially the same. Cooling ponds may add to the beauty of an area and provide recreational opportunities, but the addition of heat to the water in cooling ponds accelerates the rate of evaporation which has some effect on the humidity of the local area, possibly increasing the amount of fogging. Also, favorable sites for the ponds may not be available, particularly in urban areas.

Cooling towers, whether of the mechanical or natural draft type, are large structures which have a great visual and environmental impact. Because of the sizes of structures, cooling towers are difficult to treat esthetically. However, good structural design of facilities and landscaping of the site are helping in maintaining an acceptable appearance and minimizing disharmonies with the natural environment. Also, the plumes of visible water vapor that vary in volume and length with changing weather conditions must be considered and evaluated in selecting the type of cooling tower and the site location. Ground-level effects upon local weather, including fogging or icing, tend to be greater for the mechanical draft type owing to its lower height.[33]

The publication, *Considerations Affecting Steam Power Plant Site Selection,* makes the following statements concerning the physical requirements of the site of thermal electric generation plants:

> The siting of thermal electric generation plants has grown more complex with each passing decade. As the sizes of generators, boilers, and associated equipment have grown the problems of site location have increased.
>
> The future requirements for electrical energy and the trends in recent years with regard to power plant sizes places greater emphasis on the problems of plant location.
>
> Generating plants must be developed in such a manner as to make them good neighbors in all respects—worthy additions to the landscape and the environment as they produce the power so essential to continued economic growth and national progress.[34]

33. Ibid., p. I-12-10.
34. The Energy Policy Staff, Office of Science and Technology, *Considerations Affecting Steam Power Plant Selection,* (Washington, D.C., U.S. Government Printing Office, Superintendent of Documents, 1968), p. 7.

Site Location

Proximity to Load Centers

The economics of locating thermal electric generating plants near load centers is undergoing very rapid change. The requirements with regard to air quality control and aesthetics of the plant and the surrounding area place greater restraints on the development of plant sites located near load centers.

Highway, Rail, Water Access

Good highway access is mandatory for large modern stations. The highway will provide access for plant construction and operation, and depending on development of rail and/or water access, may serve for delivery of all or part of construction and operation materials and equipment and fuel.

Rail or water access is very desirable for delivery of heavy equipment and for fuel (except at some nuclear plants, coal-fired mine-mouth plants and at gas- and oil-fired plants served by pipeline) and where feasible, use of both usually is economical.

Foundations

One of the most important factors in choosing a site is to determine that geological conditions are such that a satisfactory foundation for the plant structure is available.

Meteorology

The relationship of meteorology to the physical requirements of siting an electric generating plant is an important consideration, especially in designing the air pollution control features of the plant.

Hydrology

Flooding—Unless the site area is diked for protection, plant grade for all plants should be set above the elevation of the greatest flood that may reasonably be expected based on actual storm and flood records and observations.

Security of cooling water source—The cooling water source for all types of thermal plants should be made dependable for all conditions in which the plant is expected to continue operation.

Thermal quality of water source—Data on stream flows, temperatures, stratification, and depths must be obtained and evaluated with an economical condenser size and pumping facilities to ensure that the temperature of the body of water receiving the cooling water discharge will not exceed the thermal criteria adopted by the governing regulatory authority.

Fuel Supply

An essential item to be considered in selecting a site for a generating plant is the availability of an adequate supply of competitively priced fuel for the life of the plant.

Effect of Plant and Transmission Line Appearance on Surrounding Area

In site selection it is essential that proper consideration be given to the impact of the plant on the appearance of the surrounding area as well as the impact of the transmission lines that must radiate from the plant.

Amenities for Employees

Power plant personnel have increased desires for comforts and conveniences with each passing decade.

Taxes

While taxes may not directly affect the physical requirements of a site, the taxing policies of the State and local governments have considerable influence on the economics of building a generating plant in one location as compared to another.

Land Requirements

Land requirements for electric power generating plants depend upon (1) the type of plant; (2) the ultimate generating plant capacity; (3) whether urban, suburban, or rural location; (4) on-site needs for fuel storage and handling facilities; (5) the methods of disposal of waste products such as ash from coal-burning plants; and (6) the exclusion areas required for nuclear power generating plants.

Powerhouse needs—The land area required for a powerhouse for a plant containing three 500-mw units is in the range of 3 to 4 acres. For a plant containing three 1,000-mw units the range would be from 6 to 7 acres. These figures include the area needed for a service bay, but do not include space for equipment and facilities which might be located outside and adjacent to the powerhouse. The location and type of generating plant will determine the kinds of such equipment. For a three 500-mw unit coal-fired plant requiring the installation of electrostatic precipitators, the area outside the powerhouse needed for the precipitators, the stacks, walkways, drives, and parking areas immediately adjacent to the powerhouse would be about 2 to 3 acres. For a three 1,000-mw-unit plant this outside area would range from 6 to 7 acres.

Space for SO_2 removal facilities—With the emphasis now being placed on air quality control, it is possible that in the future SO_2 recovery equipment plus a sulphuric acid plant could add considerably to space requirements. Should it be desirable to provide space for SO_2 removal equipment, roughly as much as 2 to 4 acres would be needed in addition.

Land Required for Fuel

The primary fuel to be utilized will influence plant arrangement and space depending on the storage requirements for onsite reserves.

Coal—A typical coal-storage yard to provide 90 days supply at a 3,000-mw coal-fired plant could require some 40 acres and the coal pile would be 40 feet high.

Gas and oil—Gas, as such, demands little storage area except for onsite holdup tanks, although it does need a backup supply of oil that results in one or more oil storage tanks. Oil storage tanks should be located as close to the delivery point as possible, whether delivery is by barge, truck, or railway, and each tank must be provided with its own dike that is sized to retain the full tank capacity.

Nuclear fuel—Space requirements for the storage of new nuclear fuel are small.

Rail terminal for coal-fired plants—If coal is to be delivered primarily by rail, planning the rail facilities is a major item, especially if a large, expensive rail storage yard is to be avoided.

Rail terminal for nuclear plants—The space required for any rail terminal will, of course, vary with the topography of the site.

Barge terminal for coal—A modern 1,000-mw power plant operating at full output will burn between 8,000 and 10,000 tons of coal a day. Therefore, for every 1,000 mw's of plant capacity, harbor, docking, and unloading facilities capable of handling three to seven barges a day must be provided.

Barge terminal for nuclear—Facilities for handling one barge at a time would be sufficient for receiving and shipping the majority of the required material, supplies, and fuel for a nuclear plant.

Barge terminal for oil—Facilities for handling barges of oil would not be very different from those for coal except a pumping station would replace the unloader.

Truck terminal for coal—Usually receipt of coal by truck, except at mine-mouth plants, will supplement rail receipts and in such cases receiving hoppers can usually be combined.

Truck terminal for nuclear—Delivery of fuel by truck does not require an extensive or special handling system.

Coal pipeline terminal—For this type of installation space must be provided for facilities to dewater the slurry at the generating plant.

Ash Disposal Areas

A 3,000-mw plant with a 35-year useful life and assuming an over-all average capacity factor of about 50 percent would require an ash disposal area of 300 to 400 acres with ash piled to an average depth of 25 feet.

Switchyard

The switchyard area requirements for a typical 3,000-mw plant with 500-ky-transmission voltage would be in the range of 10 to 15 acres.

Transmission Access

The transmission lines connecting a typical 3,000-mw generating plant into the existing transmission system at 500 kilovolts would occupy rights-of-way totaling from 100 to 150 acres for each mile over which the lines are built.

Condenser Cooling Requirements

Water in sufficient quantities and with low enough temperature must be available to absorb the maximum plant heat release without raising the temperature of the receiving waters above satisfactory levels.[35]

Air Quality

The major environmental problems caused by electric power generating facilities have to do with air pollution, disturbance of the natural ecology of a region and hence the damage or deterioration of fish and wildlife population, and the deleterious effects on the esthetics of a particular site, area or location.

The document, *Electric Power and the Environment,* states rather succinctly both the problem of air quality and water quality in the following words:

Air Quality

Fossil-fueled electric plants discharge almost 50% of the sulfur oxide pollutants, 25% of the particulate and approximately 25% of the nitrogen oxide emissions emitted in the United States. Although motor vehicles discharge nearly 60% of the total pollutants into the atmosphere, they are not a significant source of sulfur oxides.

Power plants emitted 12 million tons of sulfur oxides into the atmosphere in the U.S. in 1966. Control technology is not yet commercially available but is being developed. It is estimated that by 1980, unless control measures are applied, some 50 million tons of sulfur dioxide would be released annually from all sources and 36 million tons of the total would be discharged from power plants. The potential for pollution is large. A 3,000 MW electrical station burning coal with 3 1/2% sulfur would emit 500,000 tons of sulphur oxide per year, 100,000 tons of oxides of nitrogen, and even with 99% efficient precipitation it will emit about 7,500 tons of fine particulate matter.

Water Quality

The discharge of heated waters poses a major problem of increasing importance in connection with siting of power plants. Each kilowatt hour of electric energy generated by a modern, large fossil fuel plant requires the equivalent of slightly more than a kilowatt hour of heat to be rejected at the condenser. Present generation nuclear plants present thermal pollution problems some 50% greater. The biological damage involved in the discharge of waste heat is obvious when the temperature change is large, but in most cases it is quite subtle. What is often not recognized is that in many of our waterways the waste head may be imposed upon an environment which is already near a critical point for certain segments of aquatic life we seek to protect, for the processes we hope to limit and for certain uses we propose to make of the water resources.

During the period 1967 to 1973, 138 electrical generating units totaling 94,000 MW have been or are scheduled to go into service. The cooling water discharge from these units will make a substantial addition to the waste heat discharged to our nation's streams.[36]

35. The Energy Policy Staff, Office of Science and Technology, *Considerations Affecting Steam Power Plant Selection,* (Washington, D.C., U.S. Government Printing Office, Superintendent of Documents, 1968), p. 7-16.

36. The Energy Policy Staff, Office of Science and Technology, *Electric Power and the Environment,* (Washington, D.C., U.S. Government Printing Office, Superintendent of Documents, 1970), p. 3.

The best coverage on explanations of the problems of fish and wildlife, aesthetic recreation considerations in planning for electric power facilities are covered in the report, *Considerations Affecting Steam Power Plant Selection.*

Preserving and improving the quality of our environment has emerged as a major public goal. It is a goal which has special relevance in the selection of sites for the giant power plants of the future. A proper concern of our environment requires that particular attention be given to fish and wildlife, esthetic and recreational values in the selection and development of sites for power plants and related facilities.

Fish and wildlife resources are an important part of our natural heritage and should receive adequate consideration in planning for power developments. The resources are a part of our total socioeconomic complex and specific commercial, recreational, scientific, biological, and aesthetic values can be related to their presence. Such values may include the worth of an individual animal sold for meat, of hunting and fishing for wildlife, of watching wildlife, or using wildlife to understand man's complicated biological processes, of using wildlife as guinea pigs in man's search for technological improvement, and of knowing that there are things wild and free living somewhere within our immensely complicated ecosystem.

Other features of our natural resources are important yet defy attempts at monetary evaluation. Many areas of great scenic and historical interest have been set aside from the inevitable rush of commercial development and preserved for public use and enjoyment in national parks, monuments, and wilderness areas, as well as other similar classifications. But many other areas remain in private ownership which if lost to development would destroy scenic and historical values which more and more people believe should be preserved for future generations of Americans to enjoy.

Outdoor recreation opportunities, once plentiful, now are needed in increasing amounts especially near large population centers. Planning for future power plant development should recognize the need for expanding outdoor recreation opportunities as well as the necessity for preserving those opportunities now available.

The burden is upon both public and private agencies as well as on each individual to retain and improve upon the desirable features of the environment. Even with farsighted planning in selecting and developing generating plant sites, with existing technology these plants are bound to degrade the quality of the environment to some extent. Hopefully we can minimize the impact through planning and controls and thus provide breathing space for the industry to develop long-range plans. But we should not forget that the water, the land, and the air can only accommodate so much waste. Research must be a continuing activity if future power needs are to be satisfied without major impairment of the quality of our environment.

Fish and wildlife—Power plants can affect fish and wildlife either directly or indirectly. Direct effects include actual displacement in inundation or other habitat changes, and mortality from generating facilities or intake structures. Other direct effects are the effluent discharged into streams, the clearing of rights-of-way for transmission lines, and the formulation of reservoirs of cooling water. Indirect effects can include changes in habitat (physically and chemically) blockage of migration routes, interruption of necessary life cycles of organizations or food supply, physiological changes effecting an organism's resistance to disease and predators, etc. Long-range effects are usually indirect and include many of the aforementioned items. Also to be considered are the possible genetic effects produced by accumulation and/or reconcentration of radionuclides by aquatic organisms, especially shellfish.

Aesthetics—Power plants and transmission facilities are not welcomed, to say the least, in a natural or historic setting. While proper design and architectural treatment can make a difference there is nothing, short perhaps of undergrounding the facilities, which could eliminate the adverse encroachment of a generating station upon an important historic setting.

Transmission lines raise much of the same problems since they require clearing of the natural vegetation on the right-of-way, construction of large steel towers and access and maintenance roads which so change the natural character of the landscape that scenic and other resources can be virtually destroyed. And even undergrounding is not a complete solution, aside from the cost, because clearing of vegetation and access roads would still be required.[37]

How then, in a comprehensive way, may these problems created by the electric power industry be solved. In keeping with the general purpose of this publication, the question may be asked—How have these problems been solved in a positive way by electric utility companies and their consultants?

It has been shown that even though electric power generating facilities of any type cause varying degrees of alteration of the natural environment, there is a continuum of solutions to this infinitely complex problem. The complexity of the problem basically does not lend itself either to easy solutions or to easy explanations of solutions.

At least eight methods of approach to arriving at solutions have been explored. These are, in outline form, as follows:

1. The development of environmental impact statements required by the Federal government.
2. The development of standards and guidelines for electrical energy and power generation and development while insuring environmental enhancement or preservation.
3. Broad scale land analyses to determine optimum location and orientation of various aspects of power generation and transmission facilities and accoutrements.
4. Site selection studies to determine the precise optimum location for power generating facilities.
5. Studies for the site development of areas surrounding power generation facilities in order to make them more acceptable and credible through reduction of the potential environmental destruction inherent in the placement.
6. Cosmetic treatment of power generation facilities to either hide or diminish their apparent environmental impact.
7. Studies in the land analysis or land planning or multi-purpose plant siting.
8. Reservoir utilization and development around the areas immediately adjacent to water bodies used in hydroelectric power generation facilities.

The following section is concerned with more complete explanations and illustrations of examples of some of these approaches to the problems of electric power generation and environmental preservation and enhancement.

Nearly all of the environmental design professions, including landscape architecture, have been involved in the development of environmental impact studies and statements for electrical generation facilities utilized by the electric industry. The procedure and impact of environmental impact statements on the electric power industry, including certain aspects of electrical generation are best described in the following:

> On balance, and with the exception of esthetic factors, concern about the impact of electric power on the quality of the outdoor environment stems more from consideration of the industry's future growth requirements than from its effects to date. Relatedly, this discussion would be incomplete without reference to an environmental control procedure that has recently been instituted which provides the nation with a major line of defense against unwarranted future environmental impact. This procedure stems from a requirement of the National Environmental Policy Act of 1969 that each federal agency, when authorizing any major undertaking which might significantly affect the environment, prepare, release publicly and circulate to interested federal and state level agencies a report assessing the expected environmental impact and evaluating all practical alternative courses of action. Under the procedure established by the President's Council on Environmental Quality to implement this statutory requirement, the agency primarily concerned must first circulate a draft report or comparable information. Later, after detailed comments have been received, a final report is pre-

37. The Energy Policy Staff, Office of Science and Technology, *Considerations Affecting Steam Power Plant Selection,* (Washington, D.C., U.S. Government Printing Office, Superintendent of Documents, 1968), pp. 49-50.

pared. These reports are regularly received and reviewed by the Council on Environmental Quality and the Environmental Protection Agency.

The potential value of this procedure as an environmental control can perhaps best be conveyed by listing the basic areas these environmental reports are required to cover. They are as follows:

1. The environmental impact of the proposed action;
2. Any adverse environmental effects which cannot be avoided should the proposal be implemented;
3. Alternatives to the proposed action;
4. The relationship between local short-term uses of man's environment and the maintenance and enhancement of long-term productivity;
5. Any irreversible commitments of resources which would be involved in the proposed action should it be implemented.[38]

(A more complete delineation of the purposes of environmental impact statements is contained in Appendix A.)

In the preparation of these environmental impact statements many disciplines and professionals are represented in order to gain the necessary input. At these early stages in preparing such statements it is not always possible to have the "state of the art" at an advanced enough stage to develop optimum answers as to the environmental impact of many actions and activities. In time to come, as the procedure is developed and refined, the research and development branches will be able to develop some of the necessary information to provide more accurate information on environmental impact statements.

Assistance with the development of guidelines and standards has been an emerging role of environmental design professionals and of natural scientists in working with the electric utility industry in reducing the deleterious environmental impact of power generating facilities.

A number of excellent environmental protection checklists and guidelines have been developed. One of the most complete has to do with the siting of electrical transmission structures and will be mentioned later upon discussion that particular subject.

The August, 1970 report, "Electric Power and the Environment" sponsored by the Energy Policy Staff, Office of Science and Technology contained an environmental protection checklist and guidelines for site selection. This is included in Appendix B of this publication.

Increasingly, the electric utility industry is realizing at the same time, as environmental design professionals, that there is a greater interactive role possible in the development of such standards or guidelines. Both the Federal Power Commission and the Atomic Energy Commission are working to develop more precise standards and guidelines for the siting and site development of electrical power generating facilities in order to minimize environmental damage, destruction or disruption.

Since there currently exists so little basic essential information concerning large land areas of the nation, it is not always possible to conduct complete broad scale land analyses prior to any contemplated land development. This is especially so in the case of utilization of natural resources, including land, for electrical power generation.

The landscape architectural firm of Eckbo, Dean, Austin & Williams conducted such a land use study for the Pacific Gas and Electric Company under proposed Davenport power plant site south of San Francisco on the Pacific Coast of California. This study was characterized as "A study of 7,000 acres of land near the town of Davenport in Santa Cruz County to develop a general land use program which would be compatible with the county's overall land use plan in the north coast area. The investigation into the potential for and consequences of the element in this area has been conducted mindful of Pacific Gas and Electric Company's indication of a need to locate a power plant in this area."

The final report on this study contains some very significant statements in regard to both the processes in conducting such a

38. Federal Power Commission, *The 1970 National Power Survey: Guidelines for Growth of the Electric Power Industry,* (Washington, D.C., U.S. Government Printing Office, Superintendent of Documents, 1971), p. I-1-27.

study and the responsibility of electric companies for environmental matters.

In marshalling our forces to deal with these diverse issues, our efforts have been coordinated throughout both within the team and with all levels of operation of the Pacific Gas and Electric Company. Our work has been concerned with the process as well as the product so that each step, as described herein, is clearly visible. The resultant recommendations are based in fact and derived by logic, intuition and experience. Their accuracy and validity can now be examined by all those interested.

It is often felt that, since the Pacific Gas and Electric Company is a public utility, it bears a greater than usual responsibility to keep its public clients informed of its own activities. For this reason our study has included the accumulation and organization of data in such a way that it should serve the company well in making decisions relative to the Davenport property and also clearly indicate to other interested parties, such as local public agencies, how the information was obtained and conclusions reached.

Eckbo, Dean Austin & Williams are convinced that this process is both productive and creative and are confident that it can be used as a guide to answer similar questions on other sites. Another advantage of the process we have used, and the form it is now presented in, is that it is flexible. The changing needs and desires of both private persons and large corporate bodies are often in the same direction and these changes are now occurring with increasing frequency. Such changes should not be thwarted by inflexible plans and therefore our process permits this flexibility.[39]

The following chart then shows the landscape causality levels as referred to by the consultants on the Pacific Gas and Electric Davenport land use studies.

The consultants then began to explain more fully their process and the lines of approach to developing information upon which decision making may be based.

Landscape Causality Levels[40]

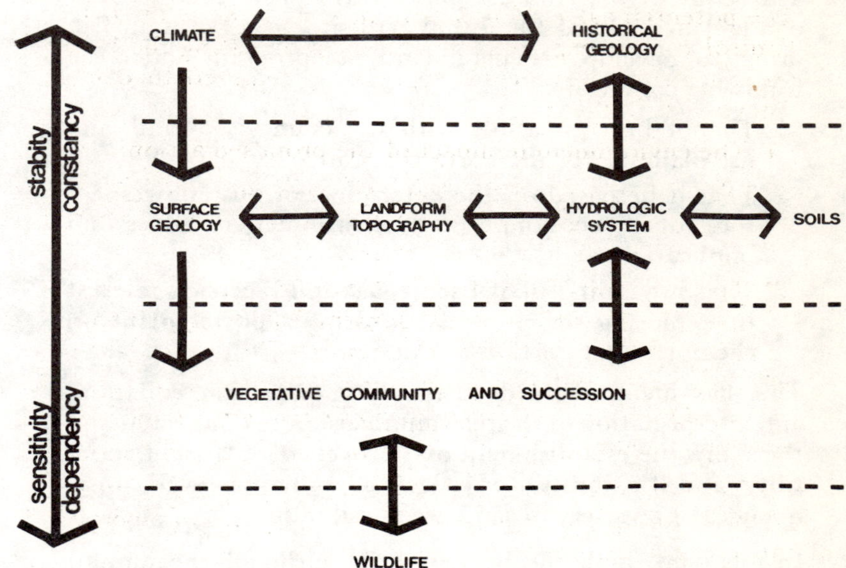

We have indicated a need to make the process timeless, and flexible, quickly responsive at any time and at any level both now and to the extent practicable, in the future. What is clearly needed is an organization of data into meaningful categories, lines of approach which sort data of like kind and of like implications into meaningful groupings and patterns.[41]

39. Garrett Eckbo, Francis H. Dean, Donald B. Austin and Edward A. Williams, *Proposed Davenport Power Plant Site Land Use Study,* (San Francisco, California, Eckbo, Dean, Austin and Williams for Pacific Gas and Electric Company, 1970), p. 3.
40. Ibid., p. 7.
41. Ibid., p. 8.

The scope of the Davenport land use study was characterized by the consultants in the following words.

> The scope of the land use phase of the project has been to conduct a study of the 7,000 acre site near Davenport in order to develop a general land use program under the following conditions:
>
> 1. That it be compatible with the county's overall land use planning for the north coast area.
> 2. That it be based on the determination of, and respect for, social, economic, and environmental values and implications.
> 3. That it be mindful of Pacific Gas and Electric's interest in exploring the potential locating a power plant on the northwest portion of the site.
>
> This has involved the determination, collection, collation, and interpretation of the relevant decision making data, and therefore the establishment of a process of sufficient versatility and flexibility to judge the potential and consequences of a variety of land use possibilities.
>
> In this light, it should be stated that although the amount of data gathered and contained herein is very thorough when measured against the ultimate land use recommendations, the consultants have been working under several basic assumptions:
>
> 1. That Pacific Gas and Electric's actions at Davenport would be subject to considerable scrutiny, and therefore should have thorough documentation.
> 2. That Pacific Gas and Electric might reject the consultants recommendations and thus should have the data with which to support another course of action.
> 3. That if the process is to be valid and viable, the basic data must be available and adaptable in the face of changing time; technologies, and values, as will be discussed below under "process."
> 4. That no preliminary assumptions or intuitions regarding land use should alter the process, which is equipped to handle the range from heavy development to open space.[42]

Certainly these processes, approaches, assumptions and suppositions are an adequate basis for all environmental studies for electrical generating facilities. This statement on scope could be utilized as a guide by either other utility industry representatives or by other environmental design professionals in undertaking similar studies. It is exemplary and unique thus far in the environmental design utility industry interaction. Probably the most unique aspect however of this overall study has to do with the basic process utilized in the ultimate decision making.

The consultants in their report introduced their process in the following way.

> The establishment and maintenance of a system to guide long-range management and development of large land areas has become imperative for all public and private authorities.
>
> The complexities and difficulties of relating day-to-day decisions concerning land use and resource allocation to long-term objectives have been brought into sharp focus by the rapidly increasing and often conflicting growth demands being placed upon limited available land, human, and capital resources.
>
> With the acceleration of resource utilization and the subsequent acceleration of modification to land and social structures a new awareness of the inter-relationships of all actions has evolved.
>
> It is no longer desirable or efficient to consider each demand on its own short-term merits. The overall implications which may reverberate through the entire land and social structures also affect decisions.
>
> As it is not possible to preconceive all the demands on resources which may materialize as a result of shifting or changing social values or capital and technical capabilities, traditional and accepted methods of decision making which rely on piecemeal projections and static plans are no longer adequate. Both projections and plans may be obsolete before implementation and neither can be considered as final guidelines.

42. Ibid., p. 3.

In the conduct of this study, Eckbo, Dean, Austin & Williams has recognized these problems and formulated concepts appropriate to their solution. Of fundamental significance is the development of rational process to assist decision making which can operate in perpetuity where each action taken becomes an element of consideration for the next decision. This 'feedback' characteristic allows refinement and maintains continuity. Thus, at any given time the decision maker has access to information, relationships, and potential consequences of future actions, limited only by accuracy and completeness, which is constantly improved without redoing everything.

The general process consists of four basic parts, presented graphically in the Introduction to this report, and is discussed here as it applies to Land Use in:

(a) Information gathering,
(b) Interpretation,
(c) Planning and design development, and
(d) Evaluation

As the process is cyclic in nature, each part of the process affects and conditions the others to produce a dynamic system; because of the accumulative nature of the process, it is not necessary to have all the information before beginning. Decisions will in any case be based on available information which increases with time and action and provides an increasingly accurate base for decision making. Implemented actions become information to generate added depth and accuracy in the entire system.

The essence of the process is that it concerns itself with constants, those criteria which affect land 'usability' rather than with unpredictable variables such as land 'use.' Resource characteristics become the primary guidelines for decisions related to resource utilization.

Thus while the more temporal aspects are critical to the fulfillment of this contract, the emphasis here will be on the timeless constants of the land.[43]

The basic planning process utilized by the consultants on this study is shown on the following chart.

Planning Process[44]

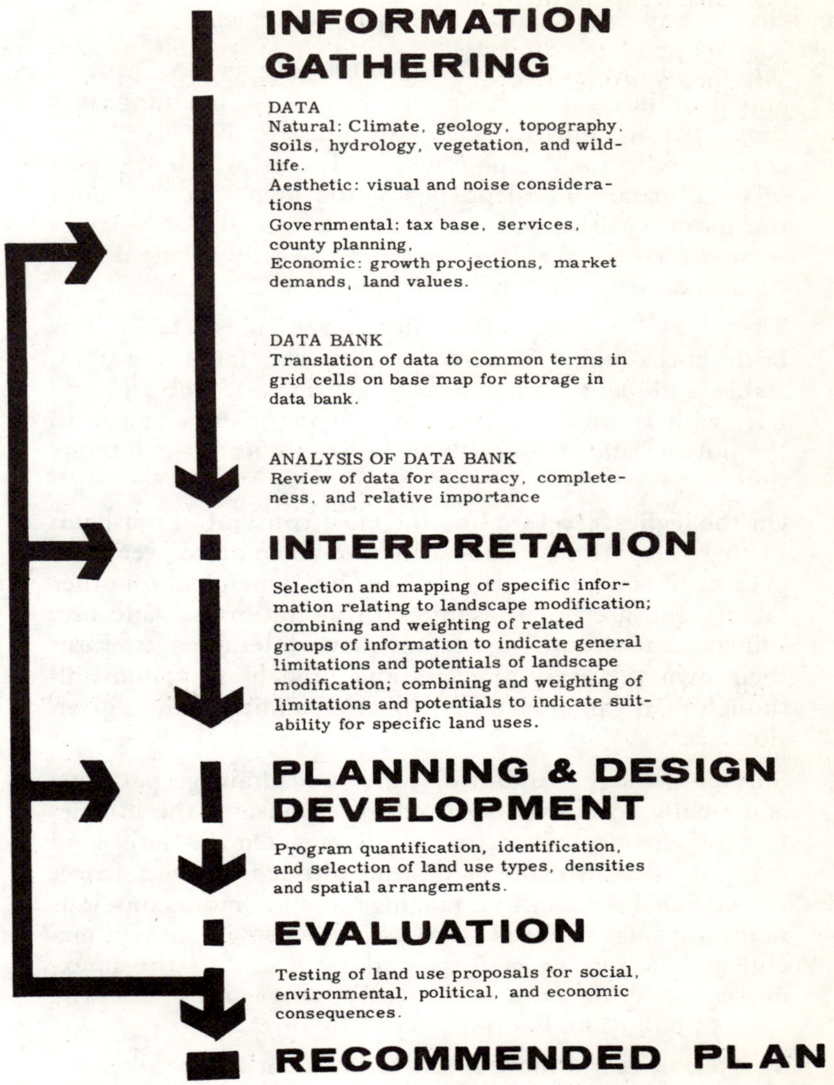

INFORMATION GATHERING

DATA
Natural: Climate, geology, topography, soils, hydrology, vegetation, and wildlife.
Aesthetic: visual and noise considerations
Governmental: tax base, services, county planning.
Economic: growth projections, market demands, land values.

DATA BANK
Translation of data to common terms in grid cells on base map for storage in data bank.

ANALYSIS OF DATA BANK
Review of data for accuracy, completeness, and relative importance

INTERPRETATION

Selection and mapping of specific information relating to landscape modification; combining and weighting of related groups of information to indicate general limitations and potentials of landscape modification; combining and weighting of limitations and potentials to indicate suitability for specific land uses.

PLANNING & DESIGN DEVELOPMENT

Program quantification, identification, and selection of land use types, densities and spatial arrangements.

EVALUATION

Testing of land use proposals for social, environmental, political, and economic consequences.

RECOMMENDED PLAN

43. Ibid., pp. 4-6.
44. Ibid., p. 5.

The Davenport study is conceived by the consultants more as a process than as an entity. In their description of their apprehension in this way they make the following statement.

> Indeed, just as we have discussed the four levels of process and their interaction in terms of land use planning, it is instructive to think of the landscape as four levels of process as well. 'Landscape Causality Levels' which also possess a dynamic interdependency and help to reveal both the more sensitive and the more constant of the variables upon which we are depending most heavily, is the phrase used to describe the landscape processes.
>
> The chart 'Landscape Causality Levels' illustrates a flow both horizontal and vertical, with the more constant, stable and determining factors generally to the top and left, with the more dependent and sensitive generally to the bottom and right, with arrows to indicate the interaction.
>
> On the highest level we find the most constant of our data. In that the patterns established have taken place over great periods of time; they are relatively less dependent on other factors and are not so subject to alteration by land use. Climate and historical geology have interacted to form their own relatively timeless and unyielding system, although man can alter microclimatic conditions on a given site.
>
> Surface geology, landform, water and drainage patterns, and finally soils have occurred in response to the interaction of climate and historical geology. On this level land use policies can make significant alternations and wreak considerable devastation, ranging from the more conscious modifications required through the use of heavy machinery on surface geology and landform to the unexpected havoc that can occur within water and soils in the form of pollution, flooding, depletion or erosion.
>
> Vegetation is still more sensitive in that it is dependent on the level above and is thus subject to ready alteration in both a direct and indirect way, and is additionally beginning to show a sensitivity to air pollution in many cases. The system is reactive moreover in that disturbance of vegetation can revert back to unfavorable alterations to the water and soils systems.
>
> Most sensitive and dependent of all is wildlife, whose dependencies run back through all levels of landscape causality, in addition to being in many cases subject to disturbance and alteration by noise and even the mere presence of man's activities. This is particularly true in the case of carnivores, at the very head of the food chain. Wildlife change also has its effect in that many vegetation types depend upon it for propagation.
>
> We have looked briefly at the landscape as a source of determinants on a lasting scale, but we must draw also from probable and potential land uses with their attendant demands and potential for land modification. Finally, a decision on approach requires a close look at man-made features and determinants to see what opportunities and limitations they present in a physical or aesthetic way.[45]

Six general levels of approach were derived for the Davenport site:

1. Environmental impact
2. Land productivity
3. Generalized engineering suitability
4. Man-made features
5. Climatic considerations
6. Aesthetic considerations[46]

The E.D.A.W. staff then go on to explain the process chart and the graphic summary which it contains.

> If the six lines of approach of the process are to be flexible and responsive, they must also be graphic—and in such a way that the piecing together of data and interpretation forms a clear and visible pattern. The boxes on the accompanying chart entitled 'Land Usability: Flow Diagram of Process' represent blocks of information and are maps which appear later in the text, if they are numbered.

45. Ibid., pp. 6, 8.
46. Ibid., pp. 8-9.

LAND USABILITY[47]

FLOW DIAGRAM OF PROCESS

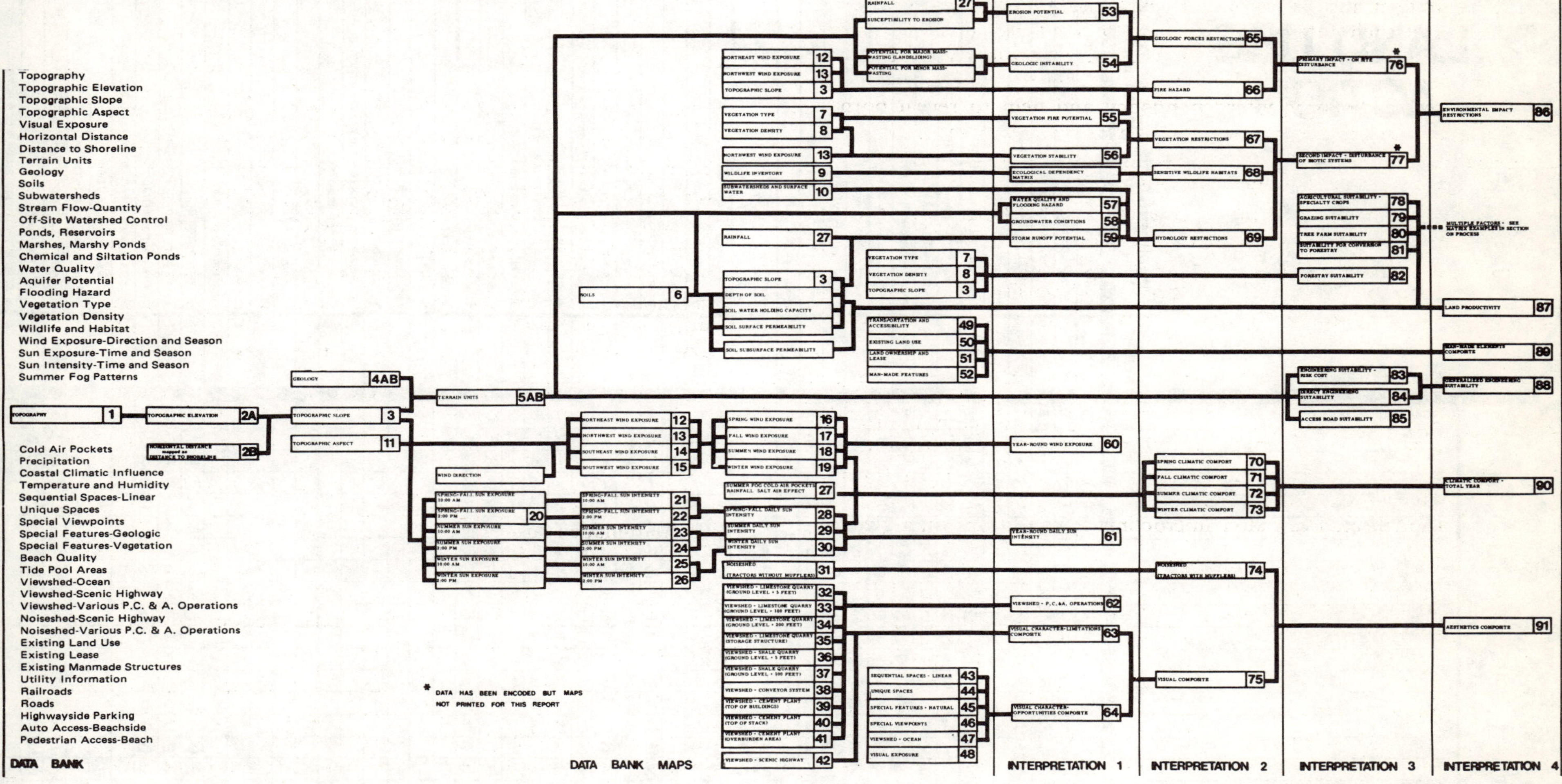

Ibid., p. 10.

LAND USE OR LAND USE ACTION[48]

ENVIRONMENTAL CONDITIONS, PROPERTIES, ACTIONS, IMPLICATIONS

ENGINEERING

DATA		EXCAVABILITY-MASS GRADING	TRENCHABILITY	TRENCH WALL STABILITY	EROSION POTENTIAL- CUT SLOPE	STABILITY-CUT SLOPE	COMPROMISE ∠-CUT SLOPE	EROSION POTENTIAL-FILL SLOPE	STABILITY-FILL SLOPE	COMPROMISE ∠-FILL SLOPES	SUITABILITY FOR FILL	TUNNELING CONDITIONS	BEARING CAPACITY-LIGHT LOADS	BEARING CAPACITY-HEAVY LOADS	SHRINK-SWELL POTENTIAL	CORROSIVITY
RELATIVE WEIGHT																
TERRAIN UNITS																
EMBANKMENT FILL	F	8	8	5	5	5	67 to 50	4	6	67 to 80	8	4	6	3	1	UNRATED--NO APPARENT PROBLEMS ON SITE
QUARRY SPOIL	SPOIL	4	4 to 1	2	8 to 1	1	67 to 50	3 to 1	7	100 to 67	5	1	4	2	1	
TAILINGS	TAILINGS	2	10	1	4	1	<20	5	2	50 to 20	2	1	2	1	5	
BEACH SAND	BS	8	10	2	10	1	67 to 50	10	6	67 to 50	4	1	5	3	1	
WIND-BLOWN SAND	WBS	9	10	2	9	2	67 to 50	10	6	67 to 50	5	1	7	3	1	
CANYON BOTTOM SOIL	CBS	8	8	5	8	5	<50	7	6	50	5 to 6	2	9	5	2	
SLOPE WASH	SW	8	8	7	4	4	50	3	4	50 to 40	7	3	8	5	3	
EARTH FLOWS	EF	8	9	4	5	1	<38	3	4	50 to 40	6	1	6	3	3	
LANDSLIDE: RELATIVELY FLAT BENCH	LSb	5	5	4 to 9	3	3	33 to 67	3	7	50	9	3	9	4	2	
LANDSLIDE: CLOSED DEPRESSION	LSd	7	7	3 to 8	4	2	38 to 67	3	5	50	6	2	6	2	4	
LANDSLIDE: REMAINDER OF SLIDE (include face and toe)	LS	5	4	2 to 6	3	2	33 to 67	2	7	50	9	3	8	2	2	
TERRACE DEPOSITS	TD	10	8	8	7	8	200 to 50	7	7	50	8	5	9	7	2	
MUDSTONE-FLAT BENCH (less than 20%)	MSb	4	2	10	1	8	VERT. to 50	2	8	67 to 50	9	10	10	9	1	
MUDSTONE-PREDOMINANTLY STEEP CANYON WALLS (more than 20%)	MSc	5	4	9	2	7	200 to 67	2	8	67 to 50	9	10	9	7	1	
MUDSTONE-SAME AS ABOVE BUT HEAVILY FORESTED	MScf	6	6	6 to 8	3	6	100 to 50	3	7	50	8	9	8	5	1	
SANDSTONE-CANYON WALL EXPOSURE	SSc	8	7	7	8	6	200 to 67	9	7	50	8	5	8	6	1	
SANDSTONE-CANYON BOTTOM AREAS	SScb	8	9	3	8	5	100 to 50	9	8	50	7	3	7	6	2	
SANDSTONE-RIDGE CAPPING EXPOSURE	SSr	8	8	8	8	6	200 to 67	9	7	50	8	5	10	7	1	
DECOMPOSED GRANITE	DG	9	8	9	6	7	200 to 67	9	7	67 to 60	8	8	10	8	1	
MARBLE	M	2	1	9	1	9	VERT. to 300	2	10	67	2	7	10	10 to 1	1	

LAND PRODUCTIVITY

DATA		SEPTIC TANK SUITABILITY	DUSTINESS	TRAFFICABILITY-MUDDINESS	HARDNESS WHEN DRY	DEPTH TO WATER TABLE	AQUIFER RECHARGE POTENTIAL	GROUND WATER AVAILABILITY (major source)	GROUND WATER AVAILABILITY (minor source)	SPRING LOCATION POTENTIAL	SOIL PERMEABILITY-SUBSURFACE	SOIL PERMEABILITY-SURFACE	SOIL WATER HOLDING CAPACITY	SOIL DEPTH
RELATIVE WEIGHT														
TERRAIN UNITS														
EMBANKMENT FILL	F	8	6	6	4	9	SUMMARIZED UNDER GROUNDWATER AVAILABILITY	1	1	1	8	5	6	10
QUARRY SPOIL	SPOIL	9	6 to 3	10 to 3	7	10 to 4		1	3	3	9	10 to 8	5 to 1	8 to 10
TAILINGS	TAILINGS	1	10	10	9	1		1	1	1	1	1	10	8 to 10
BEACH SAND	BS	8	1	7	1	1		1	1	1	10	9	3	1 to 8
WIND-BLOWN SAND	WBS	10	3	6	1	8		1	2	2	10	9	4	10
CANYON BOTTOM SOIL	CBS	6	8	9	3	2		7	10	7	8	7	10	10
SLOPE WASH	SW	3	3	7	8	4		1	6	6	3	2	8	10
EARTH FLOWS	EF	4	3	8	7	3		2	8	8	4	2	8	10
LANDSLIDE: RELATIVELY FLAT BENCH	LSb	6	3	5	8	5		3	5	4	9	6	3	8
LANDSLIDE: CLOSED DEPRESSION	LSd	6	7	10	7	4		3	5	5	8	5	7	10
LANDSLIDE: REMAINDER OF SLIDE (include face and toe)	LS	6	3	6	8	5		5	10	8	9	6	3	8
TERRACE DEPOSITS	TD	9	6	8	2	7		3	8	1 to 10	10 to 2	7	8	3
MUDSTONE-FLAT BENCH (less than 20%)	MSb	1	2	1	10	10		2	3	2	3	1	1	1
MUDSTONE-PREDOMINANTLY STEEP CANYON WALLS (more than 20%)	MSc	2	3	3	10	8		2	4	5	3	3	3	3
MUDSTONE-SAME AS ABOVE BUT HEAVILY FORESTED	MScf	4	2	6	9	6		2	6	5	3	5	8	8
SANDSTONE-CANYON WALL EXPOSURE	SSc	7	3	3	2	8		4	8	8	8	8	8	8
SANDSTONE-CANYON BOTTOM AREAS	SScb	4	5	7	4	2		9	10	8	8	7	10	10
SANDSTONE-RIDGE CAPPING EXPOSURE	SSr	9	3	3	2	9		1	2	2	8	8	7	3
DECOMPOSED GRANITE	DG	5	2	2	2	8		2	4	0	3	5	7	3
MARBLE	M	8	1	1 to 7	10	8 to 1		10	10	10	10	3	1	3 to 1

ENVIRONMENTAL IMPACT

DATA		ACTION: CUTTING-UNLOADING	FILLING-LOADING	SPRAYING-CHEMICALS, OILS	FERTILIZING	SURFACE WATER ADDITION-IRRIGATION, SEPTIC, POND	SURFACE WATER REMOVAL	PLANTING	GRAZING	CLEARING	BURNING	REACTION: RUNOFF POTENTIAL	EROSION POTENTIAL	MINOR MASS WASTING POTENTIAL	LANDSLIDE POTENTIAL	VEGETATION STABILITY-ELIMINATION POTENTIAL	STREAM FLOW REDUCTION	STREAM POLLUTION POTENTIAL	GROUND WATER POLLUTION POTENTIAL	FLOOD POTENTIAL	WILDLIFE SENSITIVITY-DESTRUCTION POTENTIAL
RELATIVE WEIGHT																					
TERRAIN UNITS																					
EMBANKMENT FILL	F												4	2	2				4	0	
QUARRY SPOIL	SPOIL												0 to 10	0 to 10	2 to 10				8	0	
TAILINGS	TAILINGS												4	4	4 to 10				3	0	
BEACH SAND	BS												10	0	0				6	10	
WIND-BLOWN SAND	WBS												8	6	6				4	2 to 8	
CANYON BOTTOM SOIL	CBS												8	4	0				9	8	
SLOPE WASH	SW												4	6	4				4	4	
EARTH FLOWS	EF												4	6	10				8	2	
LANDSLIDE: RELATIVELY FLAT BENCH	LSb												2	2	4 to 10				8	2	
LANDSLIDE: CLOSED DEPRESSION	LSd												0	4	4 to 10				8	8	
LANDSLIDE: REMAINDER OF SLIDE (include face and toe)	LS												2	8	4 to 10				8	2	
TERRACE DEPOSITS	TD												6	4	2				8	2	
MUDSTONE-FLAT BENCH (less than 20%)	MSb												2	0	2				3	2	
MUDSTONE-PREDOMINANTLY STEEP CANYON WALLS (more than 20%)	MSc												2	4	2				4	2	
MUDSTONE-SAME AS ABOVE BUT HEAVILY FORESTED	MScf												2	2	4				5	2	
SANDSTONE-CANYON WALL EXPOSURE	SSc												6	6	4				9	2	
SANDSTONE-CANYON BOTTOM AREAS	SScb												6	4	0				10	8	
SANDSTONE-RIDGE CAPPING EXPOSURE	SSr												4	6	4				4	0	
DECOMPOSED GRANITE	DG												4	6	4				4	2 to 8	
MARBLE	M												0	2	2				10	2 to 8	

NUMBERS IN MATRIX BOXES REFER TO PRESENCE OF CONDITION IN TERMS OF ITS RELATED DATA. THIS SCALE IS BASED ON 0-10: 10 INDICATES A VERY HIGH DEGREE OF PRESENCE; 0 INDICATES THE CONDITION IS NOT PRESENT. NO POSITIVE OR NEGATIVE IS IMPLIED. THIS IS COVERED BY MAKING "RELATIVE WEIGHTING" + OR -

48. Ibid., p. 12.

MATRIX I

MAP TITLE & NUMBER	Maps Used	3	13	12	55
FIRE HAZARD 66[49]	Information Used	TOPOGRAPHIC SLOPE	NW WIND EXPOSURE	NE WIND EXPOSURE	VEGETATION FIRE POTENTIAL
DATA VARIABLE	Rel. Wt.	2	1	1	1
Chaparral					10
Knobcone Pine					9
Coastal Scrub					8
Douglas-Fir					7
Live Oak/Douglas-Fir					6
Tanoak - Redwood					5
Redwood, Dense					4
Redwood. Nearly Pure					3
Grass					2
Wind			10	10	
100% + Slope		10			
80-100 "		9			
65-80 "		8			
50-65 "		7			
35-50 "		6			
20-35 "		5			
15-20 "		4			
10-15 "		3			
5-10 "		2			
0-5 "		1			

MATRIX II

49. Ibid., p. 11.

The first part is essentially a combination and sorting of raw data up to a point of departure, at which point the data is interpreted, combined, and recombined along the six lines of approach discussed earlier. The first maps are specific; the combination maps become increasingly general. Map graphics vary and 'positive' and 'negative' considerations may alternate depending on graphic intent and the point to be made.

The chart is both an outgrowth from and expression of the process.[50]

The consultants then explain more fully the use of the matrix, which they describe in the following way.

The primary tool in the process, not only for storage and retrieval of information and interpretation, but in the production of maps and the flow chart as well, has been the matrix. Its several forms have proved useful as visual aids, a fast cross-reference for weights and values of information, and as a first step for making use of the computer . . .

The first matrix example shown (Matrix I) is a portion of what would become a 'total matrix' if completed. It is infinite in its expandability because it can be used to describe, define or locate virtually anything pertaining to the Davenport site, as long as the information can be traced ultimately back to the original data bank, or battery of information collected. As such, it is never truly 'complete.'

Its use is as follows: All information collected appears under 'data' on the vertical scale. As will be recalled from the process chart, the number of data categories is considerable, and the divisions within are in some cases rather large. Opposed to that on the horizontal scale is a list of 'environmental conditions, properties, actions, and implications' of the data relative to virtually any land use action or modification of the land one might have in mind. The numbers in the squares refer to the degree of presence of that property or condition within the particular data unit cross-referenced.[51]

50. Ibid., p. 9.
51. Ibid., p. 13.

Map Title: AGRICULTURAL SUITABILITY SPECIALTY CROPS

NUMERICAL IMPORTANCE OF DATA VARIABLES SUBCATEGORIES

Map No. **78**

Maps Used	DATA VARIABLES	Relative Wt. of Data Variables	10	9	8	7	6	5	4	3	2	1	0	OFF	Notes
2A	COASTAL INFLUENCE (TOPOGRAPHIC ELEVATION)	10		0-400'			400' +								(1) Soils are used below 400' elevation on 0-20% slopes; terrain units used elsewhere.
3	SLOPE	8	0-5		5-10			10-15				15-20		20 +	
5A, B 6	DEPTH OF SOIL (1)	5	F, WBS, CBS, SW, EF, LSD, SSCB	Spoil, Tailings	EF, LS, MSCF, SSC	WS, WL			BS	TD, MSC, SSR, DG, LL, PY, SO, ES	M	MSB, WLS			(2) "Intensive agriculture" receives an automatic "10" on graphics scale. Remainder of site recombined into 6 graphic levels.
5A, B 6	SOIL SURFACE PERMEABILITY (1)	3	BS, WBS	Spoil, WS, SO, ES	SSC, SSR	CBS, TD, SSCB	LSB, LS, WLS, WL, LL, PY	F, LSD, MSCF, DG			SW, EF	Tailings, MSB			
5A, B 6	SOIL SUBSURFACE PERMEABILITY (1)	2	BS, WBS, M	Spoil, LSB, LS, SO	F, CBS, LSD, SSC, SSCB, SSR		TD		EF			Tailings, PY, LL, WLS, WS, WL			
50	EXISTING LAND USE (2)													Res'd Inst. Comm. Ind. Quarry	

Map Title: GRAZING SUITABILITY

NUMERICAL IMPORTANCE OF DATA VARIABLES SUBCATEGORIES

Map No. **79**

Maps Used	DATA VARIABLES	Relative Wt. of Data Variables	10	9	8	7	6	5	4	3	2	1	0	OFF	Notes
3	SLOPE	10	0-5 0-10	10-15	15-20				20-35				35-50	50%+	(1) Soils are used below 400' elevation on 0-20% slope: terrain units used elsewhere.
5 A, B 6	DEPTH OF SOIL (1)	6	F, WBS, CBS, SW, EF, LSD, SSCB	Spoil, Tailings	LSB, LS				BS	TD, MSC SSR, DG, LL, PY, SO, ES	M	MSB, WLS		BS WBS	
5 A, B 6	SOIL WATER-HOLDING CAPACITY (1)	4	Tailings, CBS, SSCB	SO	SW, EF, TD, MSCF, SSC	LSD, SSR, DG	F, WLS, WS, WL, LL, PY		WBS	SPOIL, BS, LSB, LS, MSC, ES		MSB, M			(2) Although not currently economically feasible to clear land, no penalty is accorded other vegetation on the assumption that subsidies could permit.
5 A, B 6	SOIL SURFACE PERMEABILITY (1)	4	BS, WBS	Spoil, WS, SO, ES	SSC, SSR	CBS, TD, SSCB	LSB, LS, WLS, WL LL, PY	F, LSD, MSCF, DG		MSC, M	SW, EF	Tailings MSB			
5 A, B 6	SOIL SUBSURFACE PERMEABILITY (1)	4	BS, WBS, M	Spoil LSB, LS, SO	F, CBS, LSD, SSC, SSCB, SSR		TD		EF	SW, MSB, MSC, MSCF, DG, ES		Trailings WLS, WS, WL, LL, PY			
50	EXISTING LAND-USE (2)													Resid. Inst. Comm. Ind. Quarry	

Map Title: TREE FARM SUITABILITY

NUMERICAL IMPORTANCE OF DATA VARIABLES SUBCATEGORIES

Map No. **80**

Maps Used	DATA VARIABLES	Relative Wt. of Data Variables	10	9	8	7	6	5	4	3	2	1	0	OFF	Notes
7	VEGETATION TYPE (Existing)	10	Grass, Bare Ground, Cultiv.						Coastal Scrub, Chaparal				Rest of Veg. Types	Ponds	
3	SLOPE	9	0-5 5-10	10-15	15-20			20-35					35-50		(1) Soils are used below 400' elevation on 0-20% slope: terrain units are used elsewhere.
5A, B 6	SOIL SUBSURFACE PERMEABILITY	7	BS, WBS, M	Spoil, LSB, LS, SO	F, CBS, LSD, SSC, SSCB, SSR		TD		EF	SW, MSB MSC, MSCF, DG, ES	Tailings WLS, WS, WL, LL, PY				
27	RAINFALL	5				35" +	30-35	under 30"							
5A, B 6	DEPTH OF SOIL (1)	4	F, WBS, CBS, SN, EF, LSD, SSCB	Spoil Tailings	LSB, LS, MSCF, SSC	WS, WL			BS	TD, MSC, SSR, DG, LL, PY, SO, ES	M	MSB, WLS			
50	EXISTING LAND USE													Resid. Inst. Comm. Ind. Quarry	
2A	COASTAL INFLUENCE (TOPOGRAPHIC ELEVATION)	'												0-150'	
															7 levels of graphics as selected

Map Title: SUITABILITY FOR CONVERSION TO FORESTRY

NUMERICAL IMPORTANCE OF DATA VARIABLES SUBCATEGORIES

Map No. **81**

Maps Used	DATA VARIABLES	Relative Wt. of Data Variables	10	9	8	7	6	5	4	3	2	1	0	OFF	Notes
3	SLOPE	4	0-5 5-10	10-5	15-20				20-35			35-50		50% +	
5A, B 6	DEPTH OF SOIL (1)	4	F, WBS, CBS, SW, EF, LSD, SSCB	Spoil, BS	LSB, LS MSCF SSC	WS, WL			BS	TD, MSC, SSR, DG, LL, PY SO, ES	M	MSB, WLS			(1) Soils are used below 400' elevation on 0-20% slope; terrain units are used elsewhere
5A, B 6	SOIL WATER HOLDING CAPACITY (1)	4	Tailings, CBS, SSCB	SO	SN, EF, TD, SSC, MSCF	LSD, SSR, DG	F, WLS WS, WL, LL, PY		WBS	Spoil, BS, LSB, LS, MSC, ES		MSB, M			
5A, B 6	SOIL SUBSURFACE PERMEABILITY (1)	4	BS, WBS M	Spoil, LS, SO, LSB	F, CBS, LSD, SSR, SSC, SSCB		TD		EF	SW, MSB MSC, DC MSCF, ES		Tailings, WLS, WS WL, LL, PY			
5A, B 6	SOIL SURFACE PERMEABILITY (1)	2	BS, WBS	Spoil, WS, SO, ES	SSC, SSR	CBS, TD, SSCB	LSB, LS, WLS, WL LL, PY	F, LSD, MSCF, DG		MSC, M	SW, EF	Tailings, MSB			
27	RAINFALL	2				35" +	30-35	under 30							
13	NW WIND EXPOSURE	2	not exposed										exposed		
5A, B	SUSCEPTIBILITY TO EROSION	-3	Spoil BS		WBS CBS		TD, SSC SSCB		F, SSR Trailings SW, EF, DG		LSB LS MSB MSC MSGF		LSD M		7 levels of graphics as selected
2B S	DISTANCE TO SHORELINE													0-3900 feet	
7	VEGETATION TYPE	1	Grass, Bare Ground					Coastal Scrub, Chaparal					Knobcone Pine, Hard-woods	All "R" or "D" Types	

52. Ibid., p. 14.

MATRIX III[52]

The use of the computer as a means to an end in the process is described in the following quote:

> Although the matrix would have had storage and retrieval value under any circumstances, it was particularly useful in that such a numbered system was ideal for the use of the computer. It was felt at an early stage that time was too short to put together a significant process package by slow, painstaking, and repeated overlay of maps by hand and eye methods.
>
> The computer could weigh more factors faster, and more accurately in that it could maintain a hold on multiple values or ratings, and had the advantage of being able to define certain situations which would have been impossible by hand, such as sun intensity, wind exposure and view and noisesheds. Another advantage lay in its counting of 'grid cells' for each value in a range, which was readily transferable into acreage figures. One grid cell on the following maps equals about .9 acres.
>
> The grid cell system also had its liabilities, however, in that precise lines located on the ground in plan become distorted by the 'toothy' edge, and somewhat inaccurate, although not to a worrisome degree at a planning scale.
>
> The computer, although a useful tool, is by no means a designer or planner; it does only what it is told, and cannot improve upon the information given it.[53]

This process is probably one of the outstanding used in any such environmental studies. Such methodology has been widely utilized in recent years in computer mapping for land use studies leading to decision making. Undoubtedly, in time to come, further sophistication will be developed in other situations and by other consultants. But these existing methods still may yet be further sophisticated as information is interchanged and methodologies are compared and utilized in other new situations. The following illustration is an overview of the data banks (1-52) prepared by the consultants for this study.

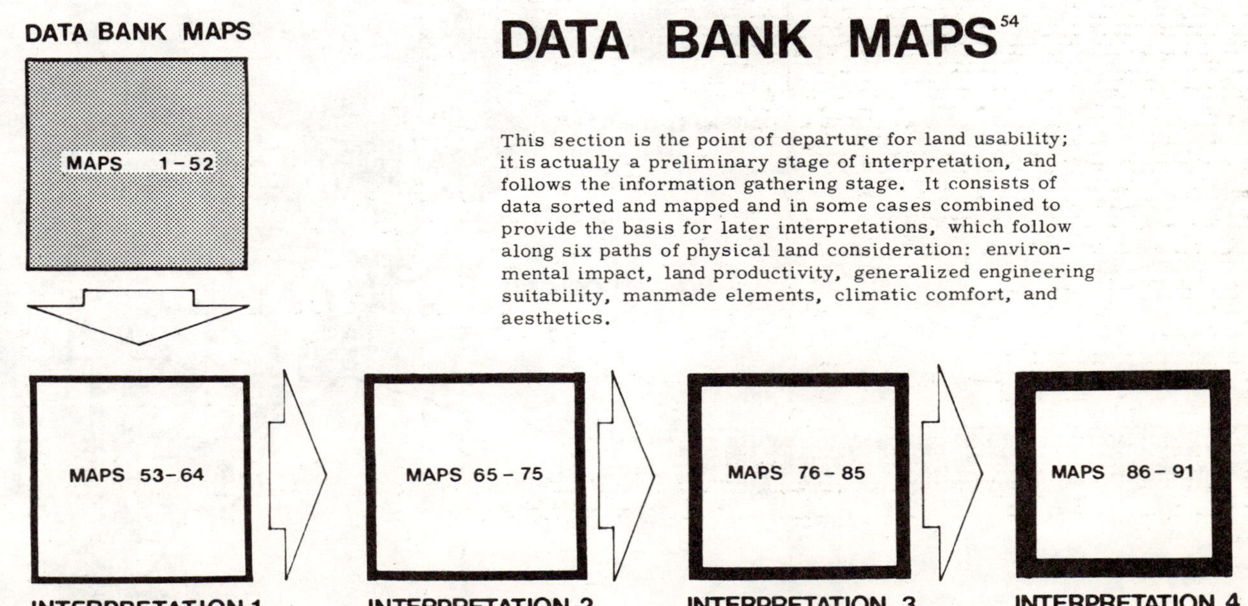

53. Ibid., pp. 13-14.
54. Ibid., p. 35.

GEOLOGY[55]

REGIONAL GEOLOGY AND GEOLOGIC STRUCTURE:

The Davenport planning study area is on the southwest flank of the Santa Cruz Mountains, the westernmost range within the Coast Range province, a complex structural unit comprising many separate ranges, coalescing mountain masses, and intervening valleys. The region is underlain by a complex assemblage of rocks ranging in age from a pre-Cretaceous "basement complex" of igneous and metamorphic origin which has only slight exposure on the subject property, to a sequence of tertiary sedimentary rocks which increases in thickness toward the coastline; terrace, landslide, and soil deposits of Pleistocene and Recent age cap some of the basement rock.

The sedimentary rocks predominate at the surface, and form a relatively undeformed homoclinal structure dipping gently seaward. The Santa Cruz mudstone, youngest of the series, predominates.

The younger rock includes slopewash, residual soils, narrow canyon-bottom strips of relatively thick alluvium, coastal dune and beach sand deposits and two landslide types -- earth flows and blockslumps, the latter occurring mainly in Santa Cruz mudstone on steep canyon wall exposures.

Source:

Appendix D

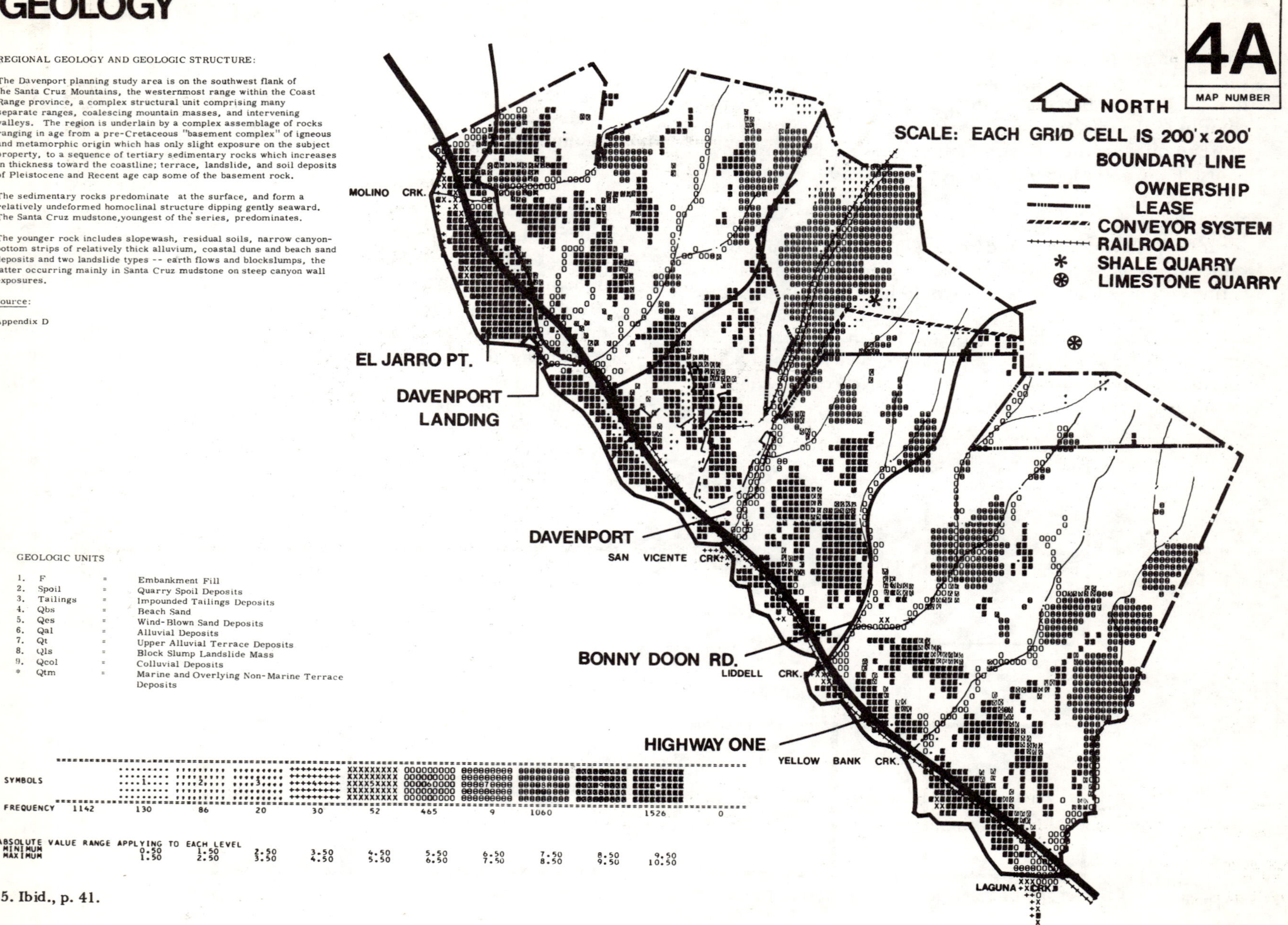

GEOLOGIC UNITS

1. F = Embankment Fill
2. Spoil = Quarry Spoil Deposits
3. Tailings = Impounded Tailings Deposits
4. Qbs = Beach Sand
5. Qes = Wind-Blown Sand Deposits
6. Qal = Alluvial Deposits
7. Qt = Upper Alluvial Terrace Deposits
8. Qls = Block Slump Landslide Mass
9. Qcol = Colluvial Deposits
* Qtm = Marine and Overlying Non-Marine Terrace Deposits

SYMBOLS		1	2	3	4	5	6	7	8	9	*	
FREQUENCY	1142	130	86	20	30	52	465	9	1060		1526	0

ABSOLUTE VALUE RANGE APPLYING TO EACH LEVEL

MINIMUM	0.50	1.50	2.50	3.50	4.50	5.50	6.50	7.50	8.50	9.50
MAXIMUM	1.50	2.50	3.50	4.50	5.50	6.50	7.50	8.50	9.50	10.50

55. Ibid., p. 41.

VEGETATION TYPE[56]

A rich diversity of vegetation exists on the Davenport site. Cwing to varied microclimatic effects and sudden topographic change, there is a highly varied juxtapositional pattern, all of which contributes to ecological health and variety, wildlife habitat, and scenic value. The site is part of the humid transition zone, a coastal strip influenced by coastal climatic patterns, primarily summer fog. This fog acts as a break on evapotranspiration during long California summer drought periods; it is this factor which allows the presence of redwoods.

1. Grassland - Not the true native grassland perennials, but a mixture of European Annuals resulting from early overgrazing and plant introduction.
2. Chaparral - Sprouting, woody plants found mainly on poor shallow soils; comes in on fire, depends on fire to continue; manzanita and chamise.
3. North Coastal Scrub: The grey textured dense shrubbery of the coast and inland; dominated early by baccharis, followed by coffeeberry; highly invasive into grasslands.
4. Knobcone Pine: Shortlived closed-cone pine coming in as pure, even-aged stands on shallow, poor soils following fire.
5. Live Oak - The hardwood forest, on extension of the understory of the conifer forest and lends character to the rural landscape; associated with madrone and Douglas fir.
6. Tanoak - Usually an associate with redwood, but in at least one case here is the dominant species; a sprouter which eventually is shaded out by redwood; also with live oak, madrone, canyon live oak, Douglas fir.
7. Douglas fir - Found with redwood but more durable on higher, drier, hotter slopes; regenerated by seed, and usually requires bare mineral soil in autumn to make a good seedbed. Under controlled management of non-burning, Douglas fir may be on its way out; associated with redwood, tanoak, live oak, madrone, bay.
8. Redwood - The dominant tree over much of the site from about 4000 feet back away from direct coastal winds and salt air. It is a sprouting tree which comes back again and again from disruption; associated mainly with tanoak and Douglas fir.
9. Riparian - A narrow band along stream courses cutting across other plant communities. Red alder, willow, bigleaf maple, and red elderberry. Actually western extension of the eastern deciduous hardwood forest.

See also maps 8, 55, 56, 66, 67.

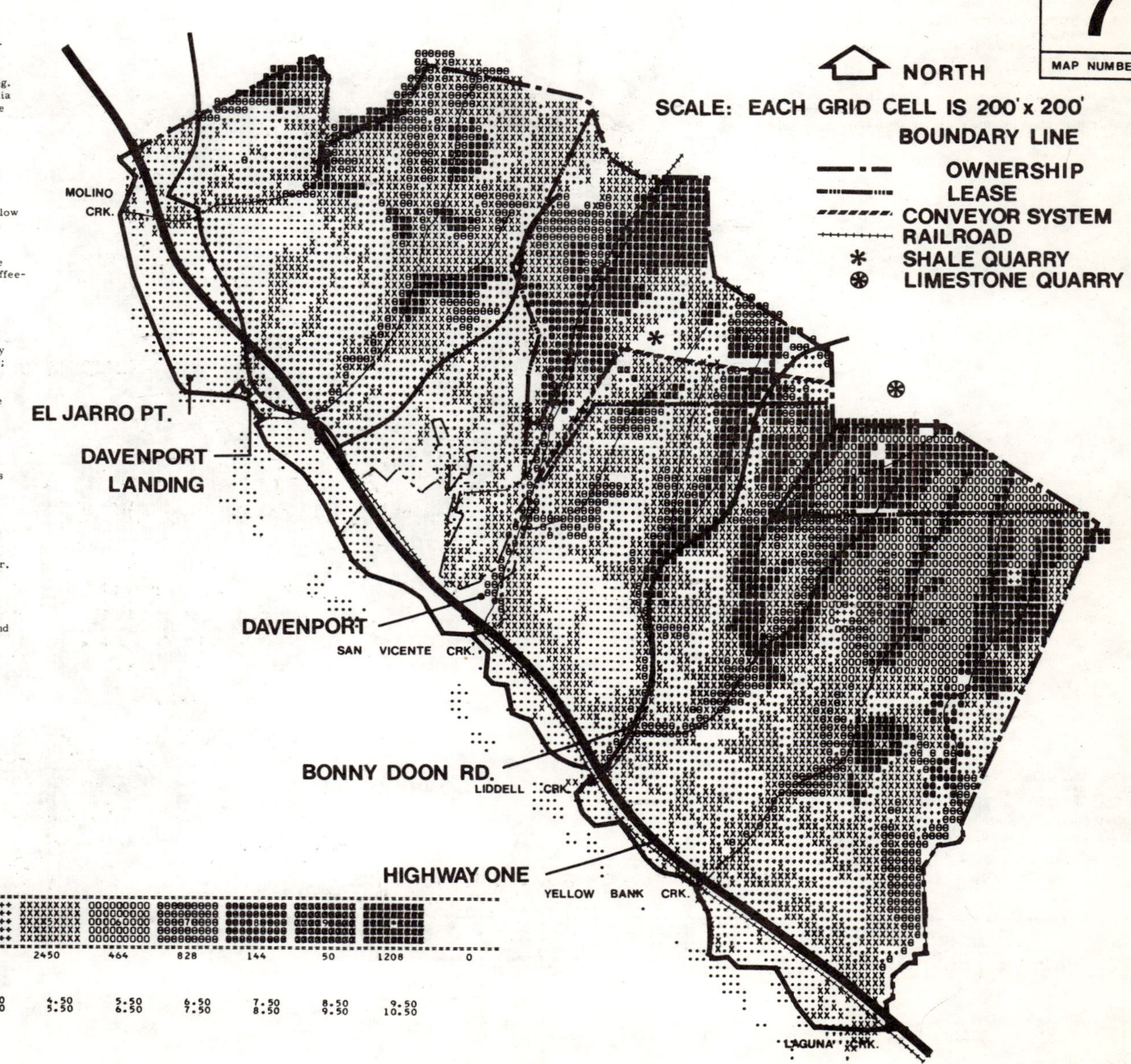

1 = CULTIVATED
2 = DUNEGRASS
3 = GRASSLAND
4 = CHAPARRAL
5 = N. COASTAL SCRUB
6 = KNOBCONE PINE
7 = LIVEOAK
8 = TAN OAK
9 = DOUGLAS FIR
10 = REDWOOD

SYMBOLS		1	2	3	4	5	6	7	8	9	10	
FREQUENCY	1326	1241	52	1523	11	2450	464	828	144	50	1208	0

ABSOLUTE VALUE RANGE APPLYING TO EACH LEVEL

MINIMUM	0.50	1.50	2.50	3.50	4.50	5.50	6.50	7.50	8.50	9.50
MAXIMUM	1.50	2.50	3.50	4.50	5.50	6.50	7.50	8.50	9.50	10.50

56. Ibid., p. 45.

WIND EXPOSURE[57]

SUMMER

Summer wind exposure is a combination of NW, NE, SW & SE wind exposure in that season. Light values on the map indicate minimum exposure. Dark values indicate maximum exposure.

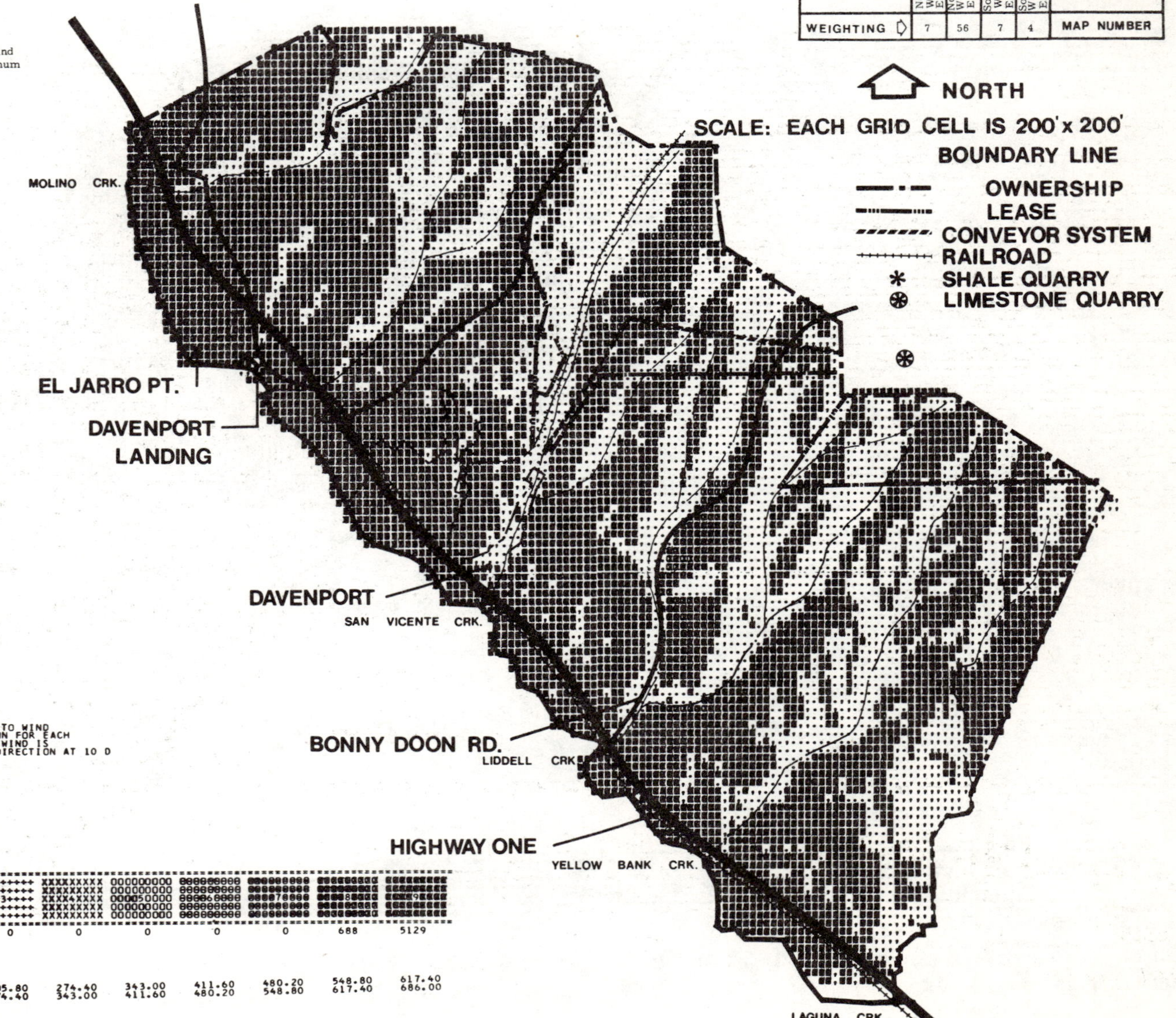

THE VALUE THAT IS ASSIGNED TO EACH CELL THAT IS EXPOSED TO WIND IS DERIVED BY COMBINING THE WIND SPEED AND WIND DIRECTION FOR EACH OF THE DOMINATE DIRECTIONS. THE DIRECT EXPOSURE TO THE WIND IS CALCULATED BY USING ASPECT, SLOPE AND PREDOMINATE WIND DIRECTION AT 10 D ABOVE HORIZONTAL TANGENT WITH EARTH'S SURFACE

1 = AREAS EXPOSED TO MINIMUM DIRECT WIND EXPOSURE
TO
9 = AREAS EXPOSED TO MAXIMUM DIRECT WIND EXPOSURE

SYMBOLS		0	1	2	3	4	5	6	7	8	9
FREQUENCY	0	1543	1046	0	0	0	0	0	0	688	5129

ABSOLUTE VALUE RANGE APPLYING TO EACH LEVEL

MINIMUM	0.0	68.60	137.20	205.80	274.40	343.00	411.60	480.20	548.80	617.40
MAXIMUM	68.60	137.20	205.80	274.40	343.00	411.60	480.20	548.80	617.40	686.00

57. Ibid., p. 56.

SUN INTENSITY[58]

10:00 a.m. WINTER

WINTER SUN INTENSITY AT 10:00 A. M.

Sun angle and sun azimuth conditions used for this map represent sun conditions at the winter solstice at 10:00 A. M. (December 21). Light tones on the map represent areas of high sun intensity relative to other areas on the site. Dark tones represent low sun intensity.

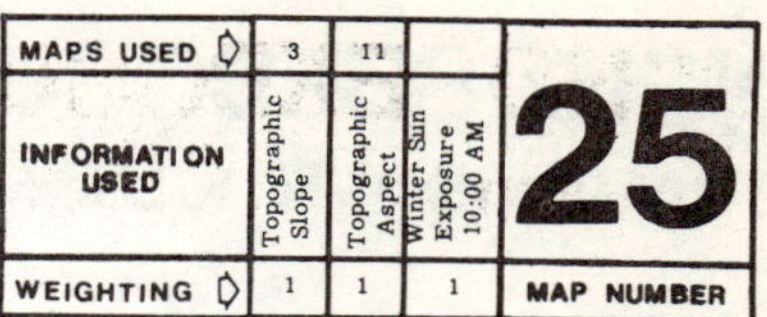

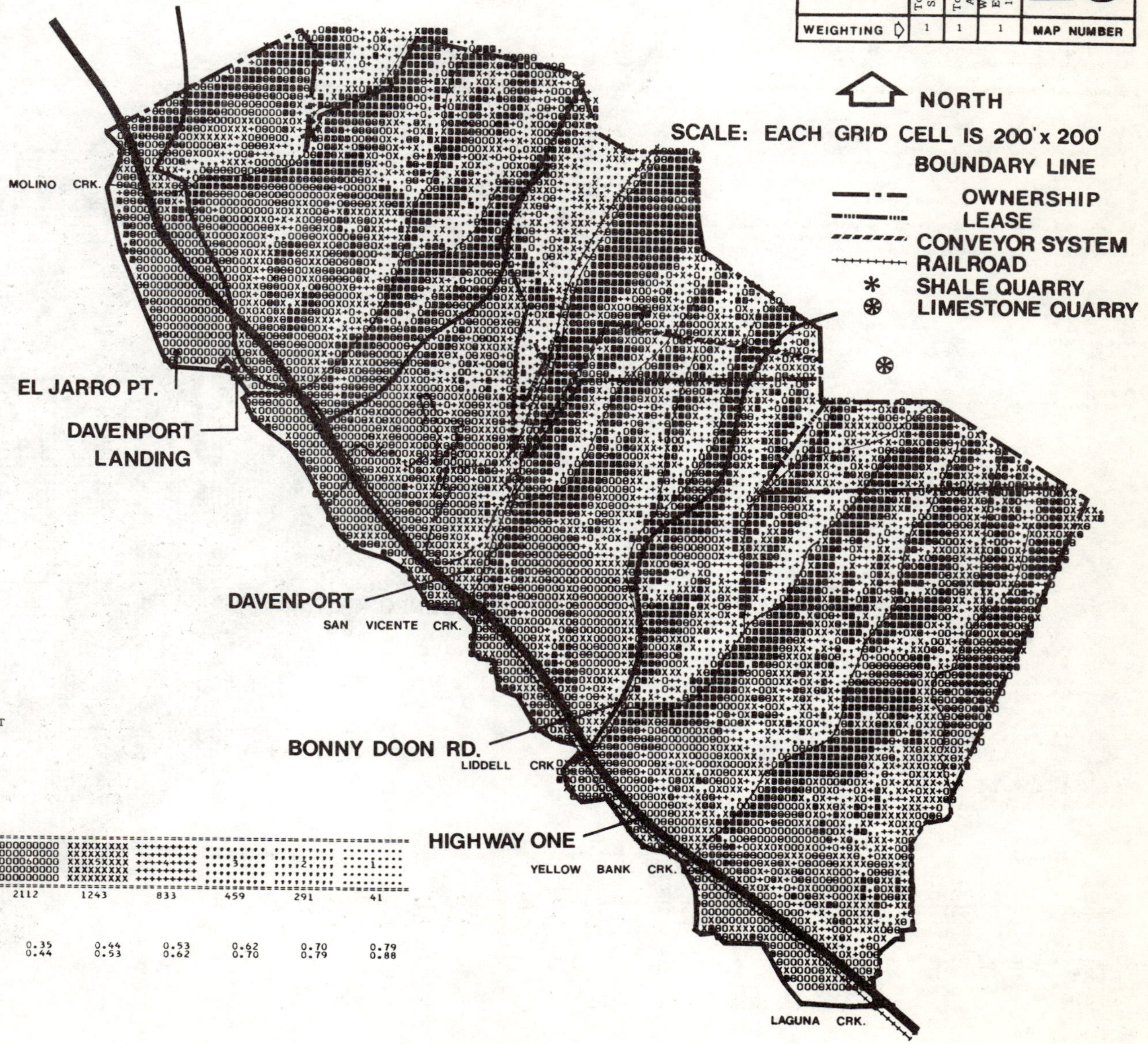

MAPS USED	3	11		25
INFORMATION USED	Topographic Slope	Topographic Aspect	Winter Sun Exposure 10:00 AM	
WEIGHTING	1	1	1	MAP NUMBER

NORTH

SCALE: EACH GRID CELL IS 200' x 200'

BOUNDARY LINE

OWNERSHIP

LEASE

CONVEYOR SYSTEM

RAILROAD

* SHALE QUARRY

⊛ LIMESTONE QUARRY

MOLINO CRK.

EL JARRO PT.

DAVENPORT LANDING

DAVENPORT

SAN VICENTE CRK.

BONNY DOON RD.

LIDDELL CRK.

HIGHWAY ONE

YELLOW BANK CRK.

LAGUNA CRK.

* = MINIMUM SUN INTENSITY
TO
1 = MAXIMUM SUN INTENSITY

THIS MAP WAS PRODUCED BY CALCULATING THE SLOPE, ASPECT AND SUN LOCATIONS FOR EACH OF A SERIES OF TIME PERIODS

SYMBOLS		*	9	8	7	6	5	4	3	2	1
FREQUENCY	0	814	945	813	854	2112	1243	833	459	291	41

ABSOLUTE VALUE RANGE APPLYING TO EACH LEVEL

MINIMUM	C.0	0.09	0.18	0.26	0.35	0.44	0.53	0.62	0.70	0.79
MAXIMUM	0.09	0.18	0.26	0.35	0.44	0.53	0.62	0.70	0.79	0.88

58. Ibid., p. 63.

NOISESHED[59]

TRACTORS WITHOUT MUFFLERS

31
MAP NUMBER

This map represents sound contours modified by line-of-sight data determined by the topography.

Every grid cell considered shows the noise exposure level existing at that point.

The principal noise sources measured include:

1. The diesel tractor operations at the quarry sites.
2. The rock crushing operations at the quarry sites and at the cement plant.
3. Rock handling conveyor system
4. Auto and truck traffic on the coast highway.

The noise zones shown on the computer map represent a summary of the several hundred measurements made of noise from the various noise sources and of the ambient noise levels in a number of locations. The levels are all expressed in terms of audibility.

The +10 DBA zones represent the maximum distance at which noise levels are 10 A-weighted decibels higher than the threshold of audibility. Audibility of these noises, from the quarry operations and from the highway, extends out to the point where they are masked by the ambient sounds from the natural environment. The ambient level, or limit of audibility for intrusive noises, is approximately 30 DBA or A-weighted decibels.

This map differs from map 74 only in the matter of tractor operations in the quarry, map 31 representing existing conditions.

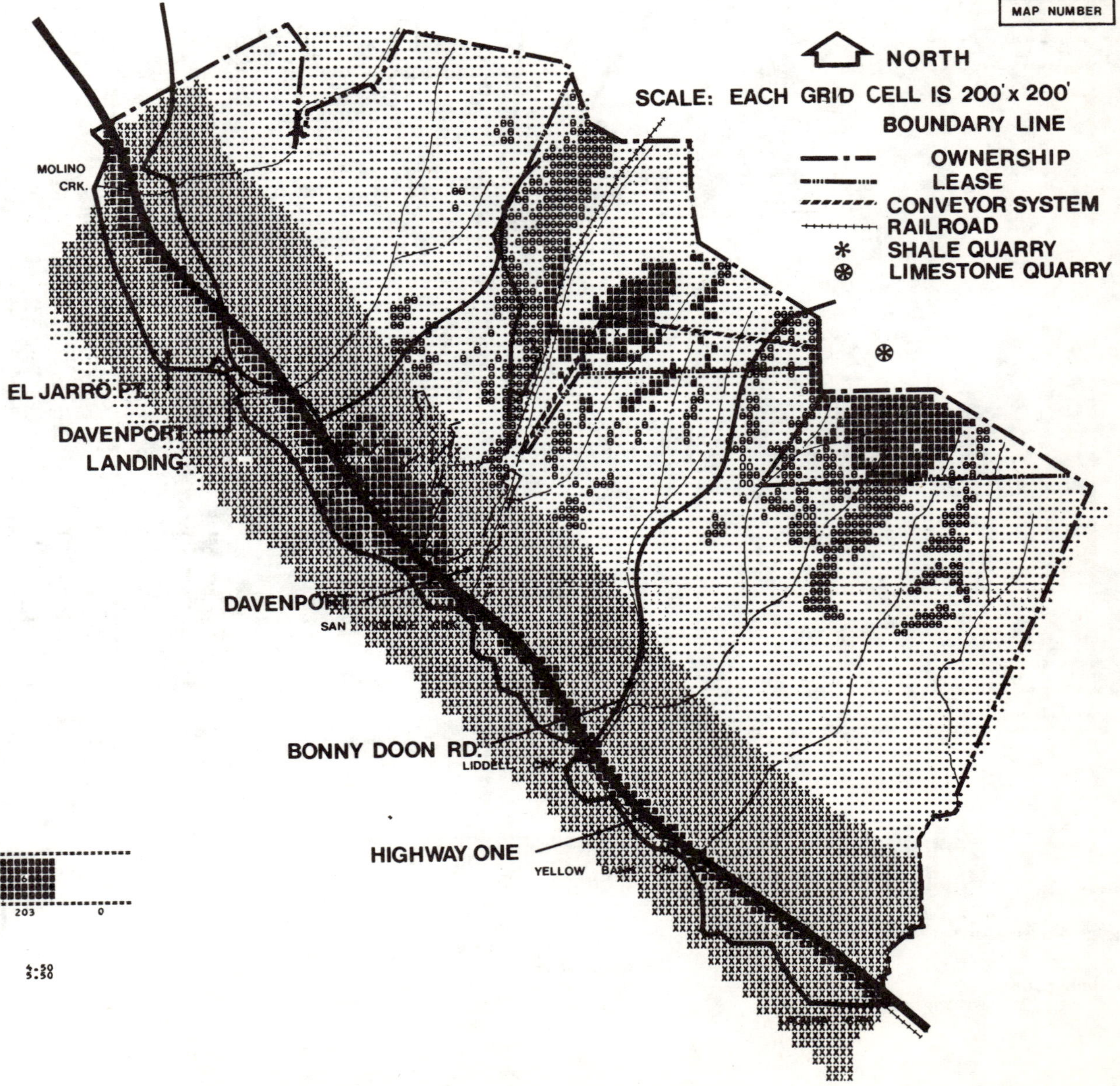

NORTH
SCALE: EACH GRID CELL IS 200' x 200'
BOUNDARY LINE
OWNERSHIP
LEASE
CONVEYOR SYSTEM
RAILROAD
* SHALE QUARRY
⊛ LIMESTONE QUARRY

MOLINO CRK.
EL JARRO PT.
DAVENPORT LANDING
DAVENPORT
SAN VICENTE CRK.
BONNY DOON RD.
LIDDELL CRK.
HIGHWAY ONE
YELLOW BANK CRK.

0 = Ambient (30 DBA)
1 = + 10 DBA
3 = Road Corridor (+30)
4 = Ambient +20 DBA
5 = Ambient +20 DBA
6 = Quarry and Rock Crusher Areas

THIS MAP WAS MADE USING EXTRAPOLATED AND INTERPOLATED DATA
USING POINT SOURCE ASSUMPTIONS AND POINT RECEPTOR DATA
THIS DATA WAS THEN SUBJECTED TO LINE OF SITE ANALYSIS
FOR DETECTION OF "AUDIBLE CONTOURS"

FREQUENCY	0	4476	307	641	3927	203	0

ABSOLUTE VALUE RANGE APPLYING TO EACH LEVEL

MINIMUM	-0.50	1.50	2.50	3.50	4.50
MAXIMUM	0.50	2.50	3.50	4.50	5.50

59. Ibid., p. 69.

VIEWSHED-CEMENT PLANT[60]

TOP OF BUILDINGS

39
MAP NUMBER

The view of the cement plant affects much of Highway One and adjacent areas, including some areas south of Bonny Doon Road.

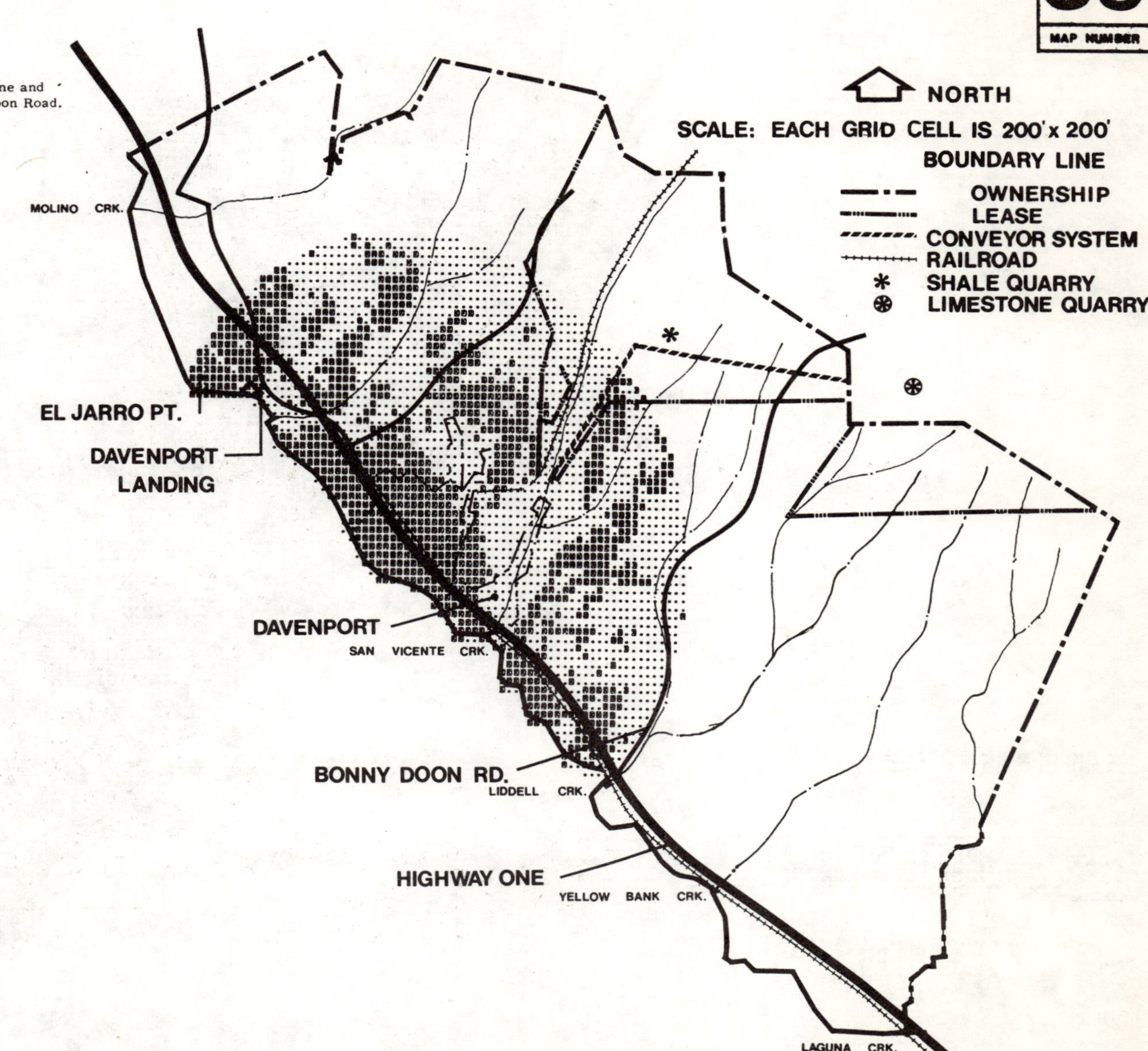

= POINT NOT TESTED
. = POINT TESTED AND NOT OBSERVED
■ = POINT TESTED AND OBSERVED
■ = OBSERVATION POINT

60. Ibid., p. 77.

OCEAN VIEWSHED[61]

MAP NUMBER 47

Ocean view considerations are very important for coastal land, as much aesthetic value must derive from this source to balance out the discomfort of strong coastal climatic influence. The black cells indicate highest value, as a view of the surf is the optimum viewing experience and indicates a view down a canyon if it is present well back into the site. A view at the 1500' mark is considered a rather good view as this allows at least a view of distant rocks and approaching waves. The ocean horizon is considered a minimal view as a result of the near perennial haze, although the sunset is a possibility if the viewing angle is right.

Less than half the site has any potential view of the ocean (tree cover is not considered, which would reduce the figure still more) as a result of the high cliffs at the ocean and the diverse topography inland.

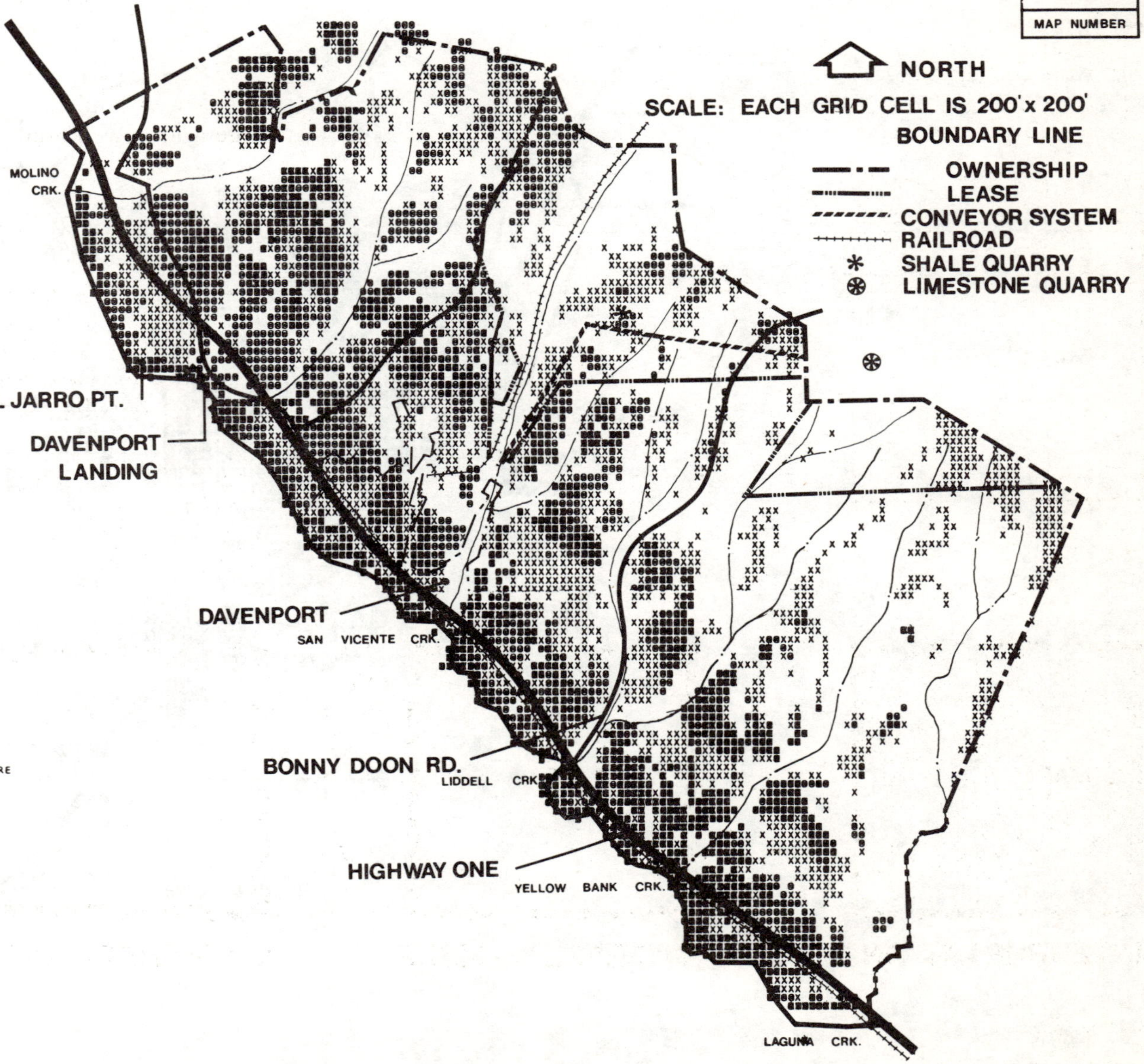

NORTH

SCALE: EACH GRID CELL IS 200' x 200'

BOUNDARY LINE

OWNERSHIP

LEASE

CONVEYOR SYSTEM

RAILROAD

* SHALE QUARRY

⊛ LIMESTONE QUARRY

MOLINO CRK.

EL JARRO PT.

DAVENPORT LANDING

DAVENPORT

SAN VICENTE CRK.

BONNY DOON RD.

LIDDELL CRK.

HIGHWAY ONE

YELLOW BANK CRK.

LAGUNA CRK.

1 = AREAS HAVING VIEW OF OCEAN HORIZON ONLY
2 = AREAS HAVING VIEW OF OCEAN FROM BEYOND 1500 FEET FROM SHORE
3 = AREAS HAVING VIEW OF SHORELINE

ABSOLUTE VALUE RANGE APPLYING TO EACH LEVEL

MINIMUM	0.50	1.50	2.50
MAXIMUM	1.50	2.50	3.50

61. Ibid., p. 85.

INTERPRETATION 1[62]

A first combining and weighting of related groups of information directed at indicating specific limitations and potential for landscape modification.

DATA BANK MAPS

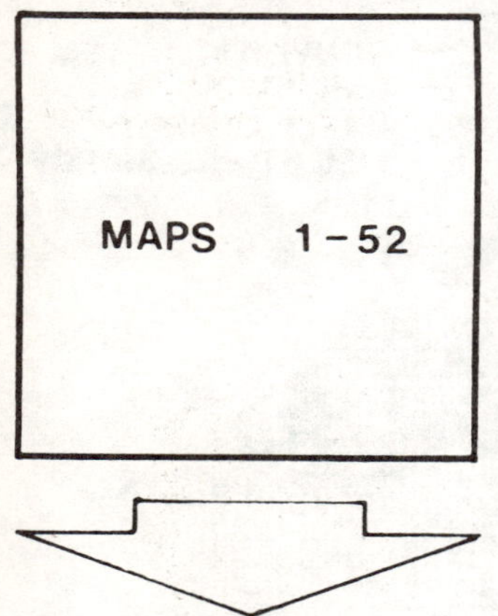

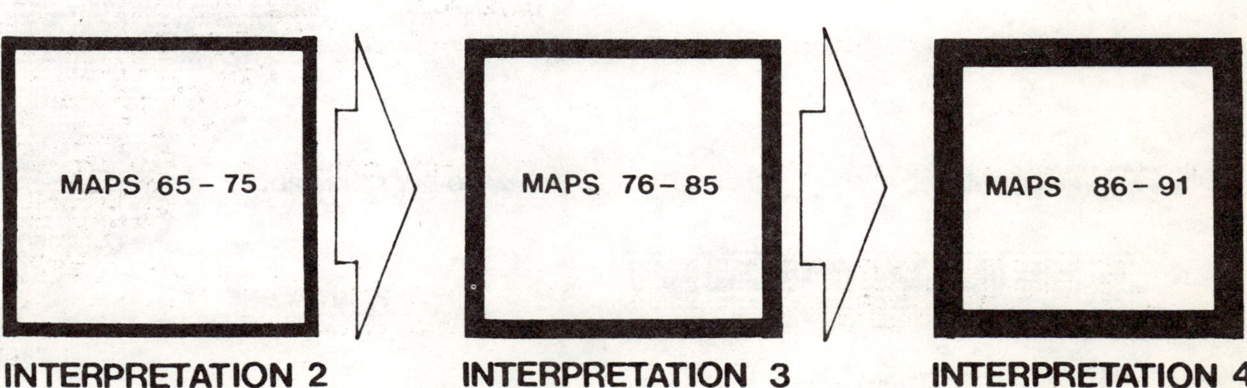

62. Ibid., p. 91.

The following then are the examples of the first stage of interpretation as depicted in computer maps 53 through 64 of the report. These are a first combining and weighing of related groups of information directed at indicating specific limitations and potential for landscape modification. These maps deal with:

Erosion Potential
Geologic Instability
Vegetation Fire Potential
Vegetation Stability
Water Quality and Flooding Hazards
Groundwater Conditions
Storm Runoff Potential
Year-Round Wind Exposure
Sun Intensity Day Long for the Total Year
View Shed in Relation to the Pacific Cement and Aggregate Quarry within the Area
Visual Characteristics
Limitation Composites
Visual Character Opportunities Composites

EROSION POTENTIAL[63]

This map is a measure of susceptibility to erosion under disturbed conditions, and combines terrain unit values (Matrix I) and rainfall. Erosion is most important as a secondary factor which can prevent successful landscaping and forestation, cause gullying and undermining of structures and tree roots, and silt up streams, thereby reducing water quality and wildlife habitat.

The site is revealed to be generally no worse than moderate in erosion potential, largely because of the influence of the dominant Santa Cruz mudstone, although the level terrace deposits do represent a somewhat more difficult problem.

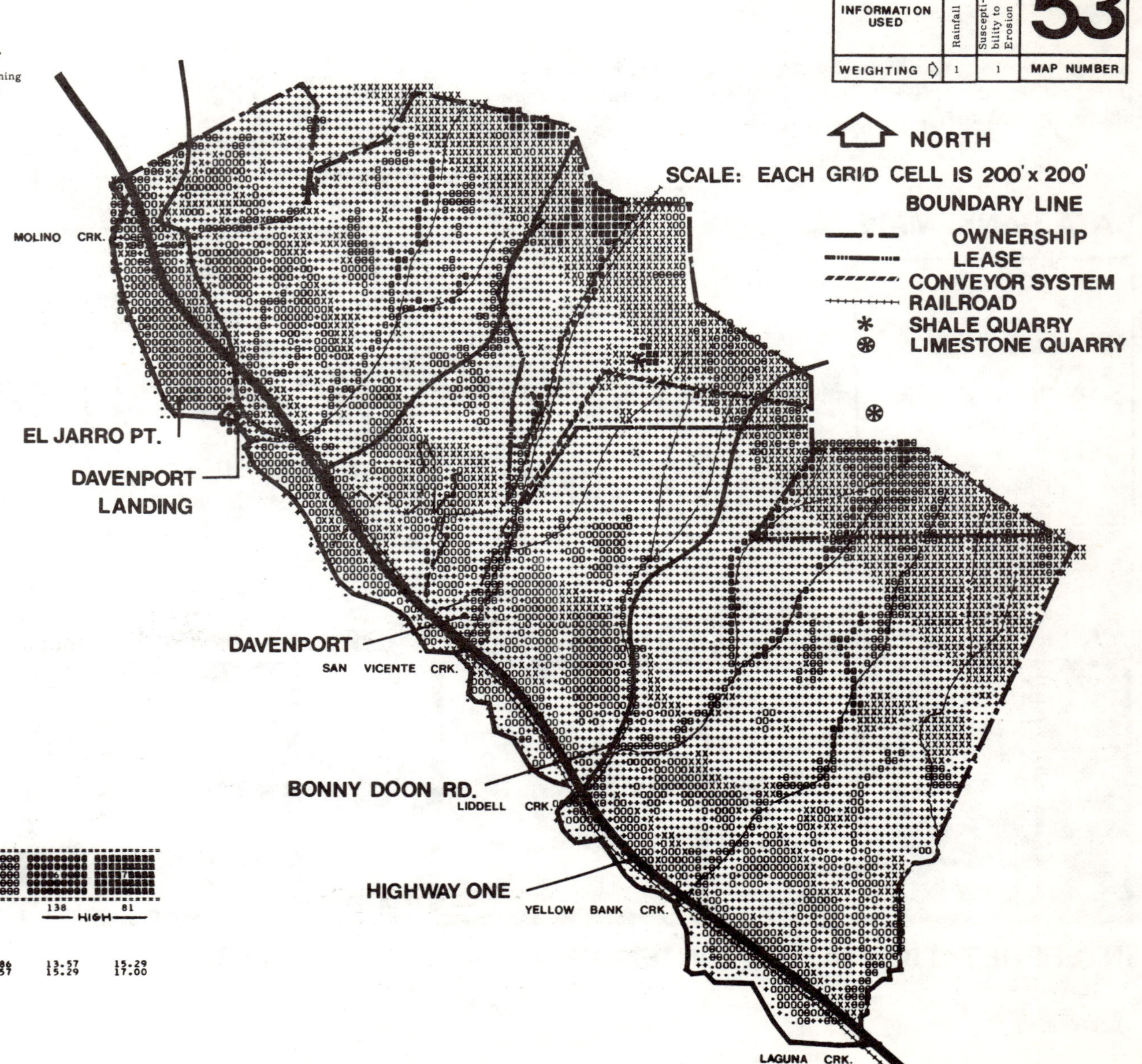

7 = HIGH EROSION POTENTIAL
TO
1 = LOW EROSION POTENTIAL

SYMBOLS	1	2	3	4	5	6	7
FREQUENCY	186	4152	1623	1504	720	138	81
	LOW		MOD.			HIGH	

ABSOLUTE VALUE RANGE APPLYING TO EACH LEVEL

MINIMUM	5.00	6.71	8.43	10.14	11.86	13.57	15.29
MAXIMUM	6.71	8.43	10.14	11.86	13.57	15.29	17.00

63. Ibid., p. 92.

VEGETATION FIRE POTENTIAL[64]

This map and the following are the key maps in the vegetation study. This map does not necessarily imply fire hazard, as matters of slope, wind, and general accessibility prevail there. It is a combination of flamability and susceptibility once a fire has begun:

The following list is valid for this map and map 56, which contains a matrix of numerical values. The information was based on discussions with Professor Ed Stone of the U.C. Berkeley School of Forestry.

1. Grassland; highly flamable if not overgrazed, very unstable; in need of constant management to prevent scrub encroachment.
2. Chaparral; a fire type; a sprouter; comes back as chaparral after fire but ultimately goes on toyon and finally oak - madrone. Under fire suppression duff buildup could lead to a very hot fire, killing off the chaparral.
3. N. Coastal Scrub - an invasive mixture of plants which stays as is near the coast, but moves on to oak - madrone inland. Usually on steeper slopes in fairly good soils. A good watershed-control plant which is highly flamable and hard to control;returns to scrub after fire.
4. Knobcone pine - a fire associate usually found on shallow, poor soils. Breaks up in 50-80 years and goes back to chaparral. If its nearby and a fire comes along, especially in fall, it will spread or return. Very prone to fire.
5. Live Oak-madrone - sprouting trees which can hold on in the face of fire and return; follows scrub in succession; creates a rather severe fire hazard. Very sensitive to environmental disruptions - subject to disease.
6. Tanoak; - Not as flamable as Live-oak; a sprouter, which will return. Rather sensitive to environmental disruptions.
7. Douglas-Fir - Mighty forest giant of the northwest which is not only quite sensitive to environmental disruption but is highly subject to fire and dependent on circumstances for survival. It requires bare mineral soil for its seeds, preferably in the fall so winter rains can help it overcome competition from scrub. Under forest management and fire exclusion, duff buildup and an ultimately large fire could kill off all the seed trees.
8. Redwood; the stable giant; keeps sprouting and coming back in the face of all manner of environmental degradation. Not flamable but once a crown fire is started, it could be bad.
9. Riparian - not notably flamable but would go with the rest; the sprouting willow would return quickly, but seed bearing alders and maple come back more slowly requiring bare mineral soil, especially in spring.

This map is included as an element in the building forward of vegetation restrictions partly because of the degradation to the landscape which would result in the event of fire. Fire tends to break down the pore structure of the soil, and the first rains knock the upper particles down into the soil to form a shallow, impervious layer which then takes the full force of the rain and results in erosion and runoff problems.

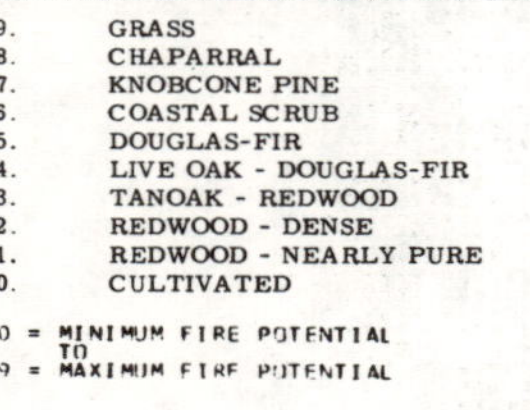

9. GRASS
8. CHAPARRAL
7. KNOBCONE PINE
6. COASTAL SCRUB
5. DOUGLAS-FIR
4. LIVE OAK - DOUGLAS-FIR
3. TANOAK - REDWOOD
2. REDWOOD - DENSE
1. REDWOOD - NEARLY PURE
0. CULTIVATED

0 = MINIMUM FIRE POTENTIAL TO 9 = MAXIMUM FIRE POTENTIAL

SYMBOLS	0	1	2	3	4	5	6	7	8	9
ABSOLUTE VALUE RANGE APPLYING TO EACH LEVEL — MINIMUM	0.50	1.50	2.50	3.50	4.50	5.50	6.50	7.50	8.50	9.50
MAXIMUM	1.50	2.50	3.50	4.50	5.50	6.50	7.50	8.50	9.50	10.50

FREQUENCY 829 1042 0 76 1104 777 99 2452 464 11 1522 0

LOW MOD LOW MOD MOD HI HI

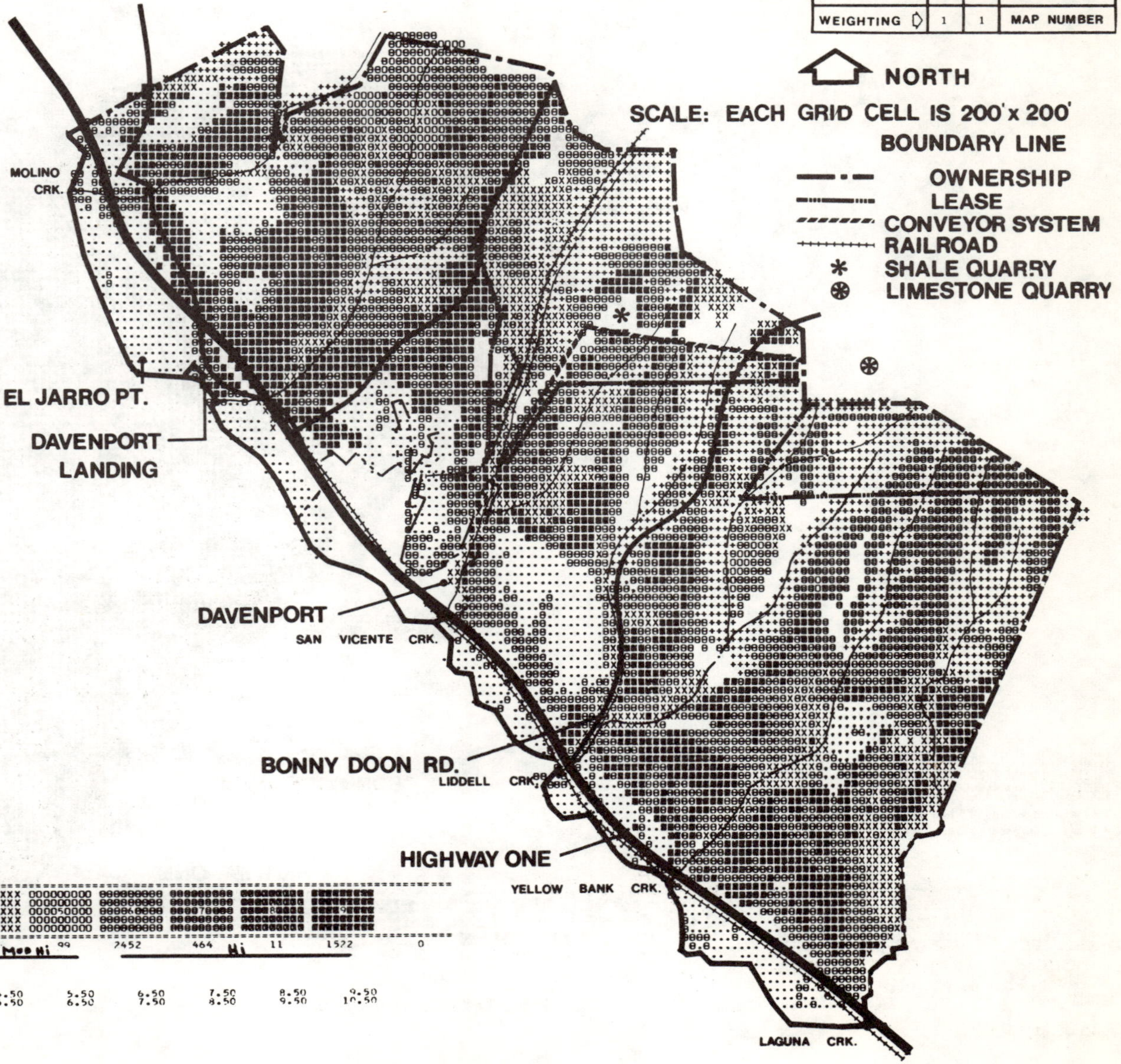

64. Ibid., p. 94.

MAPS USED ⇨	7	8	55
INFORMATION USED	Vegetation Type	Vegetation Density	
WEIGHTING ⇨	1	1	MAP NUMBER

NORTH

SCALE: EACH GRID CELL IS 200'x 200'

BOUNDARY LINE
OWNERSHIP
LEASE
CONVEYOR SYSTEM
RAILROAD
* SHALE QUARRY
⊛ LIMESTONE QUARRY

MOLINO CRK.
EL JARRO PT.
DAVENPORT LANDING
DAVENPORT
SAN VICENTE CRK.
BONNY DOON RD.
LIDDELL CRK.
HIGHWAY ONE
YELLOW BANK CRK.
LAGUNA CRK.

YR- ROUND WIND EXPOSURE[65]

Year round wind exposure is a combination of seasonal wind exposures. Light values on the map indicate minimum exposure. Dark values indicate maximum exposure.

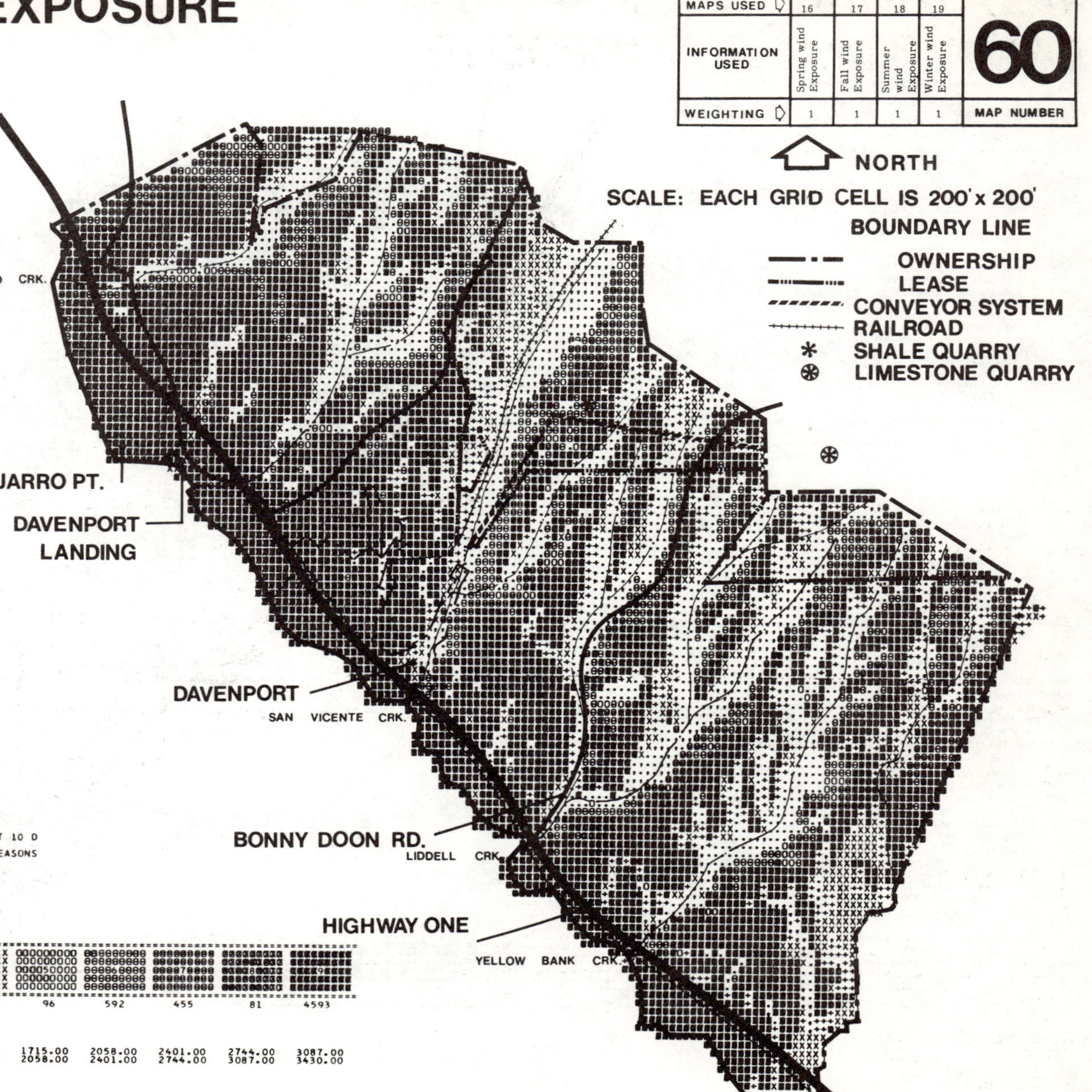

THE VALUE THAT IS ASSIGNED TO EACH CELL THAT IS EXPOSED TO WIND
IS DERIVED BY COMBINING THE WIND SPEED AND WIND DIRECTION FOR EACH
OF THE DOMINATE DIRECTIONS. THE DIRECT EXPOSURE TO THE WIND IS
CALCULATED BY USING ASPECT, SLOPE AND PREDOMINATE WIND DIRECTION AT 10 D
ABOVE HORIZONTAL TANGENT WITH EARTH'S SURFACE
THE ROUTINE USED FOR CALCULATING THE MAP COMBINES DATA FROM FOUR SEASONS

1 = AREAS EXPOSED TO MINIMUM DIRECT WIND EXPOSURE
TO
9 = AREAS EXPOSED TO MAXIMUM DIRECT WIND EXPOSURE

SYMBOLS		0	1	2	3	4	5	6	7	8	9
FREQUENCY	0	1025	366	152	545	501	96	592	455	81	4593

ABSOLUTE VALUE RANGE APPLYING TO EACH LEVEL

MINIMUM	0.0	343.00	686.00	1029.00	1372.00	1715.00	2058.00	2401.00	2744.00	3087.00
MAXIMUM	343.00	686.00	1029.00	1372.00	1715.00	2058.00	2401.00	2744.00	3087.00	3430.00

65. Ibid., p. 99.

VISUAL CHARACTER–OPPORTUNITIES COMPOSITE[66]

MAPS USED	47	46	44	43	45	64
INFORMATION USED	Viewshed - Ocean	Special Viewpoints	Unique Spaces	Sequential Spaces - Linear	Special Features - Natural	
WEIGHTING	1	1	1	1	1	MAP NUMBER

This map, a composite of identifiable and identified positive visual elements on the site, reveals several interesting points: that the southeastern portion of the site, although short of ocean views, has otherwise strong aesthetic value and obviously has the most "inland" feel on the site. Immediately to the ocean side of this area is the site's outstanding ocean view/ridgeline/special viewpoint area, with a ridge area in the far northwest corner offering another fine visual experience.

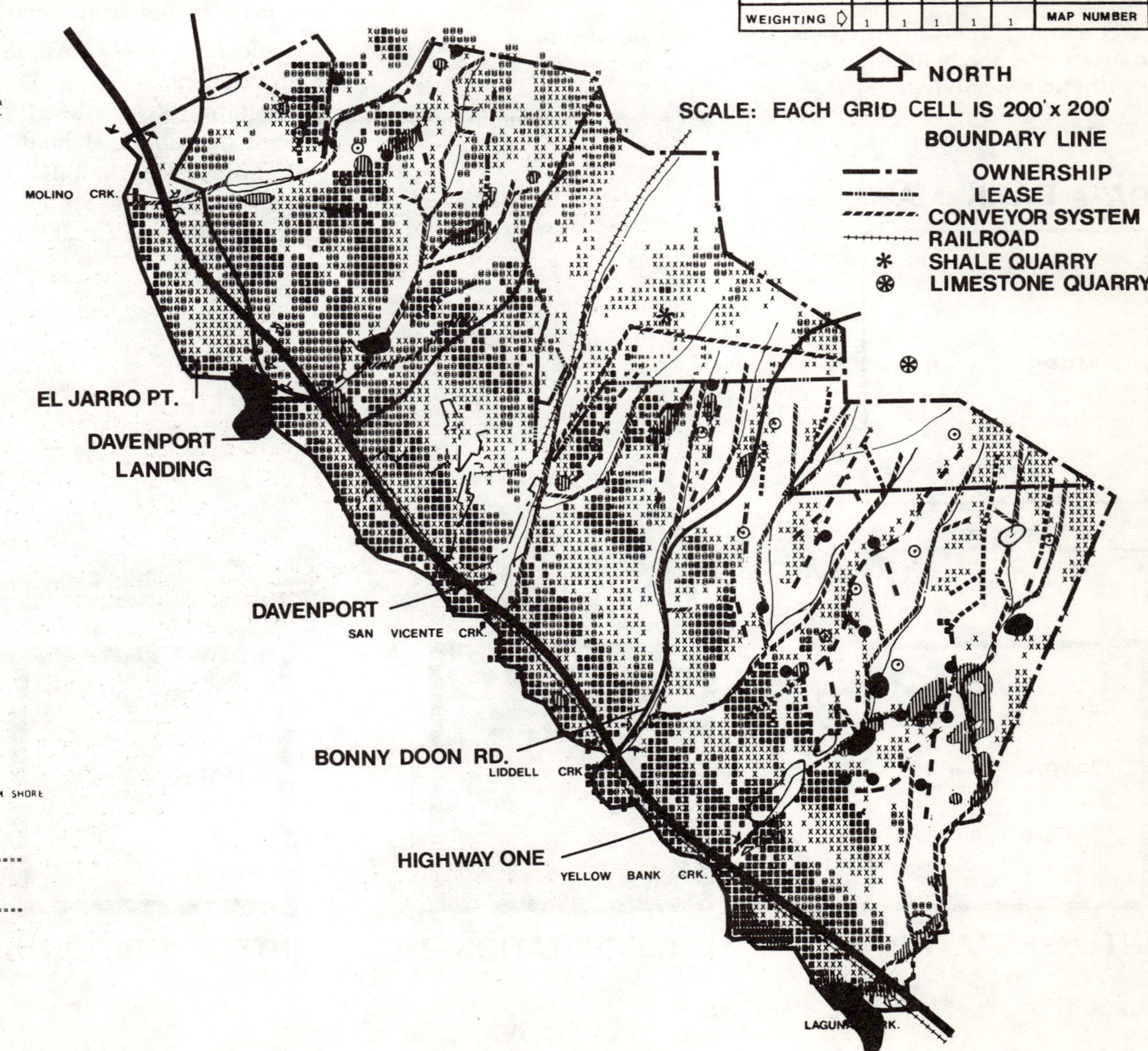

GOOD - V. GOOD RIDGE LINE SEQUENCE

MODERATELY GOOD RIDGE LINE SEQUENCE

GOOD - V. GOOD CANYON SEQUENCE

FAIR - MODERATELY GOOD CANYON SEQUENCE

GOOD - V. GOOD VIEW POINTS

MODERATELY GOOD VIEWPOINTS

GOOD - V. GOOD SPACES

FAIR - MODERATELY GOOD SPACES

SPECIAL FEATURES - NATURAL

1 = AREAS HAVING VIEW OF OCEAN HORIZON ONLY
2 = AREAS HAVING VIEW OF OCEAN FROM BEYOND 1500 FEET FROM SHORE
3 = AREAS HAVING VIEW OF SHORELINE

SYMBOLS		XXXX1XXXX	2	3	
FREQUENCY	5363	1815	1092	1284	0

ABSOLUTE VALUE RANGE APPLYING TO EACH LEVEL

MINIMUM	0.50	1.50	2.50
MAXIMUM	1.50	2.50	3.50

66. Ibid., p. 103.

INTERPRETATION 2[67]

A combining and weighting of related groups of information into composits for each of the major natural and aesthetic categories.

The following are examples of the maps 65-75, which are a combining and weighing of related groups of information into composites for each of the major natural and aesthetic categories. These maps include considerations of:

- Geologic Forces/Restrictions
- Fire Hazard
- Vegetation Restrictions
- Sensitive Wildlife Habitat
- Hydrology Restrictions
- Climate Comfort—Spring, Fall, Summer, Winter
- Noise-Shed and Visual Composite

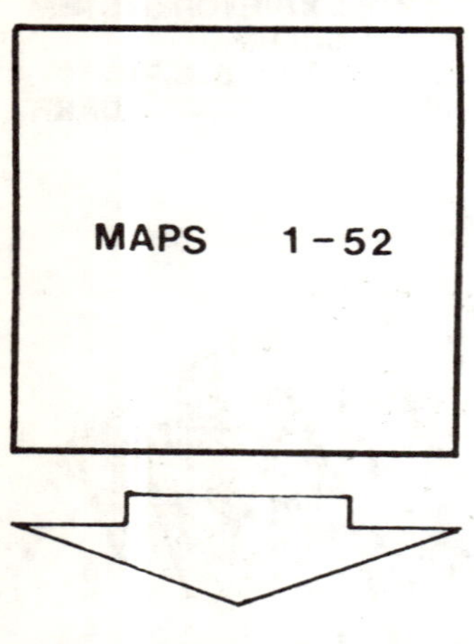

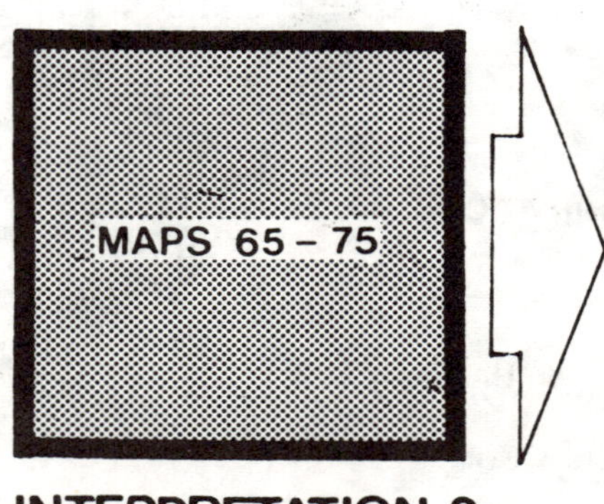

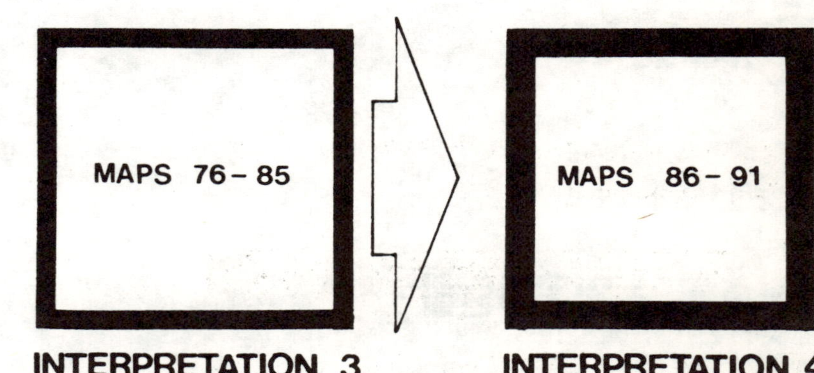

67. Ibid., p. 104.

VEGETATION RESTRICTIONS[68]

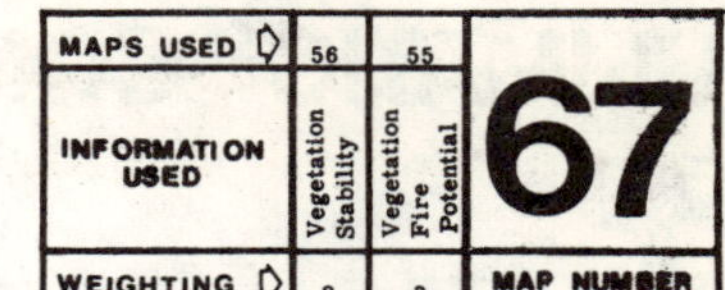

This map, a composite of vegetation stability and fire potential, indicates that only the redwood stands out as a "reliable" plant. The map is indicative of how stable and constant the redwood is, being resistive to both fire and land use alterations.

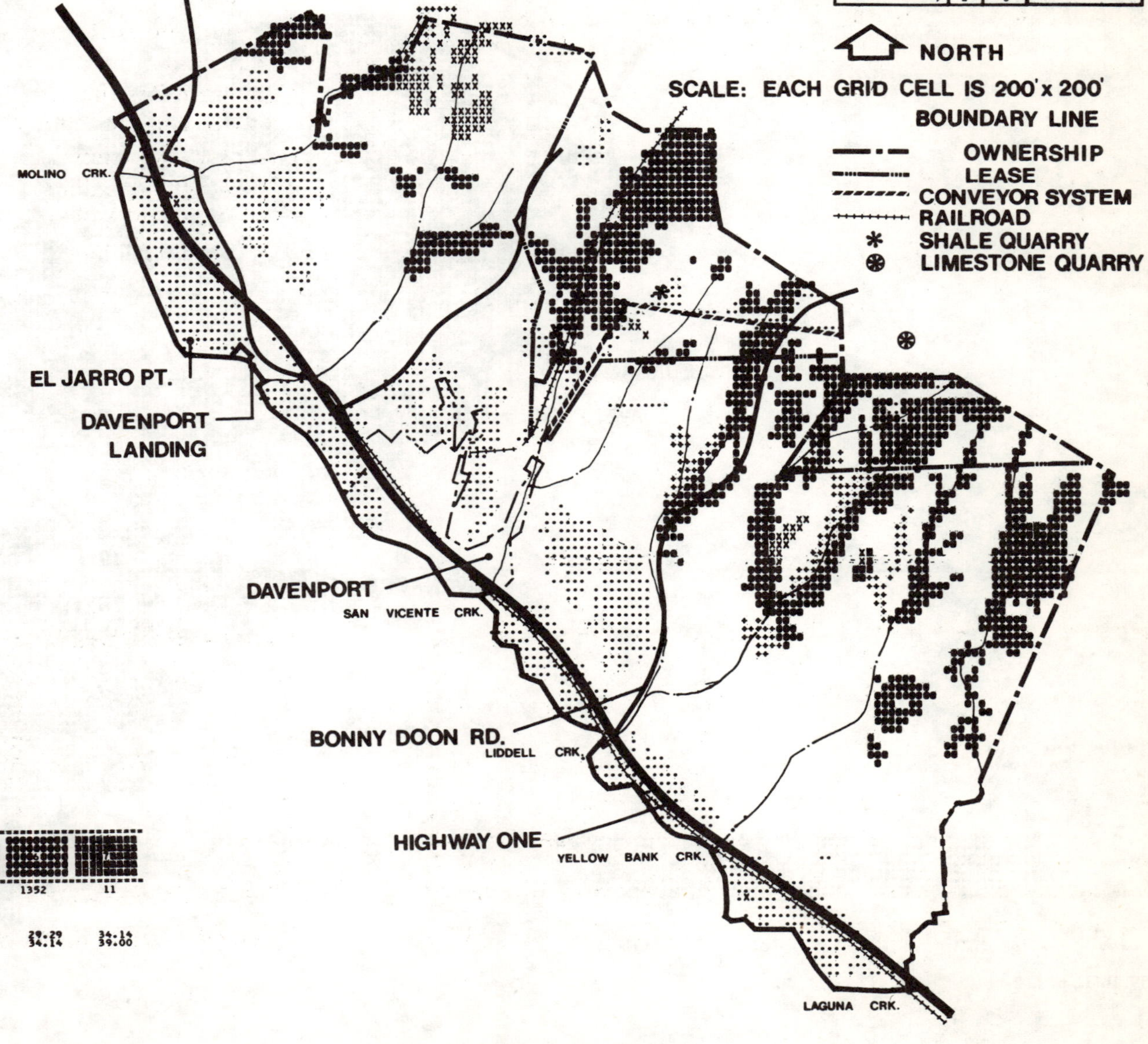

NORTH

SCALE: EACH GRID CELL IS 200' x 200'

BOUNDARY LINE

OWNERSHIP

LEASE

CONVEYOR SYSTEM

RAILROAD

* SHALE QUARRY

⊛ LIMESTONE QUARRY

MOLINO CRK.

EL JARRO PT.

DAVENPORT LANDING

DAVENPORT

SAN VICENTE CRK.

BONNY DOON RD.

LIDDELL CRK.

HIGHWAY ONE

YELLOW BANK CRK.

LAGUNA CRK.

THIS MAP COMBINES AN ANALYSIS OF VEGETATION FIRE POTENTIAL AND AN ANALYSIS OF THE VEGETATION STABILITY

7 = LOW VEGETATION RESTRICTIONS
TO
1 = HIGH VEGETATION RESTRICTIONS

SYMBOLS	1	2	3	4	5	6	7
FREQUENCY	1312	148	108	0	0	1352	11

ABSOLUTE VALUE RANGE APPLYING TO EACH LEVEL

MINIMUM	5.00	9.86	14.71	19.57	24.43	29.29	34.14
MAXIMUM	9.86	14.71	19.57	24.43	29.29	34.14	39.00

68. Ibid., p. 107.

CLIMATIC COMFORT[69]

FALL

The purpose of this map is to compare areas on the site for climatic comfort for Fall. The light areas indicate maximum comfort on the site. The dark areas indicate minimum comfort (see Climatic Comfort - Total Year for further explanation).

The maximum and minimum values are not absolute but only refer to range of climate on this site.

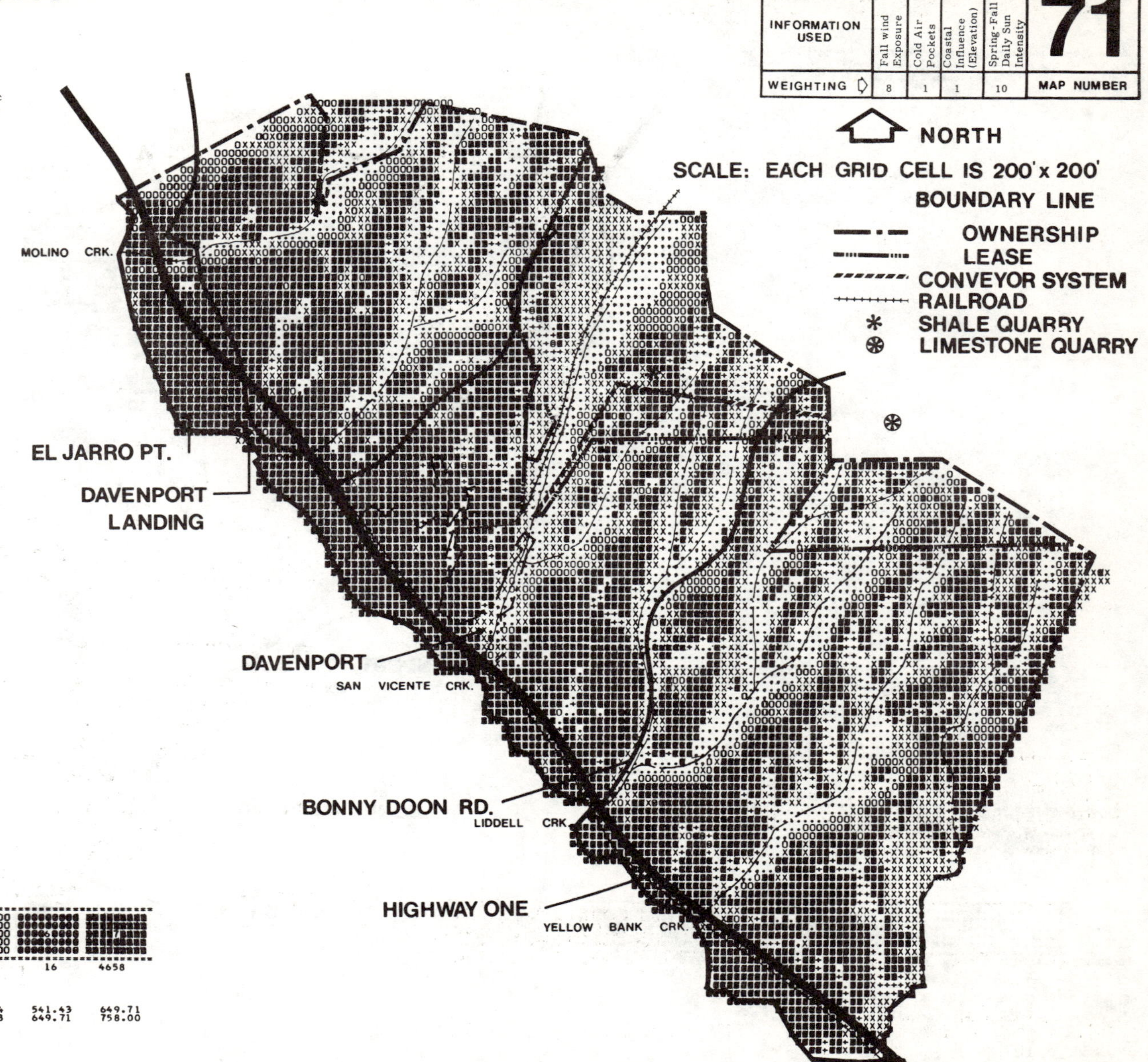

CLIMATIC COMFORT BY SEASON
1 = MOST COMFORTABLE
TO
7 = LEAST COMFORTABLE

SYMBOLS	1	2	3	4	5	6	7
FREQUENCY	1025	366	395	1373	573	16	4658

ABSOLUTE VALUE RANGE APPLYING TO EACH LEVEL

MINIMUM	0.0	108.29	216.57	324.86	433.14	541.43	649.71
MAXIMUM	108.29	216.57	324.86	433.14	541.43	649.71	758.00

69. Ibid., p. 112.

VISUAL COMPOSITE[70]

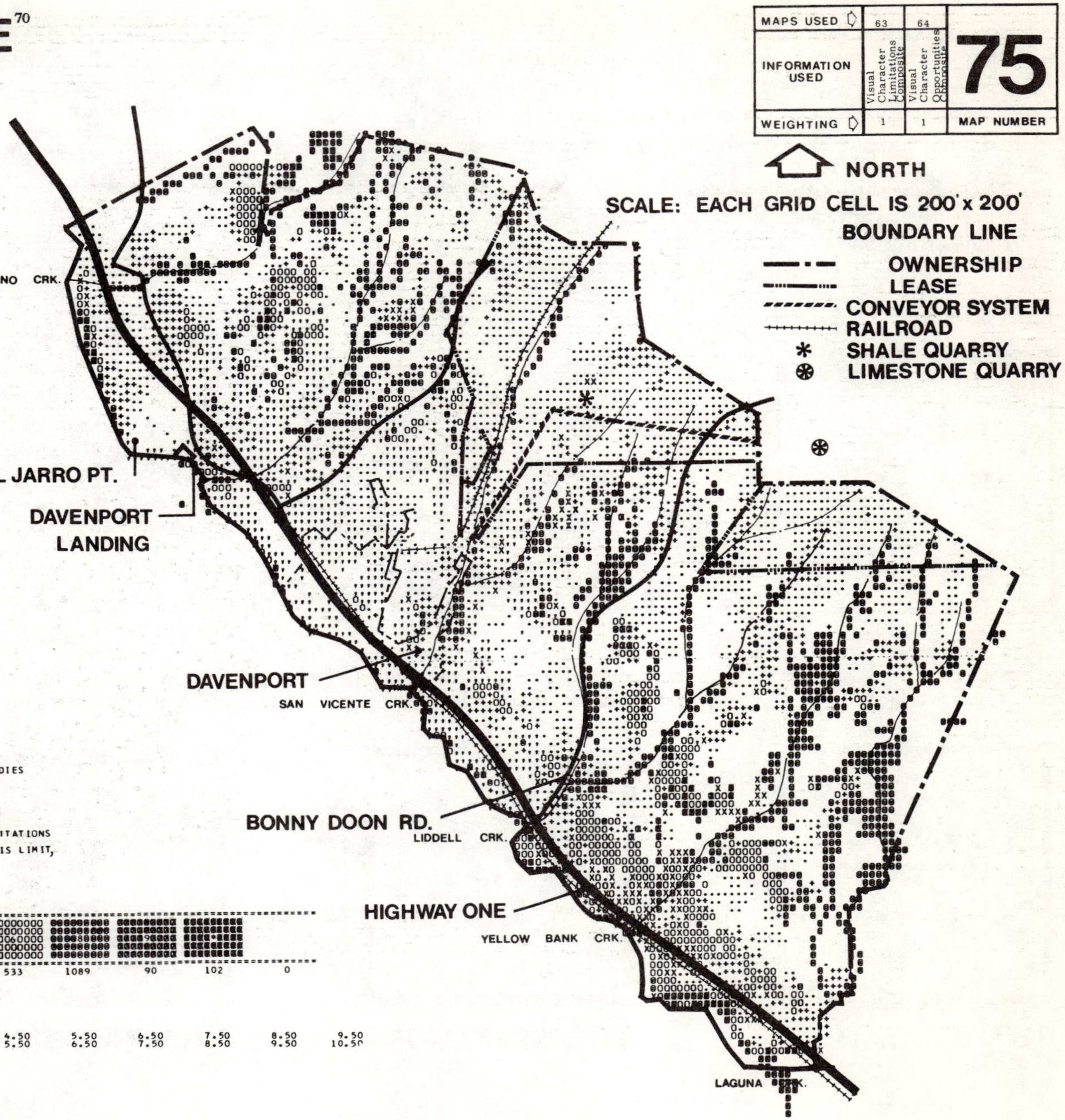

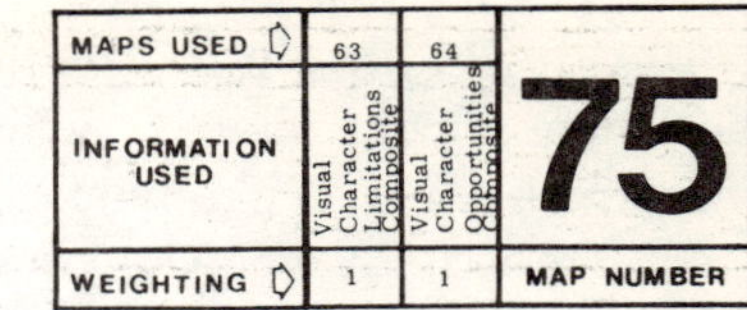

NORTH

SCALE: EACH GRID CELL IS 200' x 200'

BOUNDARY LINE
- OWNERSHIP
- LEASE
- CONVEYOR SYSTEM
- RAILROAD
- * SHALE QUARRY
- ⊛ LIMESTONE QUARRY

MOLINO CRK.

EL JARRO PT.

DAVENPORT LANDING

DAVENPORT

SAN VICENTE CRK.

BONNY DOON RD.

LIDDELL CRK.

HIGHWAY ONE

YELLOW BANK CRK.

LAGUNA CRK.

THIS MODEL COMBINES INFORMATION FROM A SERIES OF SUB-VISUAL STUDIES

* = HIGH VISUAL CHARACTER AND NO VISUAL LIMITATIONS
9 = MODERATELY HIGH VISUAL CHARACTER AND NO VISUAL LIMITATIONS
8 = FAIRLY HIGH VISUAL CHARACTER AND NO VISUAL LIMITATIONS
6 = HIGH VISUAL CHARACTER AND MILD VISUAL LIMITATIONS
5 = MODERATELY HIGH VISUAL CHARACTER AND MILD VISUAL LIMITATIONS
4 = FAIRLY HIGH VISUAL CHARACTER AND MILD VISUAL LIMITATIONS
2 = FAIRLY HIGH VISUAL CHARACTER AND MODERATE TO HIGH VISUAL LIMITATIONS
1 = NO DEFINED VISUAL CHARACTER AND MILD VISUAL LIMITATIONS
BLANK AREAS = NO DEFINED VISUAL CHARACTER AND MOD. TO HI VIS LIMIT, NO DEFINED VISUAL CHARACTER OR LIMITATIONS

SYMBOLS		1	2	4	5	6	8	9	*	
FREQUENCY	357	2077	745	873	172	533	1089	90	102	0

ABSOLUTE VALUE RANGE APPLYING TO EACH LEVEL

MINIMUM	0.50	1.50	2.50	3.50	4.50	5.50	6.50	7.50	8.50	9.50
MAXIMUM	1.50	2.50	3.50	4.50	5.50	6.50	7.50	8.50	9.50	10.50

70. Ibid., p. 116.

Also included in this section is an extremely interesting ecological dependency matrix, which is a relationship chart identifying the qualitative dependencies of various forms of wildlife, land characteristic, vegetation and climate. These relationships were identified according to the consultants in the following way:

> These relationships were identified and subjected to a computer program which analyzed the chains of interrelationship such as food chains, and determined a ranking of ecological significant areas. This information was printed in a computer map indicating the degree of ecological sensitivity. The ecological dependency matrix is shown on the following sheet and the resultant areas of sensitive wildlife habitat.[71]

The following illustrations are representative of the maps 76 through 85, which are classified as Interpretation III, or as "Land Usability" Expressed the Suitability for Certain objective land uses which fall logically within a single realm of the six paths of physical land considerations. These include:

Agricultural Suitability Specialty Crops
Grazing Suitability
Tree Farm Suitability
Suitability for Conversion to Forestry
Forestry Suitability
Engineering Suitability—Risk Cost
Direct Engineering Suitability
Access Roads Suitability

DATA BANK MAPS

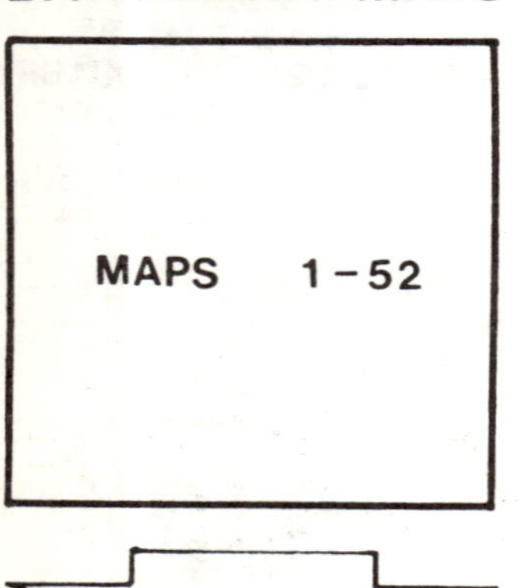

INTERPRETATION 3[72]

Land usability expressed as suitability for certain objective land uses which fall logically within a single realm of the six paths of physical land considerations.

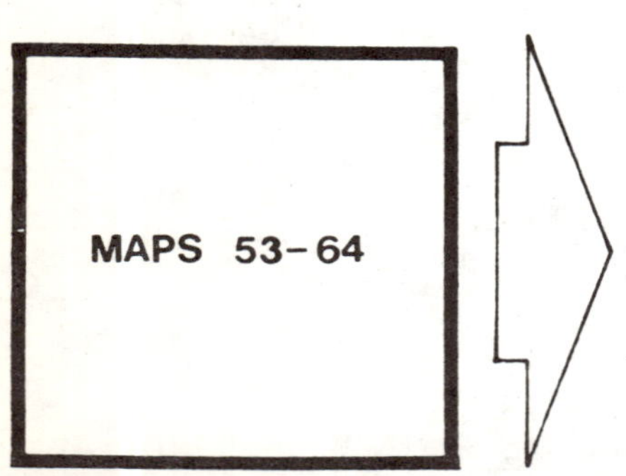

INTERPRETATION 1

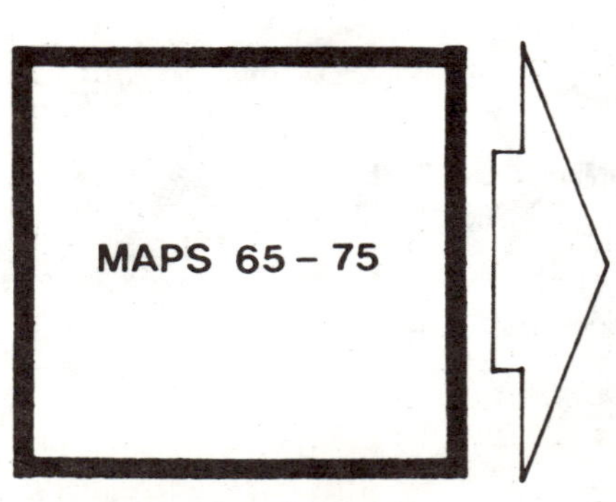

INTERPRETATION 2

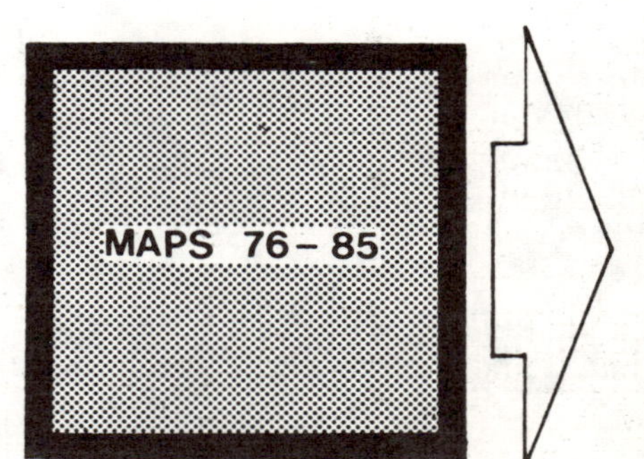

INTERPRETATION 3

INTERPRETATION 4

71. Ibid., p. 116.
72. Ibid., p. 117.

ECOLOGICAL DEPENDENCY MATRIX[73]

The matrix is a relationship matrix identifying the qualitative dependencies of various forms of wildlife, land characteristics, vegetation and climate. These relationships were identified and subjected to a computer program which analyzed the chains of interrelationships, such as food chains, and determined a ranking of ecologically significant areas. This information was printed in a computer map indicating the degree of ecological sensitivity.

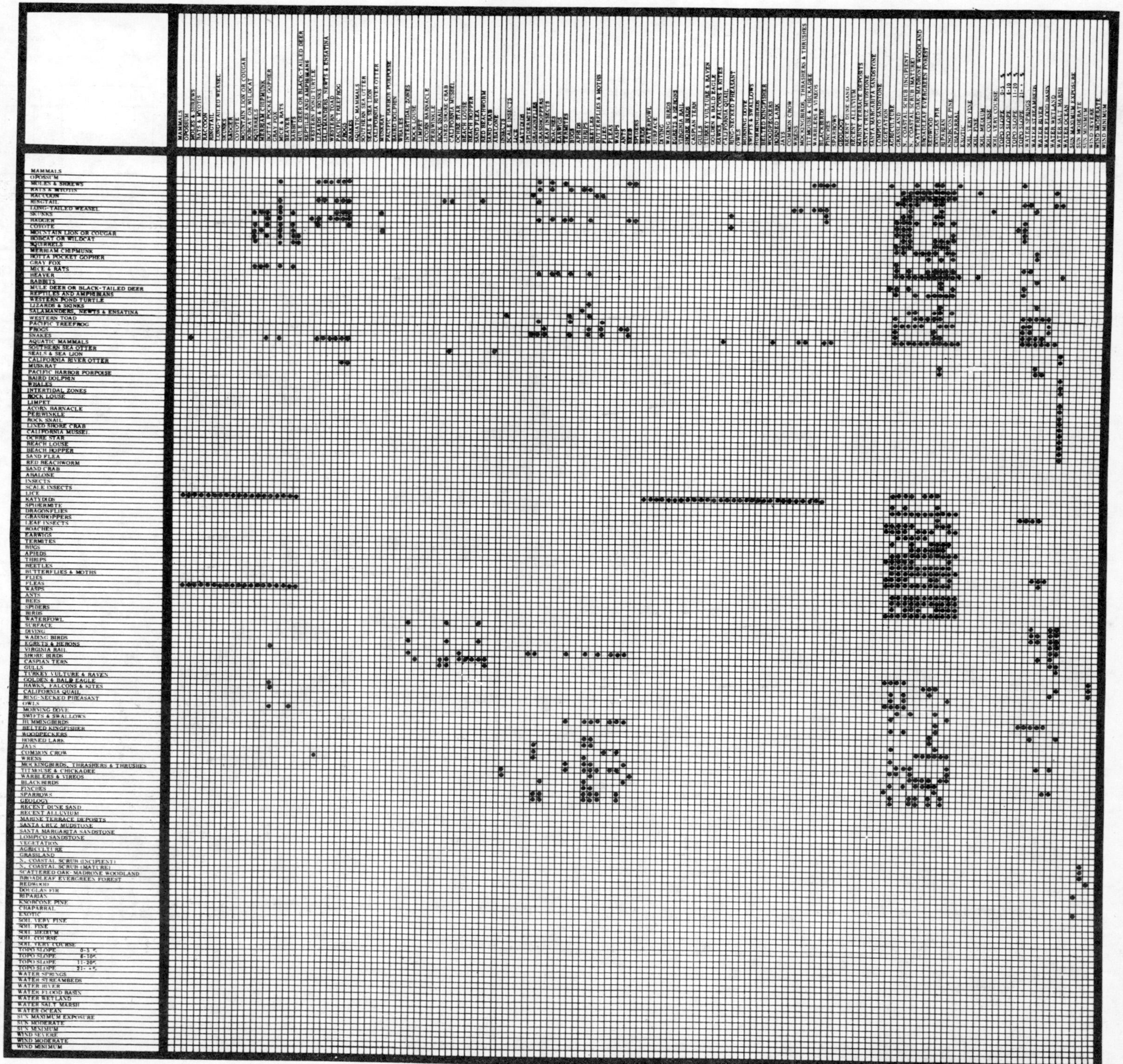

MAMMALS
OPOSSUM
MOLES & SHREWS
BATS & MYOTIS
RACCOON
RINGTAIL
LONG-TAILED WEASEL
SKUNKS
BADGER
COYOTE
MOUNTAIN LION OR COUGAR
BOBCAT OR WILDCAT
SQUIRRELS
MERRIAM CHIPMUNK
BOTTA POCKET GOPHER
GRAY FOX
MICE & RATS
BEAVER
RABBITS
MULE DEER OR BLACK-TAILED DEER
REPTILES AND AMPHIBIANS
WESTERN POND TURTLE
LIZARDS & SKINKS
SALAMANDERS, NEWTS & ENSATINA
WESTERN TOAD
PACIFIC TREEFROG
FROGS
SNAKES
AQUATIC MAMMALS
SOUTHERN SEA OTTER
SEALS & SEA LION
CALIFORNIA RIVER OTTER
MUSKRAT
PACIFIC HARBOR PORPOISE
BAIRD DOLPHIN
WHALES
INTERTIDAL ZONES
ROCK LOUSE
LIMPET
ACORN BARNACLE
PERIWINKLE
ROCK SNAIL
LINED SHORE CRAB
CALIFORNIA MUSSEL
OCHRE STAR
BEACH LOUSE
BEACH HOPPER
SAND FLEA
RED BEACHWORM
SAND CRAB
ABALONE
INSECTS
SCALE INSECTS
LICE
KATYDIDS
SPIDERMITE
DRAGONFLIES
GRASSHOPPERS
LEAF INSECTS
ROACHES
EARWIGS
TERMITES
BUGS
APHIDS
THRIPS
BEETLES
BUTTERFLIES & MOTHS
FLIES
FLEAS
WASPS
ANTS
BEES
SPIDERS
BIRDS
WATERFOWL
SURFACE
DIVING
WADING BIRDS
EGRETS & HERONS
VIRGINIA RAIL
SHORE BIRDS
CASPIAN TERN
GULLS
TURKEY VULTURE & RAVEN
GOLDEN & BALD EAGLE
HAWKS, FALCONS & KITES
CALIFORNIA QUAIL
RING-NECKED PHEASANT
OWLS
MORNING DOVE
SWIFTS & SWALLOWS
HUMMINGBIRDS
BELTED KINGFISHER
WOODPECKERS
HORNED LARK
JAYS
COMMON CROW
WRENS
MOCKINGBIRDS, THRASHERS & THRUSHES
TITMOUSE & CHICKADEE
WARBLERS & VIREOS
BLACKBIRDS
FINCHES
SPARROWS
GEOLOGY
RECENT DUNE SAND
RECENT ALLUVIUM
MARINE TERRACE DEPOSITS
SANTA CRUZ MUDSTONE
SANTA MARGARITA SANDSTONE
LOMPICO SANDSTONE
VEGETATION
AGRICULTURE
GRASSLAND
N. COASTAL SCRUB (INCIPIENT)
N. COASTAL SCRUB (MATURE)
SCATTERED OAK-MADRONE WOODLAND
BROADLEAF EVERGREEN FOREST
REDWOOD
DOUGLAS FIR
RIPARIAN
KNOBCONE PINE
CHAPARRAL
EXOTIC
SOIL VERY FINE
SOIL FINE
SOIL MEDIUM
SOIL COURSE
SOIL VERY COURSE
TOPO SLOPE 0-5 %
TOPO SLOPE 6-10%
TOPO SLOPE 11-20%
TOPO SLOPE 21- +%
WATER SPRINGS
WATER STREAMBEDS
WATER RIVER
WATER FLOOD BASIN
WATER WETLAND
WATER SALT MARSH
WATER OCEAN
SUN MAXIMUM EXPOSURE
SUN MODERATE
SUN MINIMUM
WIND SEVERE
WIND MODERATE
WIND MINIMUM

73. Ibid., p. 108.

AGRICULTURE SUITABILITY SPECIALTY CROPS[74]

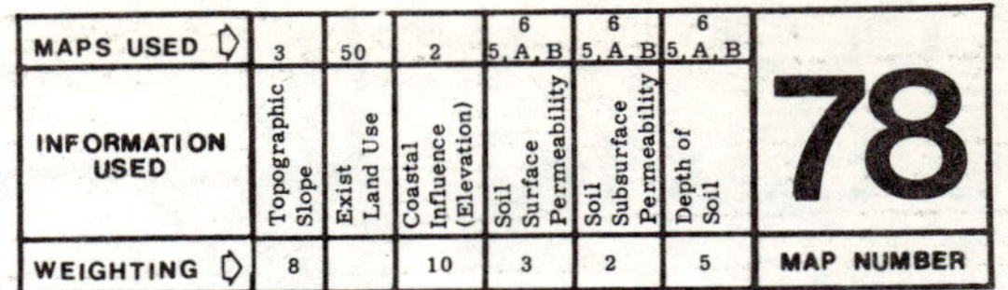

MAPS USED	3	50	2	6 5, A, B	6 5, A, B	6 5, A, B	78
INFORMATION USED	Topographic Slope	Exist Land Use	Coastal Influence (Elevation)	Soil Surface Permeability	Soil Subsurface Permeability	Depth of Soil	
WEIGHTING	8		10	3	2	5	MAP NUMBER

Although the agricultural land is currently in brussel sprouts as a result of their higher economic yield, other coastal influence crops would be suitable; broccoli, artichokes, lettuce, sugar beets and strawberries have been or could be grown, although artichokes are more deep-rooted and suffer when terrace soils become too shallow.

As the map indicates, nearly all the land of reasonable suitability exists on the first two terraces, with all cultivated lands shown, although some were not currently in crops at the time of this study.

Source: Appendix I. The Santa Cruz Mountains Regional Pilot Study Early Warning System," Patri, Streatfield, Ingmire.

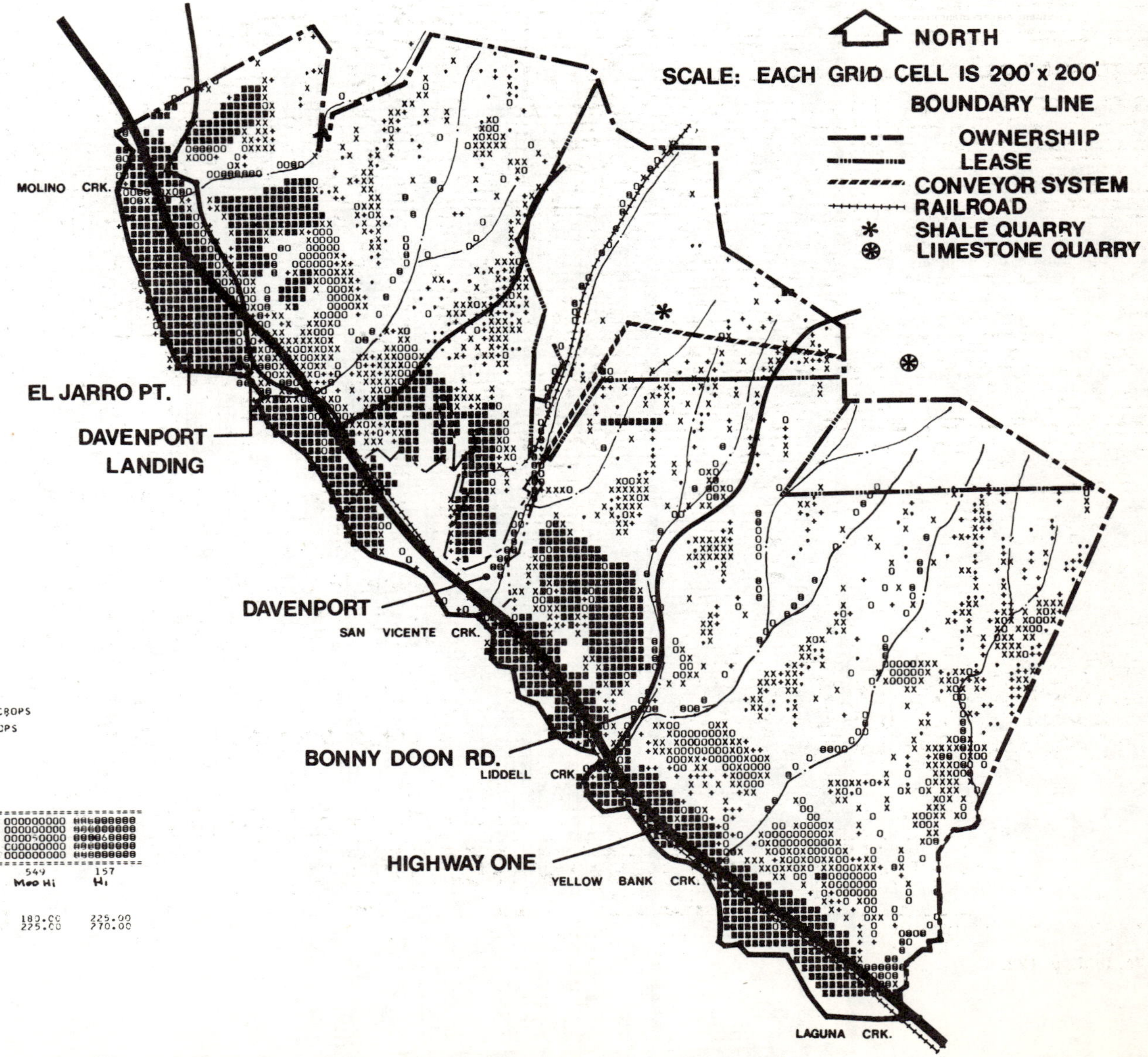

NORTH
SCALE: EACH GRID CELL IS 200' x 200'
BOUNDARY LINE
OWNERSHIP
LEASE
CONVEYOR SYSTEM
RAILROAD
SHALE QUARRY
LIMESTONE QUARRY
MOLINO CRK.
EL JARRO PT.
DAVENPORT LANDING
DAVENPORT
SAN VICENTE CRK.
BONNY DOON RD.
LIDDELL CRK.
HIGHWAY ONE
YELLOW BANK CRK.
LAGUNA CRK.

THEORETICAL MAXIMUM FOR SITE = 270

9 = EXISTING SPECIALITY CROPS
6 = HIGH POTENTIAL SUITABILITY FOR CONVERSION TO SPECIALITY CROPS
TO
1 = LOW POTENTIAL SUITABLITY FOR CONVERSION TO SPECIALITY CROPS

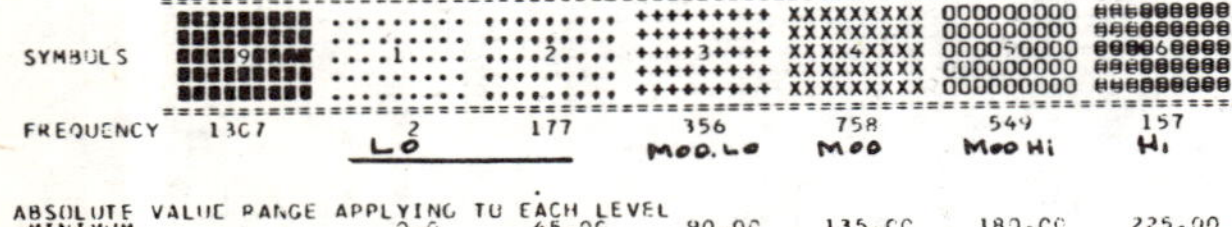

SYMBOLS	9	1	2	3	4	5	6
FREQUENCY	1307	2	177	356	758	549	157
		LO	LO	MOD. LO	MOD	MOD HI	HI
ABSOLUTE VALUE RANGE APPLYING TO EACH LEVEL							
MINIMUM		0.0	45.00	90.00	135.00	180.00	225.00
MAXIMUM		45.00	90.00	135.00	180.00	225.00	270.00

74. Ibid., p. 118.

FORESTRY SUITABILITY[75]

This map is based on a forest resource study and involves existing stands of timber trees (redwood and douglas fir) according to age and density (see map 8).

Basically, the canyons beyond the zone of high direct coastal influence are dominated by redwood, and the suitability and cost figures are based on the assumption of a Selective Thinning and Cultural improvement management program.

Recent logging has stripped the site of all merchantable timber and generally left the canyon bottoms rather torn up and littered. Many roads would be substandard under today's rules.

However, redwood and the canyons bottoms are both resilient and after some reopenings and improvements, existing roads could be utilized, giving accessibility to within 1/4 mile of all timber.

Source: Appendix K.

MAPS USED	7	8	3	82
INFORMATION USED	Vegetation Type	Vegetation Density	Topographic Slope	
WEIGHTING	1	1	1	MAP NUMBER

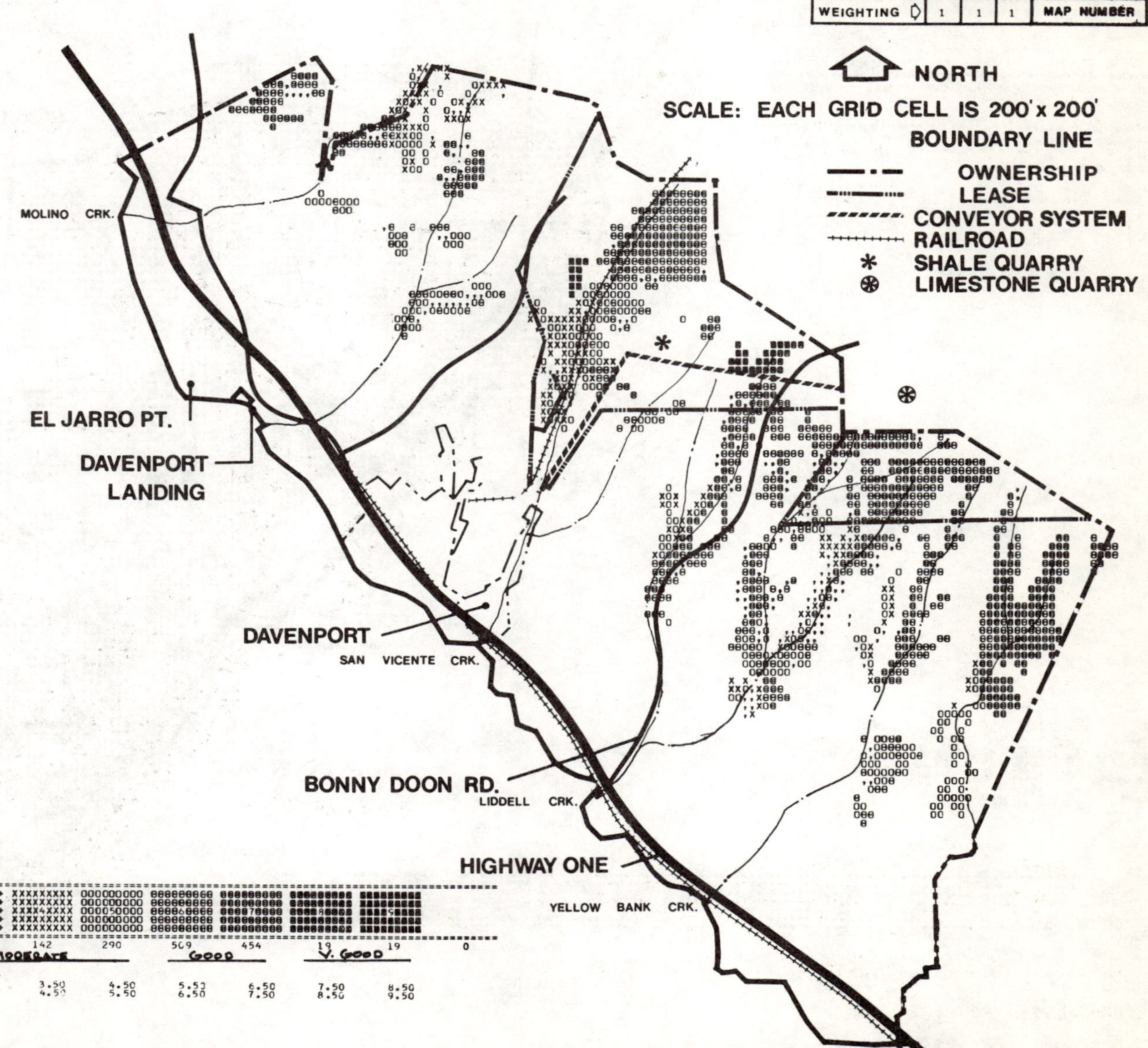

0 = NC TIMBER PCTENTIAL
1 = 100 - 150 PCTENTIAL BCARD FEET PER ACRE PER YEAR
2 = 2C0 POTENTIAL BOARC FEET PER ACRE PER YEAR
3 = 400 POTENTIAL BCARC FEET PER ACRE PER YEAR
4 = 6CC POTENTIAL BOARC FEET PER ACRE PER YEAR
5 = 750 POTENTIAL BOARC FEET PER ACRE PER YEAR
6 = 900 POTENTIAL BOARC FEET PER ACRE PER YEAR
7 = 1050 POTENTIAL BOARC FEET PER ACRE PER YEAR
8 = 1500 POTENTIAL BOARD FEET PER ACRE PER YEAR
9 = 1800 PCTENTIAL BCARC FEET PER ACRE PER YEAR

THIS MAP DISPLAYS LANDSCAPE SUITABILITY FOR FORESTRY
IT WAS PRODUCED BY COMBINING SLOPE CLASSIFICATIONS
WITH EXISTING TIMBER VEGETATICN.

SYMBOLS: ....0.... ;;;;1;;;; ••••2•••• ++++3++++ XXXX4XXXX 000050000 8888688888 888878888 ████8████ ████9████

FREQUENCY: 0 0 80 86 0 142 290 509 454 19 19 0

LOW — FR-MODERATE — GOOD — V. GOOD

ABSOLUTE VALUE RANGE APPLYING TO EACH LEVEL

	0	1	2	3	4	5	6	7	8	9
MINIMUM	-0.50	0.50	1.50	2.50	3.50	4.50	5.50	6.50	7.50	8.50
MAXIMUM	0.50	1.50	2.50	3.50	4.50	5.50	6.50	7.50	8.50	9.50

75. Ibid., p. 122.

DIRECT ENGINEERING SUITABILITY[76]

This map describes terrain units in quantitative terms of land capability as measured from the point of view of land development and use costs under conventional building standards. 1971 figures are used, and the costs are for single family detached residential housing, 4 lots/acre, site improvements, but no access road included. A minimum size developed area of 50 acres was assumed, although the essence of these prices is that their order and general ratio of suitability remains constant even as the construction type changes.

These costs tend to eliminate most of the site from consideration in terms of market demand.

Source: Appendix E

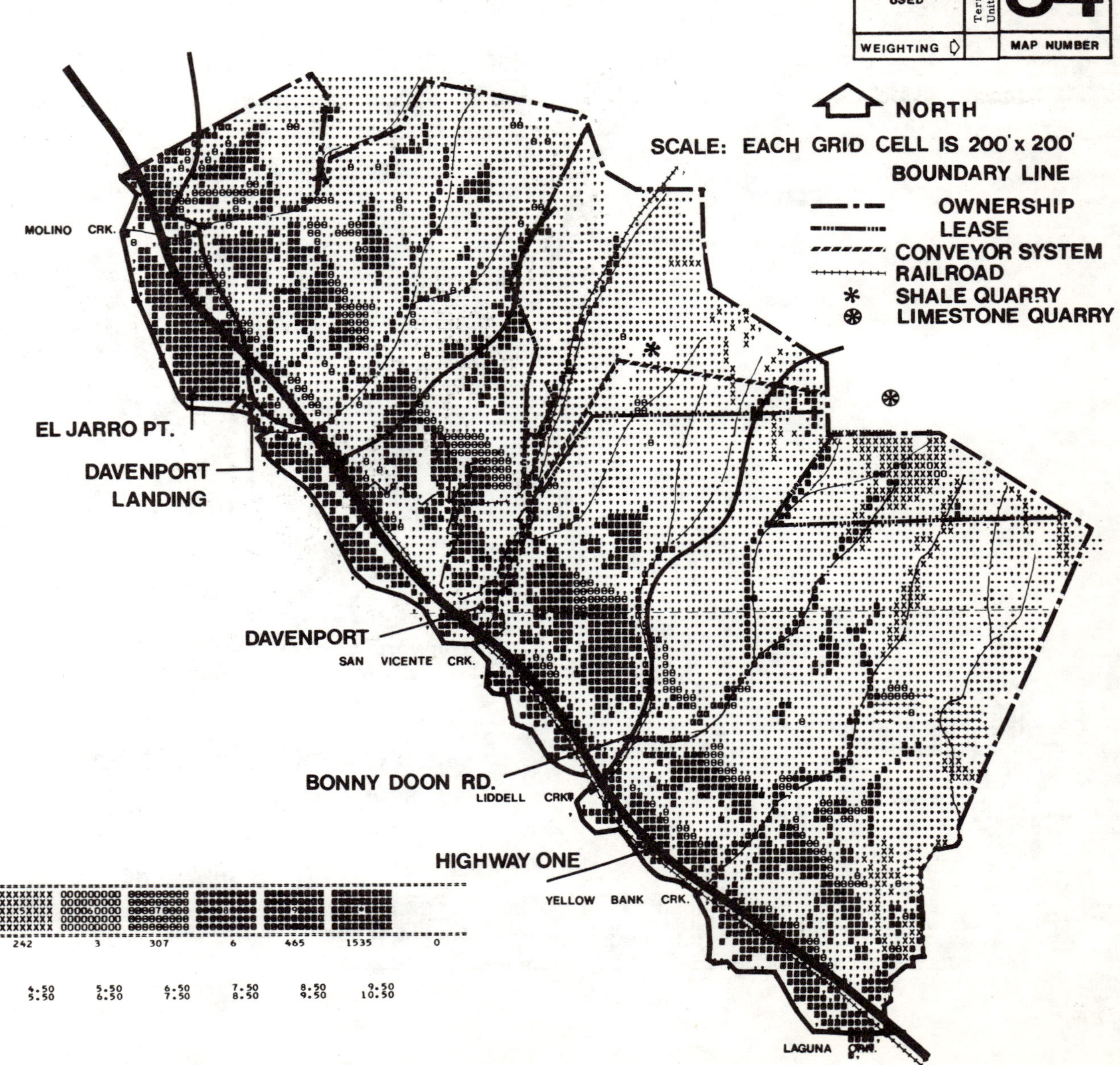

TD
MBC
MBCF

* = $4000/LOT
9 = $4130/LOT
8 = $4530/LOT
7 = $4690/LOT
6 = $5740/LOT
5 = $6430/LOT
4 = $7050/LOT
3 = $10000/LOT
2 = $13900/LOT
1 = $14200/LOT
BLANK AREAS = AREAS NOT STUDIES

SYMBOLS		1	2	3	4	5	6	7	8	9	*	
FREQUENCY	1693	636	3608	986	73	242	3	307	6	465	1535	0
ABSOLUTE VALUE RANGE APPLYING TO EACH LEVEL												
MINIMUM		0.50	1.50	2.50	3.50	4.50	5.50	6.50	7.50	8.50	9.50	
MAXIMUM		1.50	2.50	3.50	4.50	5.50	6.50	7.50	8.50	9.50	10.50	

76. Ibid., p. 124.

ACCESS ROAD SUITABILITY[77]

MAPS USED	5 A, B	85
INFORMATION USED	Terrain Units	
WEIGHTING		MAP NUMBER

This map depicts terrain units expressed in terms of accessibility; the dollar basis is a 60 foot right-of-way access road to county standards with complete utility trunk lines for a 200 unit residential development. It represents the other half of map 84, the former showing how much it costs to get there, the latter how much it costs to build once you are there. Together, they paint a bleak picture for most of the site, as the back country combines high access costs with small developable blocks of land, which means per unit costs are quite high, all things considered.

Source: Appendix E.

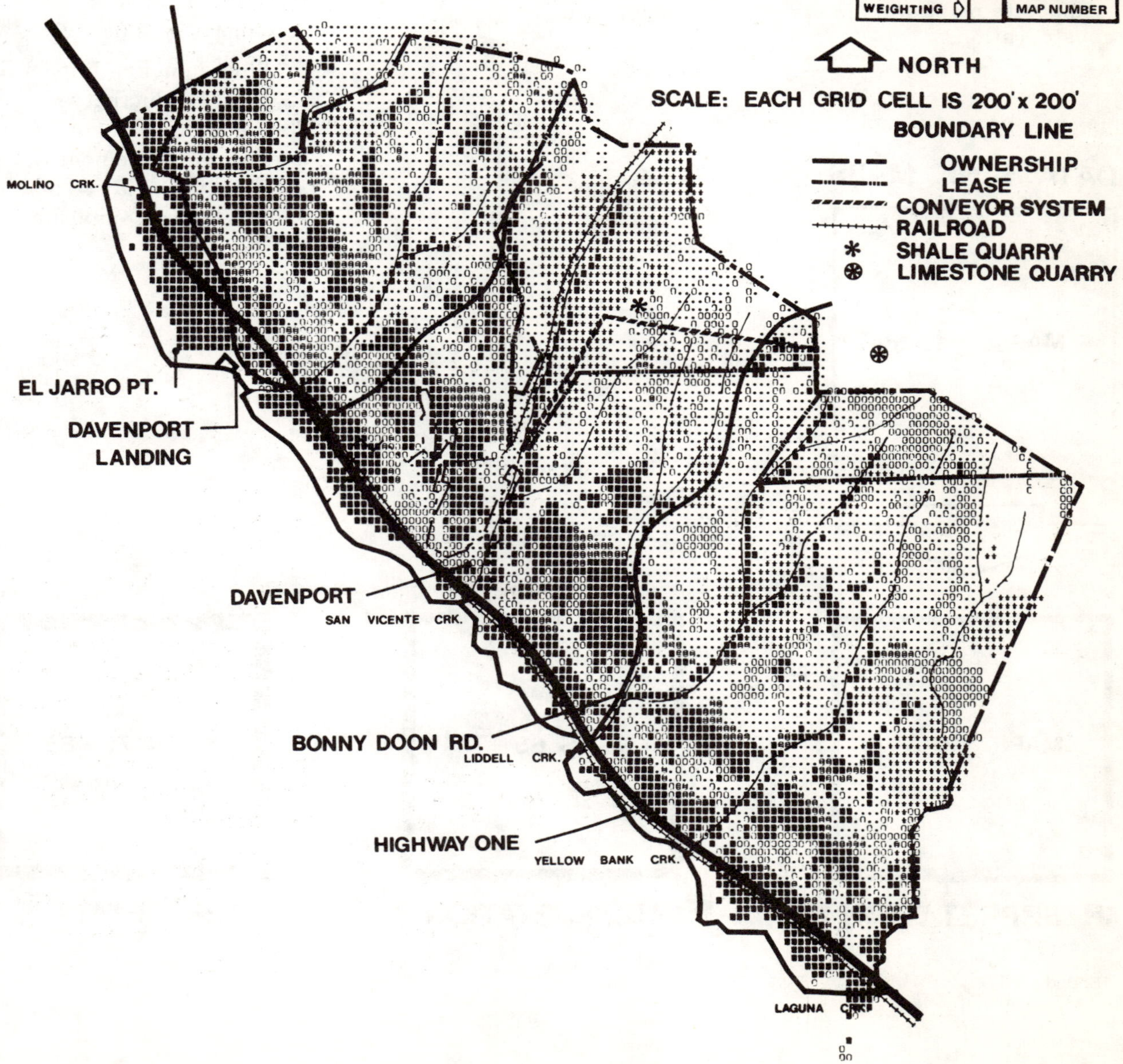

NORTH

SCALE: EACH GRID CELL IS 200' x 200'

BOUNDARY LINE

OWNERSHIP

LEASE

CONVEYOR SYSTEM

RAILROAD

* SHALE QUARRY

⊛ LIMESTONE QUARRY

MOLINO CRK.

EL JARRO PT.

DAVENPORT LANDING

DAVENPORT

SAN VICENTE CRK.

BONNY DOON RD.

LIDDELL CRK.

HIGHWAY ONE

YELLOW BANK CRK.

LAGUNA CRK.

Costs in $/Linear foot

CBS, TD	5. 61.00 - 63.00
SW	4. 71.00 ±
EF, MSB, SSC, SSR	3. 90.00 - 94.00
LSB, LSD, LS	2. 111.00 ±
MSC, MSCF	1. 141.00 - 144.00

ABSOLUTE VALUE RANGE APPLYING TO EACH LEVEL

MINIMUM	1.00	1.80	2.60	3.40	4.20
MAXIMUM	1.80	2.60	3.40	4.20	5.00

77. Ibid., p. 125.

INTERPRETATION 4[78]

Composite maps for the six paths of physical land consideration.

The following illustrations are representative of the maps numbered 86 through 91 by the consultants, and represent "composite maps for the six paths of physical land considerations."

These maps include:

- Environmental Impact Restrictions
- Land Productivity
- Generalized Engineering Suitability
- Man-Made Element Composite
- Climate Comfort Total Year
- Aesthetics Composite

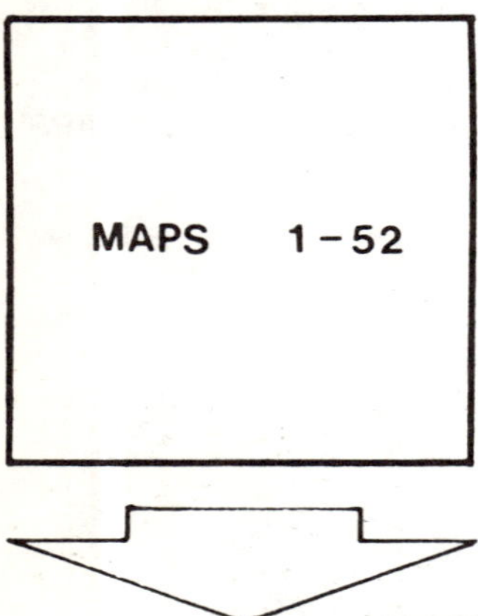

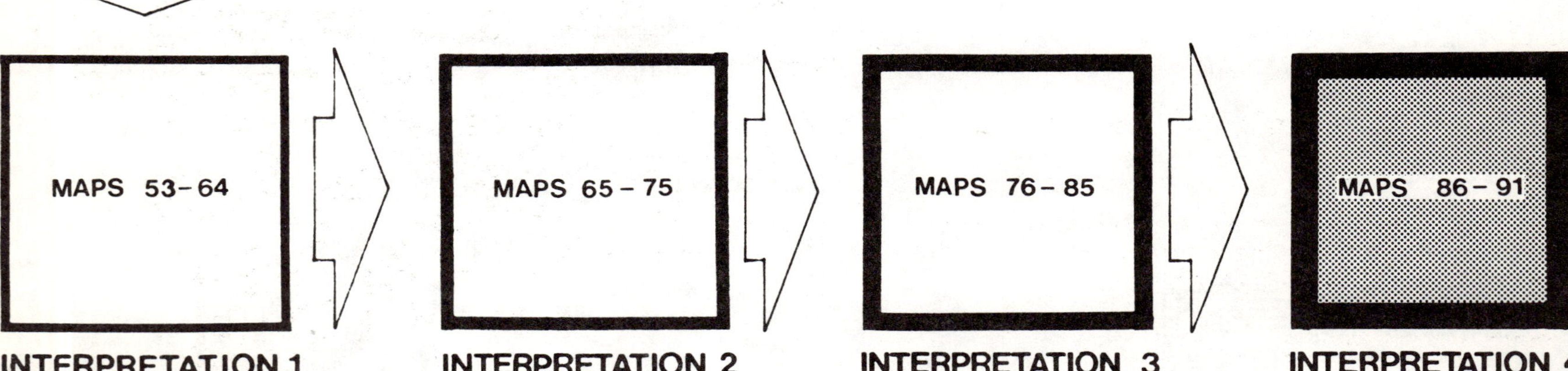

78. Ibid., p. 126.

ENVIRONMENTAL IMPACT RESTRICTIONS[79]

This map draws from land-based environmental phenomena - geology, hydrology, vegetation, and wildlife - in the expressed form of geologic forces restrictions, vegetation restrictions, hydrologic restrictions, sensitive wildlife habitat (mapped here as streams, since water and associated scrub slopes are highest in wildlife dependancy) and fire hazard, (vegetation, slope, and wind combined). The basic vehicles are thought of as Primary Impact - on site disturbance (fire geology) and 2nd Impact - Disturbance of Biotic systems (vegetation, water and wildlife. The notion is that man has a tendency to locate in ignorance of direct or primary natural forces - areas unsuited by geology or fire hazard, while his land modification frequently generates reverberations in the landed system-secondary impacts which tend to follow the landscape causality levels portrayed under Process, but at an accelerated rate.

Examples of this include the following potential chain - reactive events:

1. Groundwater removal = possible stream flow reduction in summer= vegetation disappearances = change in light quantity reaching streams = wildlife change.
2. Addition of unnatural water (irrigation, pond, septic) = destruction of native vegetation = wildlife change
   or
   = increased runoff from already - saturated soils at time of winter rain = erosion = plant destruction = stream siltation & degrading = more wildlife and vegetation destruction and flooding = more siltation.
3. Construction and development (general) = compaction = alteration of surface or groundwater flow = vegetation destruction from either water addition or removal (marked change in natural environment)
4. Logging, clearing = compaction, exposure of land to force of rain = erosion and altered runoff = siltation of stream etc.
5. Fire suppression = death of fire - dependant plants, buildup of flammable duff, etc. = severe fire when it does occur = destruction of vast acreage, unusual hazard and difficulty of fire control, destruction of many plant species, destruction of soil, and opening of lands to rainfall direct impact = heightened runoff and erosion, etc.

The site is revealed to be subject to some of these aggravations; rather high fire hazard combines with already rather unstable vegetation over much of the site along with geologic instability as the worst of the potential problems. Erosion is not great, and the scrub and much wildlife are on what should be considered "undevelopable" slopes.

VEGETATION RESTRICTIONS

Knobcone Pine; Chaparral; Hardwoods; Douglas Fir; Coastal Scrub

HYDROLOGIC RESTRICTIONS

Flooding Hazard, Recharge (Potential minor aquifer, or potential salt water intrusions)

Recharge (potential major aquifer)

Year round stream (5-6 C.F.S.)

Year round stream (2 C.F.S.)

Year round stream (1 C.F.S.)

Major intermittent stream

San Vicente Creek - Large Off-Site Watershed.

Laguna Creek East - Moderately Large Off-Site Watershed

Molino Creek - Small Watershed - But Mostly Off-Site

Liddell Creek - Small Off-Site Watershed Dominated by Quarrying Operations

FIRE HAZARD

General area of relatively high fire hazard

GEOLOGIC FORCES RESTRICTIONS

Expressed as $ 5,000.00 risk cost per 1/4 acre lot

Expressed as $ 10,000.00 risk cost per 1/4 acre lot

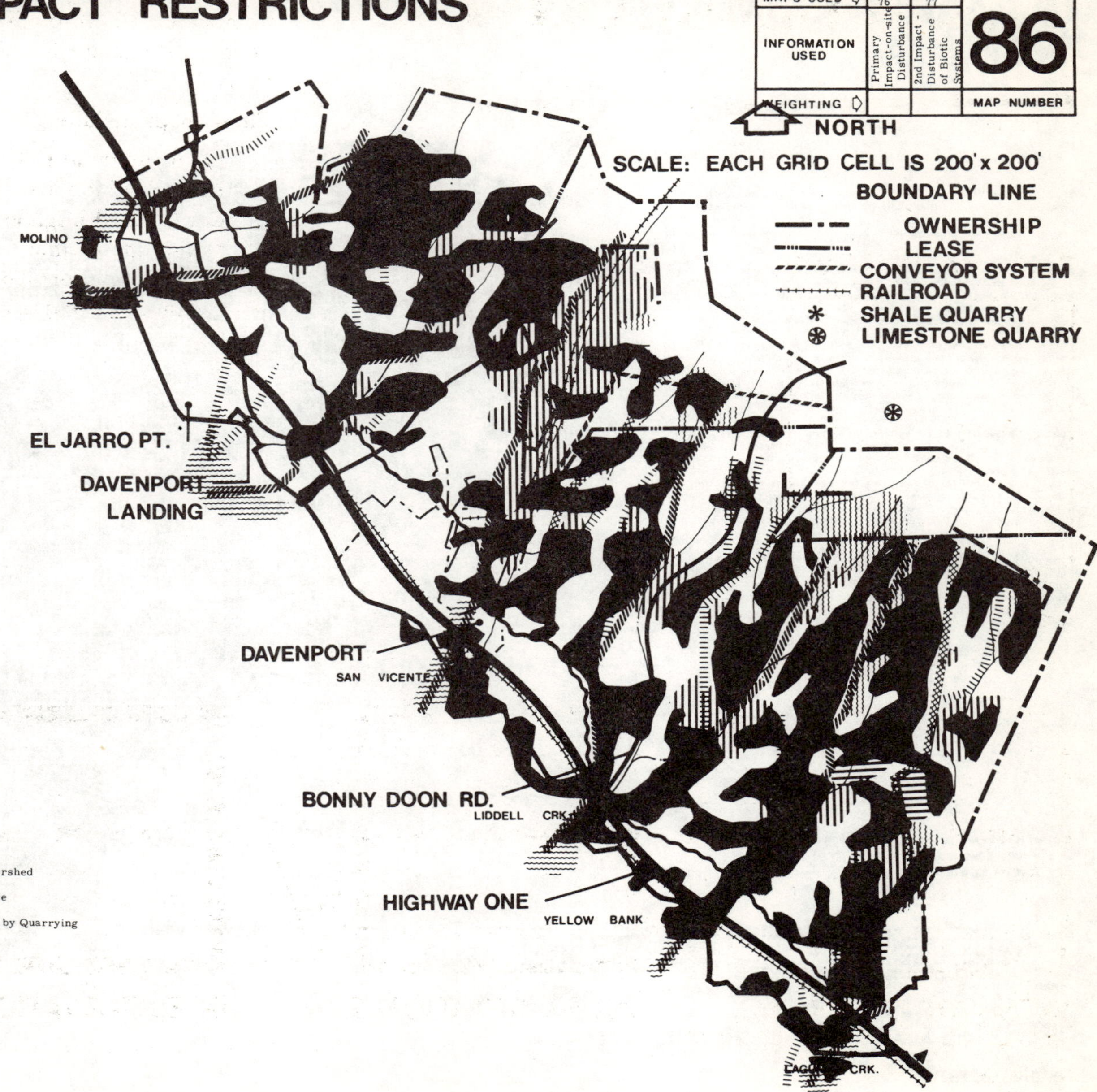

79. Ibid., p. 127.

CLIMATIC COMFORT[80]

TOTAL YEAR

MAPS USED	70	71	72	73	90
INFORMATION USED	Spring Climate Comfort	Fall Climate Comfort	Summer Climate Comfort	Winter Climate Comfort	
WEIGHTING	1	1	1	1	MAP NUMBER

The purpose of this map is to compare areas on the site for climatic comfort during the entire year. The light areas indicate maximum comfort on the site. The dark areas indicate minimum comfort. The maximum and minimum values are not absolute but only refer to the range of climate on this site.

This map allows a comparison between climatic requirement of potential land uses and a generalized picture of the climatic problems and potentials of the site. It also allows a comparison of climatic comfort with problems and potentials in other sections of the land use study.

Areas of greatest climatic comfort tend to be at stream bottoms and on the southeast facing slopes of the canyons. These areas are normally quite steep (4:1 - 2:1 slopes) and therefore present difficult problems for any kind of large scale construction. Trails however could be developed on some of the more gradual topography.

There are some areas of maximum climatic comfort which are not steep canyon slopes. The major area of this sort is near the property line at Laguna Creek extending about 1/2 the distance to the back property line. Although this area would be relatively unsuitable for large scale construction, it could support camping, picnicking and trails.

The large masses of dark tone on the map (minimum comfort) correspond to the major flat areas of the site at the first and second terrace. These areas would normally be the most highly suitable for development. However climatic conditions limit their suitability. In terms of agriculture these flat areas of the first and second terrace are somewhat limited by climatic conditions. Salt spray, wind, fog and moderate temperatures limit plant types suitable for cultivation. Grazing is not limited by the climatic conditions in the areas of minimum climatic comfort.

Other areas shown as minimum comfort are generally located on narrow ridge tops between steep canyons. They are highly suitable for large scale construction but inaccessibility and poor climate are the major factors in limiting their use for development. Grazing exists here and trails could also be developed.

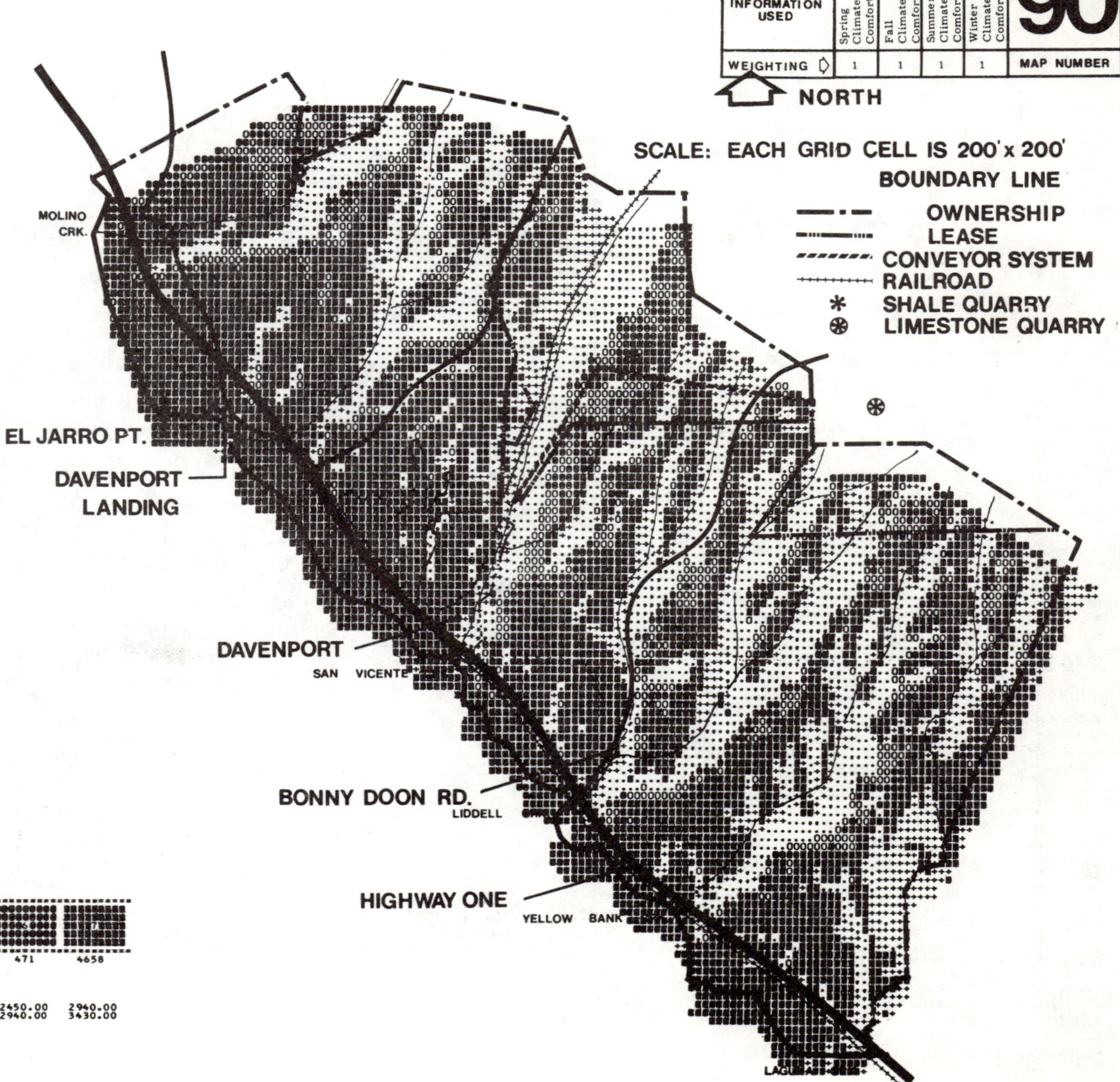

CLIMATIC COMFORT BY SEASON
1 = MOST COMFORTABLE
TO
7 = LEAST COMFORTABLE

SYMBOLS	1	2	3	4	5	6	7
FREQUENCY	1025	518	1046	0	688	471	4658

ABSOLUTE VALUE RANGE APPLYING TO EACH LEVEL

MINIMUM	0.0	490.00	980.00	1470.00	1960.00	2450.00	2940.00
MAXIMUM	490.00	980.00	1470.00	1960.00	2450.00	2940.00	3430.00

80. Ibid., p. 131.

AESTHETICS COMPOSITE[81]

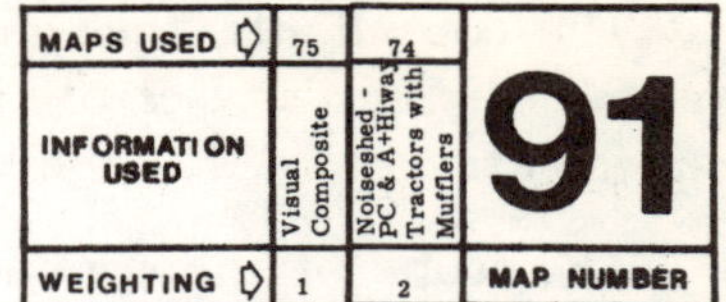

This map represents an overlaying of noise factors over the visual composite, and is indicative of the rather substantial proportion of the site which is effected by either visual or noise distractions, or both. Noise is considered the more serious of the problems, as it is less subject to screening. For purposes of this model, the noise level is considered to be generated by quarry tractors with mufflers.

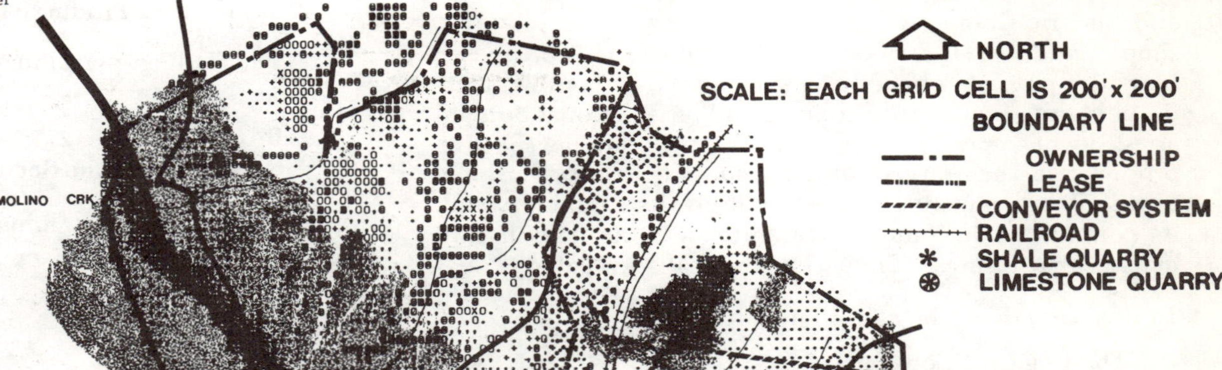

NOISE LEVELS:

- Ambient + 10 DBA
- Ambient + 20 DBA
- Road corridor or quarry and rock crushers (Ambient + 30 DBA)

THIS MODEL COMBINES INFORMATION FROM A SERIES OF SUB-VISUAL STUDIES

- * = HIGH VISUAL CHARACTER AND NO VISUAL LIMITATIONS
- 9 = MODERATELY HIGH VISUAL CHARACTER AND NO VISUAL LIMITATIONS
- 8 = FAIRLY HIGH VISUAL CHARACTER AND NO VISUAL LIMITATIONS
- 6 = HIGH VISUAL CHARACTER AND MILD VISUAL LIMITATIONS
- 5 = MODERATELY HIGH VISUAL CHARACTER AND MILD VISUAL LIMITATIONS
- 4 = FAIRLY HIGH VISUAL CHARACTER AND MILD VISUAL LIMITATIONS
- 2 = FAIRLY HIGH VISUAL CHARACTER AND MODERATE TO HIGH VISUAL LIMITATIONS
- 1 = NO DEFINED VISUAL CHARACTER AND MILD VISUAL LIMITATIONS (BLANK)
  BLANK AREAS = NO DEFINED VISUAL CHARACTER AND MOD. TO HI VIS LIMIT, NO DEFINED VISUAL CHARACTER OR LIMITATIONS

OF DATA POINT VALUES IN EACH LEVEL

ABSOLUTE VALUE RANGE APPLYING TO EACH LEVEL								
MINIMUM	0.50	1.50	2.50	3.50	4.50	5.50	6.50	7.50
MAXIMUM	1.50	2.50	3.50	4.50	5.50	6.50	7.50	8.50

81. Ibid., p. 132.

The consultants, then deal with the regional implications of various alternative uses of the 7,000 acre site with the following statement:

> The scope of the regional implication study is to set the various alternatives posed by the actions of the Pacific Gas and Electric Company into a context larger than that of a simple land-use study for a 7,000 acre parcel. To the extent that physical phenomena such as terrain and weather do not respect political property and jurisdictional boundaries, it becomes necessary always to examine specific problems in their larger contexts, so that the true meaning of possible Company choices for itself, its neighbors, and its community become visible and can be dealt with in a manner that insures effectuation and realistic goals.[82]

In this part, then, the consultants deal with:

A. The Regional Context
B. Governmental Structure which included:
   1. Public Planning and Fiscal Policy
   2. Transportation
   3. County Services
   4. School Districts
   5. Utility Districts
   6. Other Special Districts
C. Social and Economic Characteristics, under this were considerations such as:
   1. Population
   2. Housing
   3. Income and Employment
D. Historical Determinants of Land Use—The Land Itself
E. Subregional Physical Constraints

From all this Eckbo, Dean, Austin & Williams developed a series of findings and recommendations based, essentially, on two possible alternative uses of the site. These centered around a development plan, and an open-space plan. The outline of areas covered under the findings pertaining to land suitability included:

A. Aesthetics
B. Climate
C. Vegetation and Wildlife
D. Water Resources
E. Topography
F. Geology and Soils
G. Buildable Areas
H. Planned Productivity

The findings pertaining to urban development included:

A. Market
B. Economics
C. Fiscal—Public Sector

The findings pertaining to open-space uses included:

A. Market
B. Economics
C. Fiscal—Public

The conclusions and recommendations included the following categories:

A. Land-Use Suitability
B. Development
C. Open Space

The significance of this study and its process cannot be underestimated. It is probably one of the pioneering efforts by an electric utility industry representative to scientifically evaluate property utilizing environmental considerations and a natural resource base as primary determinants in decision making. Both processes and methodology are pioneering and innovative. Certainly it seems that they provide an excellent model for other sectors dealing with the entire problem of energy and environment. As mentioned at the outset, the land-use study for Pacific Gas and Electric Company for the Davenport site was only the first part of a three-part study. The second publication and part of the study dealt with the precise siting of the nuclear power plant.

This particular study, according to the consultants, "considered only aesthetics (mainly visibility) and impact on their terrestrial environment."

82. Ibid., p. 135.

The methodology of the studies included the fact that

> . . . six possible layouts representing a wide range of alternatives were prepared by P.G. & E. and provided to Eckbo, Dean, Austin & Williams for review. The objective was to develop criteria for making aesthetic and environmental evaluations to combine with the utility's cost, feasibility, geologic, seismologic, and marine ecological studies so that an optimum site layout could be developed.
>
> Closely involved in this part of the study are the character of the site itself, the nature of the power plant facilities being proposed, and the existing and potential land uses in the neighborhood.[83]

The scope, techniques and methodologies of the study, while innovative and helpful in decision-making for this particular project, provides, it seems, lessons, guidance and a starting point for further studies in other locations under similar circumstances. Certainly, few other studies of this type exist for guidance of persons faced with some of the problems of energy and environment. Therefore, it seems well worthwhile to make extensive references to that document dealing with the various aspects such as scope and process. Though the eventual findings are of less interest in this particular situation than they were in the original study, the consultants defined the scope of the project as follows:

> The assignment is to examine and make recommendations relative to the aesthetic and environmental factors involved in the siting of the power plant and its appurtenances, and within this contest to:
>
> A. Establish guidelines for the siting of power plants;
>
> B. Establish a general approach toward determining the visual and environmental acceptability of various designs;
>
> C. Determine what impact visibility will have in terms of potential and existing land uses;
>
> D. Determine the relative visibility in the study area of alternative power plant and switchyard locations;
>
> E. Recommend locations for screening with earth forms and trees;
>
> F. Determine problems and potentials of various designs in order to propose solutions to the above problems in design critiques.
>
> G. Make final recommendations regarding the specific schemes tested and from these establish general guidelines for future power plant site evaluation and selection.

The E.D.A.W. staff, then, utilized the following two units of the plant shape assumed for planning study—two units.

83. Garrett Eckbo, Francis H. Dean, Donald B. Austin and Edward A. Williams, *Power Plant Siting,* (San Francisco, California, Eckbo, Dean, Austin and Williams, for Pacific Gas and Electric Company, 1970), p. 1.

POWER PLANT SHAPE ASSUMED FOR PLANNING STUDY — TWO UNITS[84]

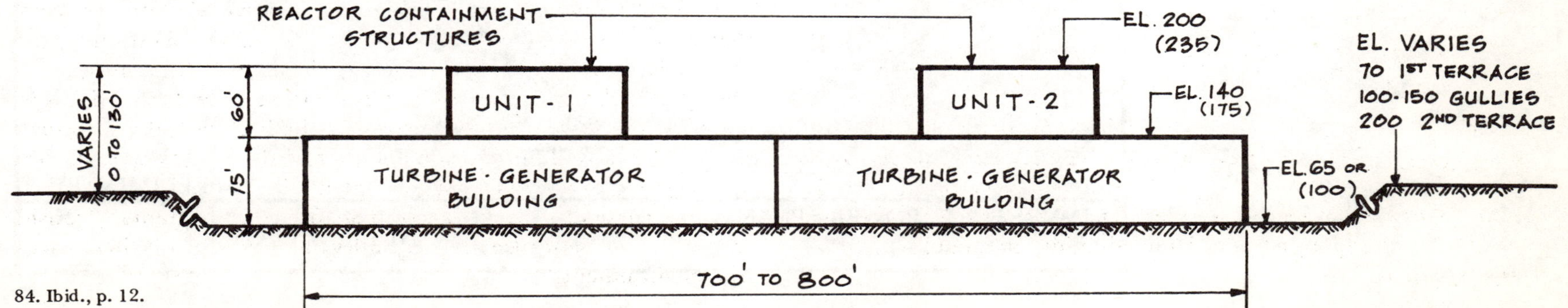

84. Ibid., p. 12.

Eckbo, Dean, Austin & Williams then prepared six alternative site plans, placing the power plant, a 230-KV Switchyard and a 115 KV Switchyard on the site. Some of these alternatives called for the relocation of existing Highway 1 along the Pacific Coast. Most called for screening with an earth form or landscaping. All of the proposed site plan alternatives contained sectional views to indicate the relationship of the various man-made accoutrements and the required alteration of the natural landscape. Two of these preliminary alternative plans are shown in the following illustrations.

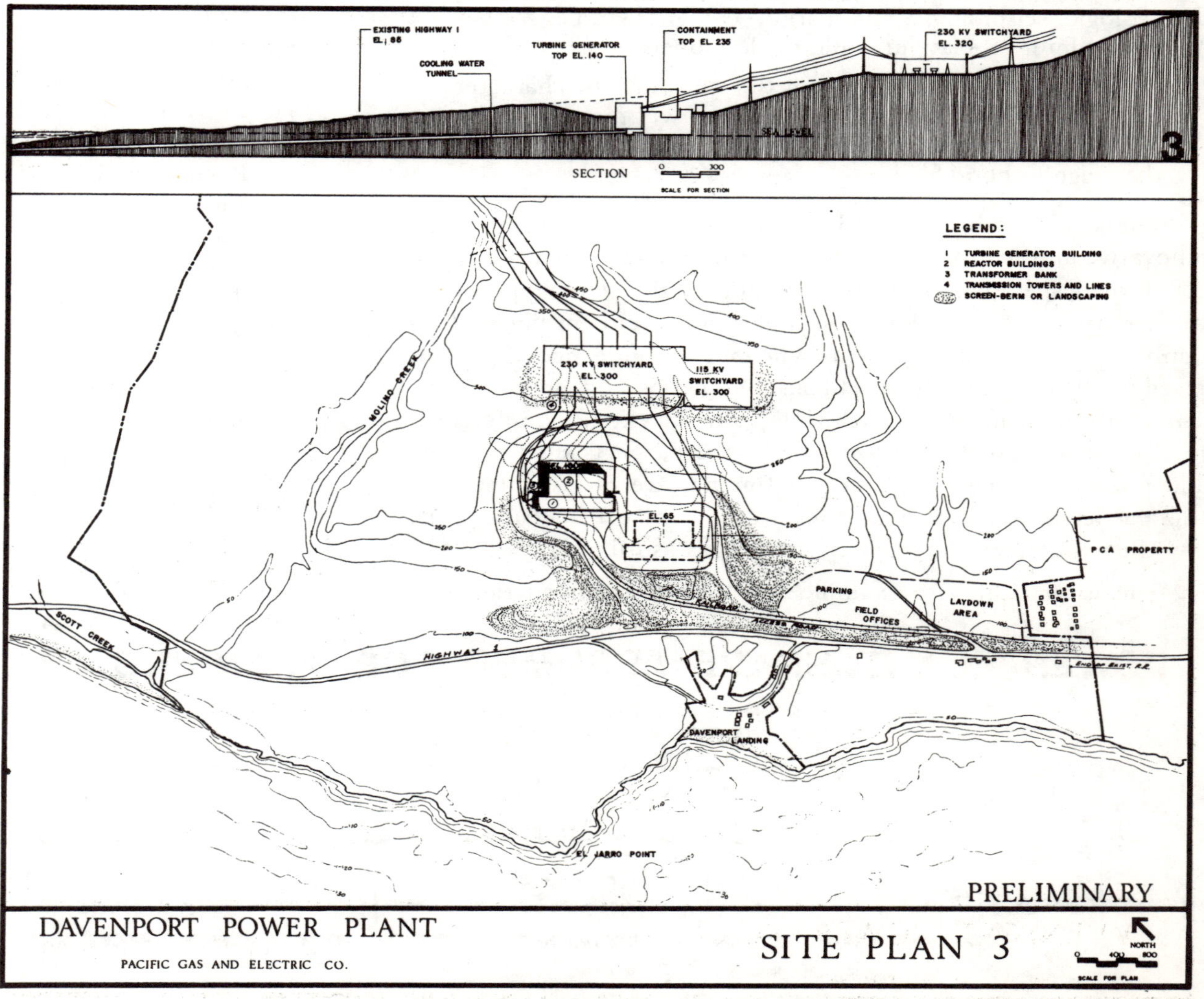

85. Ibid., p. 8.

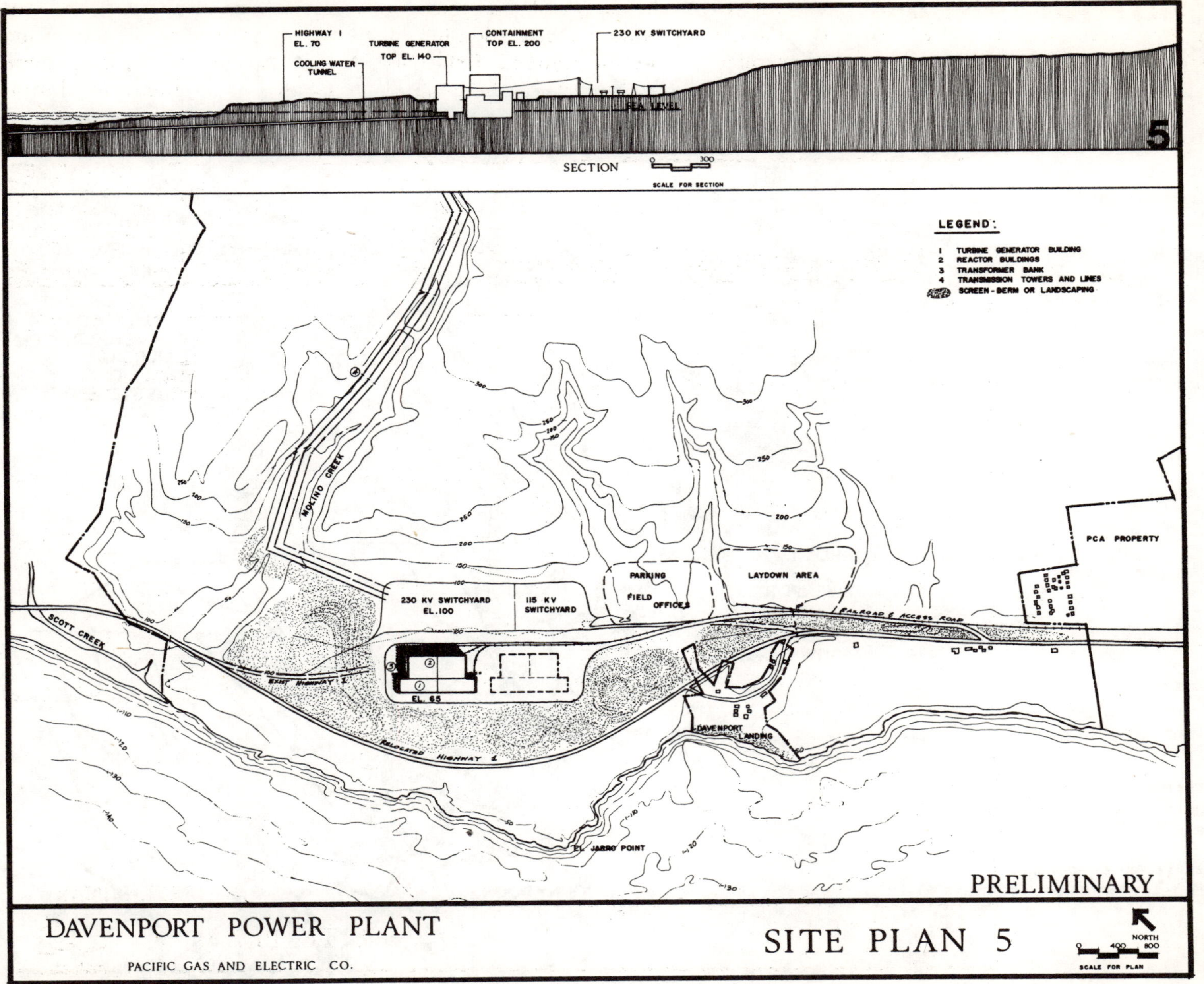

86

86. Ibid., p. 10.

86. Ibid., p. 10.

VISIBILITY STUDY AREA[87]

SCENIC HIGHWAY 1
VISUAL QUALITIES

The site itself was then analyzed in great detail to indicate the opportunities and the restraints which existed in the natural environment as it related to the possible development of the site for a power plant. One of these maps dealt with the visibility study area and is shown in the following illustration.

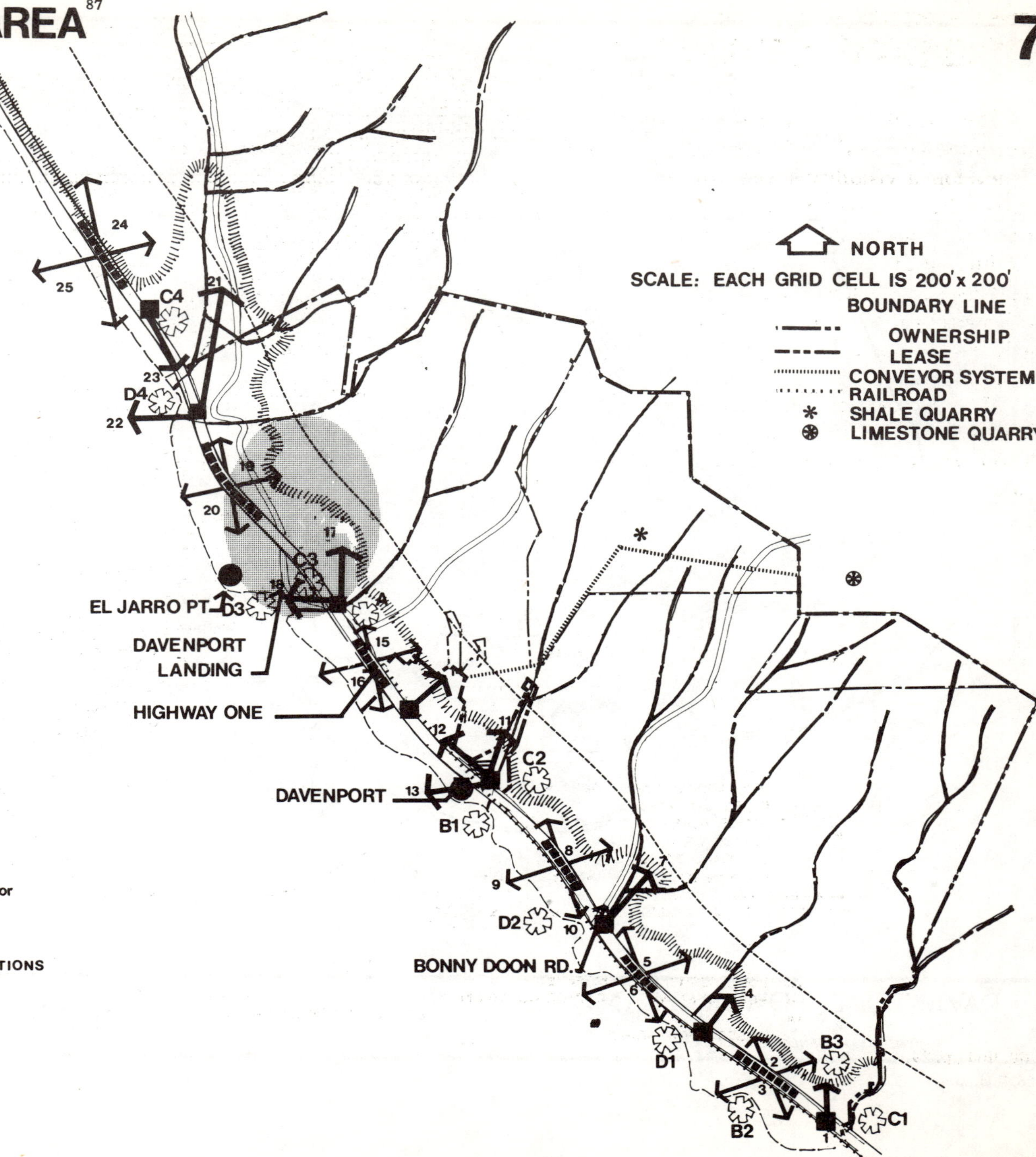

87. Ibid., p. 16.

17

POWER PLANT SITE PLAN 1 UNIT 2

VISIBILITY STUDY[88]

The consultants then utilized the computer and the grid system to develop a visibility rating for each of the proposed power plant schemes. They conducted these visibility studies for the power plant itself, as well as for the 230 KV Switchyard and the 115 KV Switchyard on each of the proposed site plans.

The following illustration, then, shows visibility study for Power Plant Site 1 and Power Plant Unit 2.

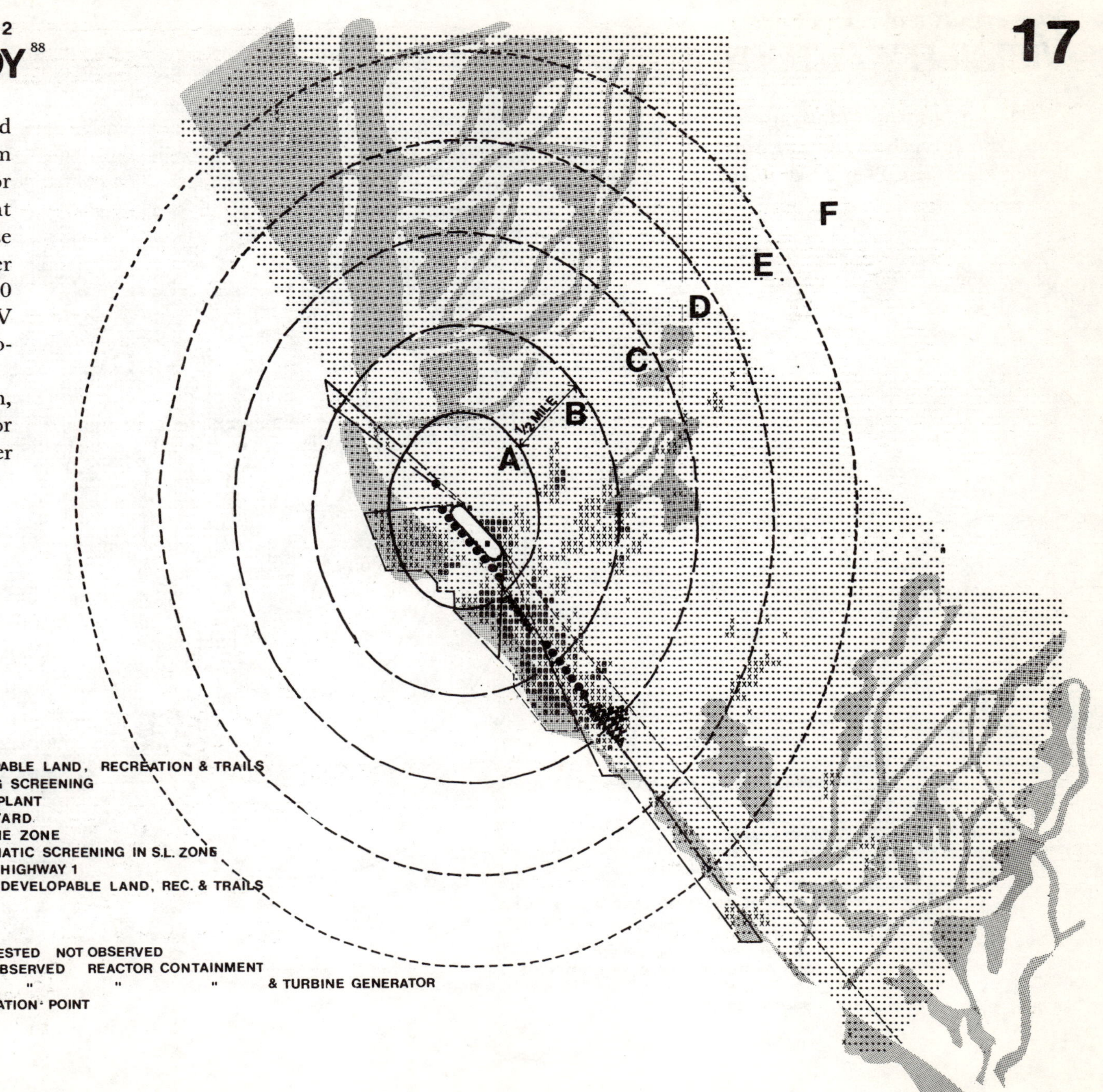

88. Ibid., p. 29.

POWER PLANT SITE PLAN 2 UNIT 1

VISIBILITY STUDY[89]

The following illustration, then, shows visibility study for Power Plant Site Plan 2—Unit 1.

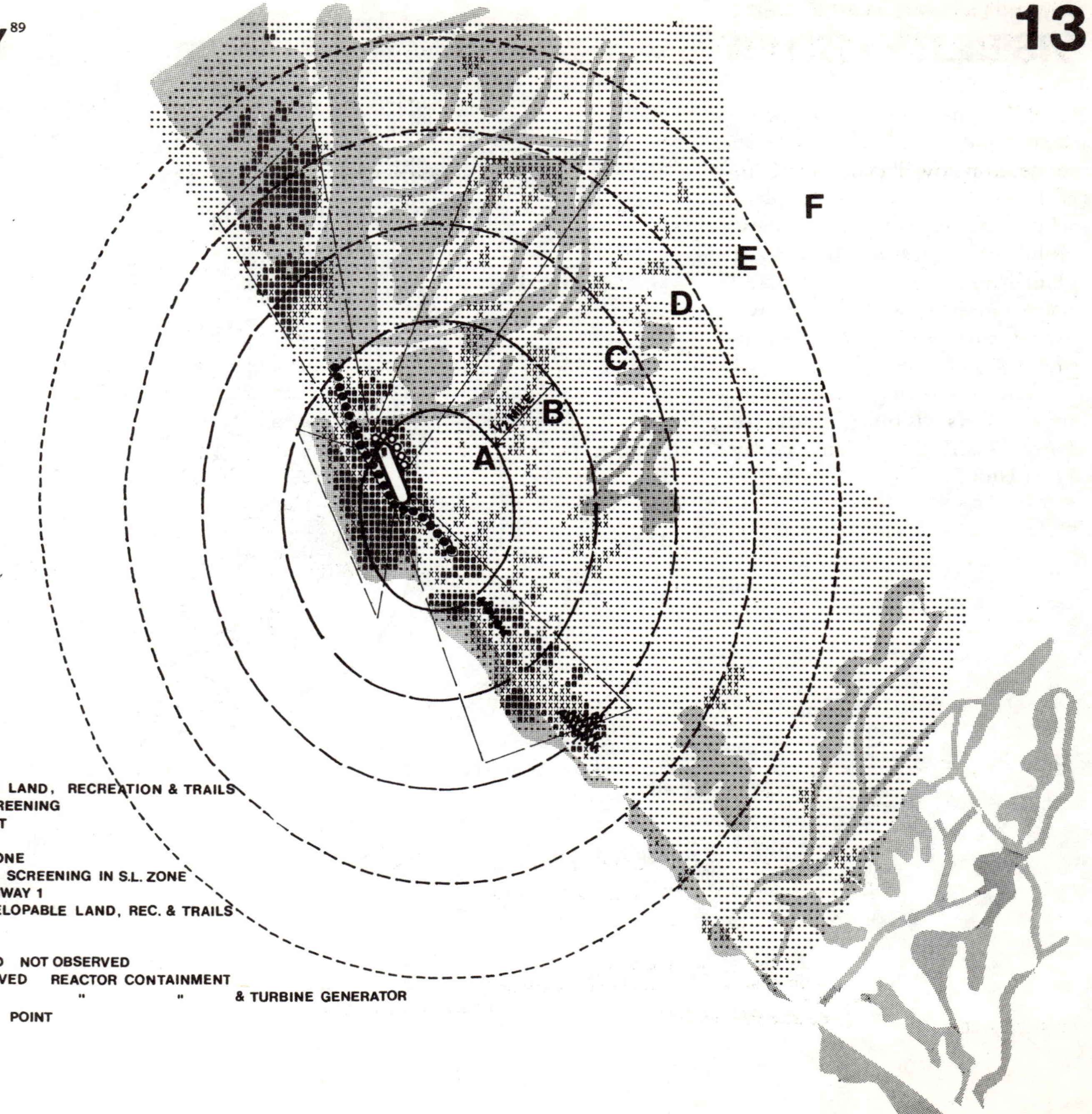

89. Ibid., p. 25.

POWER PLANT SITE PLAN 2 UNIT 2

VISIBILITY STUDY[90]

The following illustration, then, shows one of the visibility studies for Power Plant Site Plan 2—Unit 2.

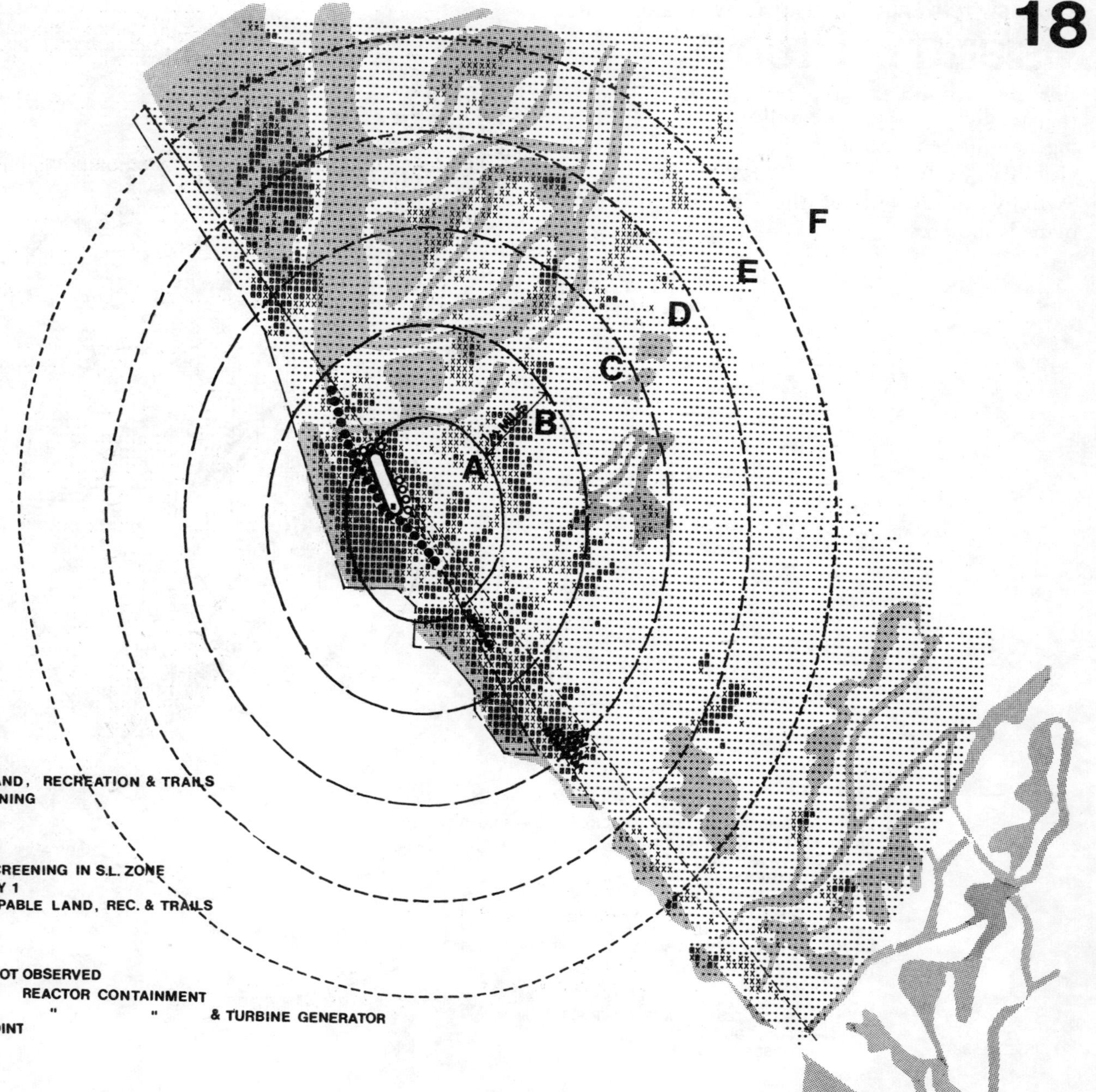

90. Ibid., p. 30.

230KV SWITCHYARD SITE PLAN 4 WEST END el. 360

26

VISIBILITY STUDY[91]

As illustrated in the following computer printout map, the visibility study for the 230 KV Switchyard on one of the site plans is shown.

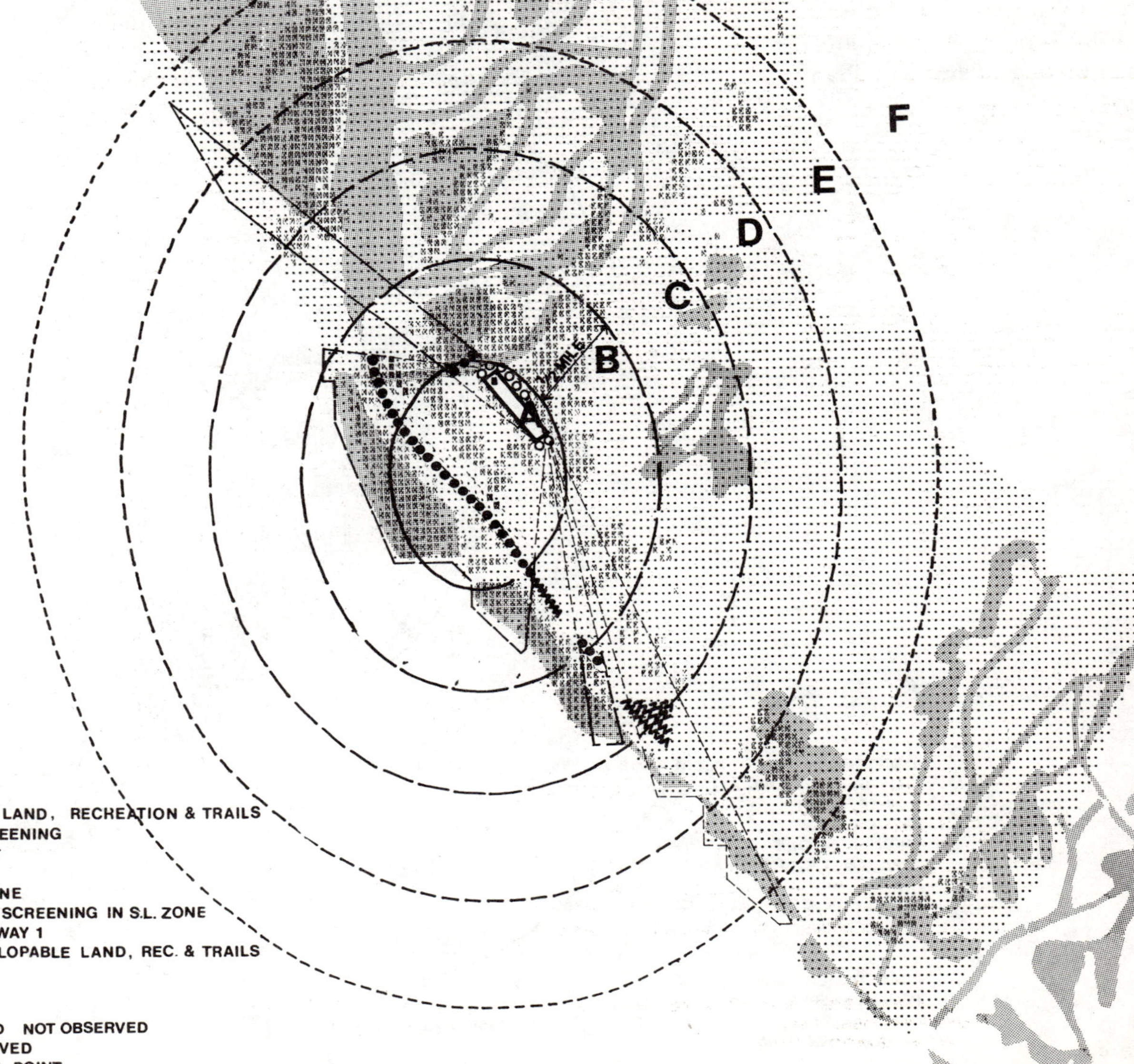

91. Ibid., p. 38.

115KV SWITCHYARD SITE PLAN 2 el. 300

VISIBILITY STUDY[92]

33

The following illustration, then, shows one of the visibility potential for one of the 115 KV Switchyard alternatives.

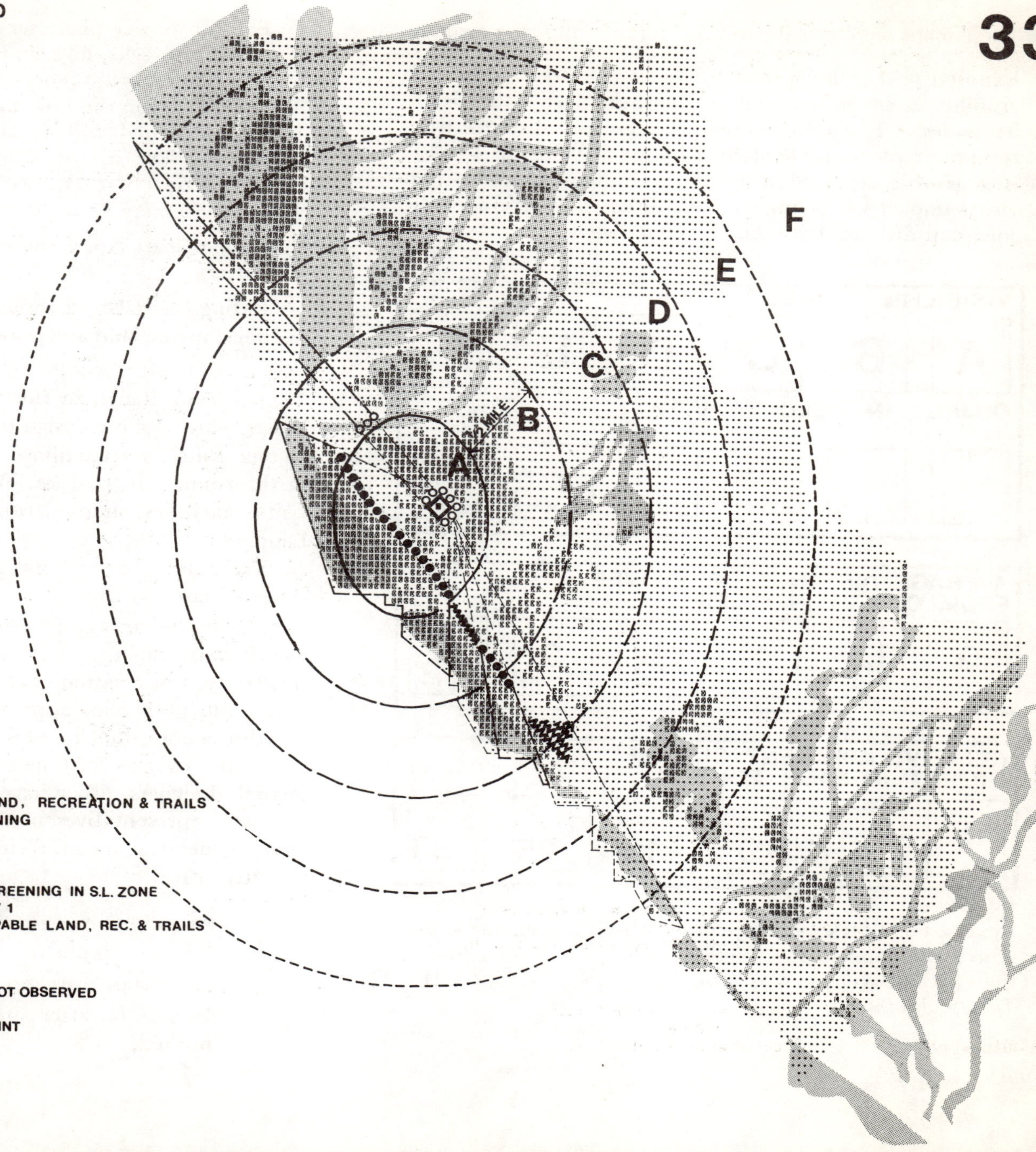

DEVELOPABLE LAND, RECREATION & TRAILS
EXISTING SCREENING
POWER PLANT
SWITCHYARD
SIGHTLINE ZONE
DIAGRAMATIC SCREENING IN S.L. ZONE
FROM HIGHWAY 1
FROM DEVELOPABLE LAND, REC. & TRAILS

POINT TESTED NOT OBSERVED
POINT OBSERVED
OBSERVATION POINT

92. Ibid., p. 45.

These maps were developed utilizing the following process:

Exposed grid cells on maps 6-24 are counted within each visibility zone and entered on the rating chart, the visibility zones A-F represent areas of increased distance from the power plant and switchyard zone (A) the chart below is a graphic representation of the basic assumption that visual impact of an object is in direct proportion to the square of distance between object and viewer.

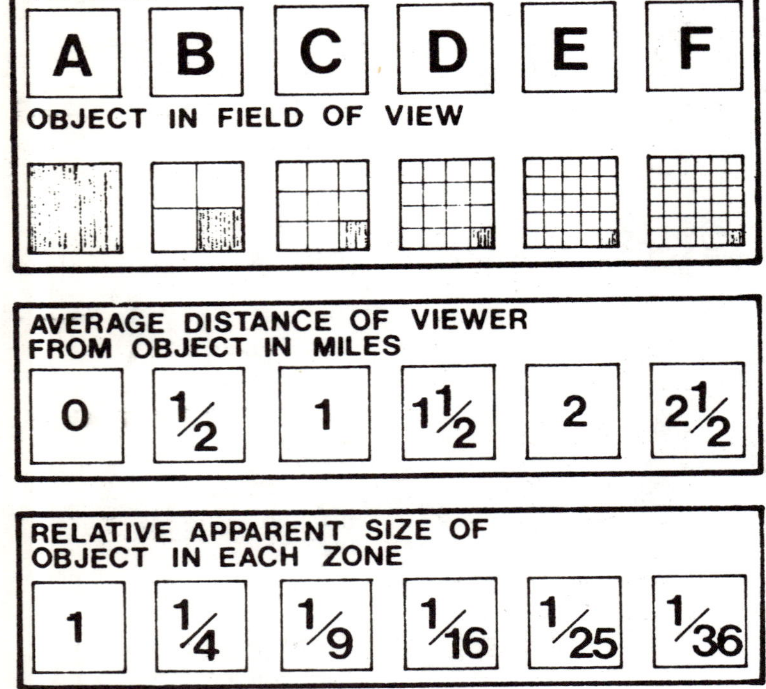

Next, a factor is established for each visibility zone which is in proportion to the fractions shown of the last line of the chart above. Such as: *100*/1, *25*/1/4, *11*/1/9, *6*/1/16, *4*/1/25, *3*/1/36

93. Ibid., p. 49.

For the power plant schemes there are two kinds of exposed grid cells, high and low. The multiplier for the high rating remains the normal one for that visibility zone. The multiplier for the low rating is 1/2 the normal. Grid cells are multiplied and a subtotal for each visibility zone is entered. The subtotals are then added to form a total. A comparison of totals will reveal the level of visibility for each scheme.

Rating Chart No. 1 shows visibility ratings for Highway 1 only.

Rating Chart No. 2 shows visibility ratings for potentially developable land and potential recreation and trail areas.

The consultants, in summary, prepared a series of design critiques which "show design problems and make recommendations among which are handling of highway and rail access, planting, earthforming, siting, location of transmission lines, location of visitor facilities, and relationship of power plant to Davenport Landing."

Two of the design critiques are shown in the illustrations on pages 87 and 88.

The consultants concluded with a series of findings and recommendations which, as mentioned previously, are not pertinent to this study, but resulted from the process utilized in this very innovative and pioneering approach to an increasingly common problem throughout the United States.

In the overall continuum of possible ways in which environmental designers either have or could work with electric utility industry representatives in the location in the design of electric power generating facilities sites, site development is the fifth possible alternative:

- After environmental impact studies and statements have been accomplished,
- After standards and guidelines have been developed,
- Hopefully, after site selection studies have been accomplished,

HIGHWAY 1

RATING CHART 1[94]

VISIBILITY ZONES	A	B	C	D	E	F	TOTAL
MULTIPLIER	100	25	11	6	4	3	

UNIT 1 POWER PLANT SCENIC HIGHWAY ONLY

	A HIGH	A LOW	A 100 HIGH	A 50 LOW	B H	B L	B 25 H	B 12 L	C H	C L	C 11 H	C 6 L	D H	D L	D 5 H	D 3 L	E H	E L	E 4 H	E 2 L	F H	F L	F 3 H	F 1.5 L	TOTAL HIGH	TOTAL LOW
SCHEME 1	8	11	800	550	0	6	0	72	0	6	0	36	0	5	0	15	0	0	0	0	0	0	0	0	800	673
SCHEME 2	13	12	1300	600	12	8	300	96	5	15	55	90	3	1	18	3	0	0	0	0	0	0	0	0	1670	789
NOT TESTED – SIMILAR TO 4 SCHEME 3																										
SCHEME 4	0	0	0	0	0	2	0	24	0	0	0	0	0	0	0	0	0	0	0	0	0	0	0	0	0	24
SCHEME 5	24	7	2400	350	23	7	575	84	23	13	263	78	10	8	50	24	0	11	0	22	3	12	9	18	3287	476
SCHEME 6	24	8	2400	400	28	5	700	60	26	10	286	60	14	3	70	9	7	9	28	18	17	4	51	6	3535	553

UNIT 2 POWER PLANT SCENIC HIGHWAY ONLY

	A HIGH	A LOW	A 100 HIGH	A 50 LOW	B H	B L	B 25 H	B 12 L	C H	C L	C 11 H	C 6 L	D H	D L	D 5 H	D 3 L	E H	E L	E 4 H	E 2 L	F H	F L	F 3 H	F 1.5 L	TOTAL HIGH	TOTAL LOW
SCHEME 1	1	1	100	50	13	5	325	60	2	12	22	72	0	1	0	3	0	0	0	0	0	0	0	0	447	185
SCHEME 2	24	2	2400	100	18	6	400	30	14	5	154	30	5	1	25	3	0	0	0	0	0	9	0	12	2979	175
SCHEME 3	11	10	1100	500	1	15	25	180	0	16	0	96	0	2	0	6	0	0	0	0	0	0	0	0	1125	782
SCHEME 4	0	1	0	50	0	2	0	24	0	0	0	0	0	0	0	0	0	0	0	0	0	0	0	0	0	74
SCHEME 5	30	7	3000	350	24	6	600	72	19	11	209	66	7	6	35	18	1	4	4	8	7	9	21	13	3869	527
SCHEME 6	29	5	2900	250	31	9	775	108	20	14	220	84	14	8	70	24	10	4	40	8	14	4	42	6	4047	480

230 KV WEST END SWITCHYARD: SCENIC HIGHWAY ONLY

	A 100		B 25		C 11		D 6		E 4		F 3		TOTAL
SCHEME 1	0	0	0	0	0	0	0	0	0	0	0	0	0
SCHEME 2	21	2100	21	525	26	286	8	48	0	0	0	0	2959
SCHEME 3	14	1400	14	350	8	88	5	30	0	0	0	0	1868
SCHEME 4	12	1200	16	400	9	99	4	24	0	0	0	0	1723
SCHEME 5	21	2100	21	525	20	220	8	48	0	0	0	0	2893
NOT TESTED SIMILAR TO 4 SCHEME 6													

230 KV EAST END SWITCHYARD: SCENIC HIGHWAY ONLY

	A 100		B 25		C 11		D 6		E 4		F 3		TOTAL
SCHEME 1	0	0	0	0	0	0	0	0	0	0	0	0	0
SCHEME 2	23	2300	17	425	15	165	10	60	0	0	0	0	2950
SCHEME 3	27	2700	18	450	11	121	7	42	0	0	0	0	3313
SCHEME 4	3	300	19	475	5	55	4	24	0	0	0	0	854
SCHEME 5													
NOT TESTED SIMILAR TO 4 SCHEME 6													

115 KV SWITCHYARD SCENIC HIGHWAY ONLY

	A GRID CELL COUNT	A SUB-TOTAL	B GRID CELL COUNT	B SUB-TOTAL	C GRID CELL COUNT	C SUB-TOTAL	D GRID CELL COUNT	D SUB-TOTAL	E GRID CELL COUNT	E SUB-TOTAL	F GRID CELL COUNT	F SUB-TOTAL	TOTAL
SCHEME 1	1	100	10	250	0	0	0	0	0	0	0	0	350
SCHEME 2	25	2500	18	450	15	165	5	30	0	0	0	0	3145
SCHEME 3	23	2300	20	500	20	220	6	36	0	0	0	0	3056
SCHEME 4	28	2800	23	575	20	220	11	66	0	0	0	0	3661
SCHEME 5	13	1300	22	550	21	231	10	60	0	0	0	0	2141
NOT TESTED SIMILAR TO 4 SCHEME 6													

94. Ibid., p. 50.

DEVELOPMENT and RECREATION

RATING CHART 2[95]

VISIBILITY ZONES	A	B	C	D	E	F	TOTAL
MULTIPLIER	100	25	11	6	4	3	

UNIT 1 POWER PLANT DEVELOPMENT AND RECREATION

		A HIGH	A LOW	A 100 H	A 50 L	B H	B L	B 25 H	B 12 L	C H	C L	C 11 H	C 6 L	D H	D L	D 5 H	D 3 L	E H	E L	E 4 H	E 2 L	F H	F L	F 3 H	F 1·5 L	TOTAL HIGH	TOTAL LOW
	SCHEME 1	36	18	3600	900	16	26	400	912	0	28	0	68	0	50	0	150	0	38	0	76	0	6	0	9	4000	2115
	SCHEME 2	68	2	6800	100	58	58	1450	696	27	11	297	66	63	32	378	96	48	16	142	32	2	13	6	19·5	9123	1009·5
NOT TESTED- SIMILAR TO 4	SCHEME 3																										
	SCHEME 4	0	0	0	0	43	11	132	0	0	23	0	138	14	18	84	54	0	0	0	0	0	0	0	0	216	192
	SCHEME 5	60	2	6000	100	95	20	2375	240	43	30	473	180	103	24	515	72	97	87	388	54	17	12	51	18	9802	664
	SCHEME 6	72	1	7200	50	100	26	2500	312	68	32	748	192	110	45	550	135	119	38	476	76	25	10	75	15	10549	780

UNIT 2 POWER PLANT DEVELOPMENT AND RECREATION

	A HIGH	A LOW	A 100 H	A 50 L	B H	B L	B 25 H	B 12 L	C H	C L	C 11 H	C 6 L	D H	D L	D 5 H	D 3 L	E H	E L	E 4 H	E 2 L	F H	F L	F 3 H	F 1·5 L	TOTAL HIGH	TOTAL LOW
SCHEME 1	1	37	100	1850	25	43	625	516	22	10	242	60	2	2	12	6	0	9	0	18	0	8	0	12	979	2462
SCHEME 2	72	2	7200	100	79	40	1975	480	36	41	396	246	77	31	462	93	64	31	256	62	9	5	27	7·5	10316	988·5
SCHEME 3	33	34	3300	1700	24	69	600	828	2	28	22	168	9	5	36	15	0	21	0	42	0	4	0	6	3958	2759
SCHEME 4	0	3	0	150	7	31	175	372	0	9	0	54	0	15	0	45	1	60	4	120	1	5	3	2·25	182	743·25
SCHEME 5	74	1	7400	50	88	18	2200	216	53	25	583	150	85	20	425	60	74	26	296	52	20	3	60	4.5	10964	532.5
SCHEME 6	65	0	6500	0	123	16	3075	192	83	37	913	222	113	23	565	69	127	47	508	94	28	11	84	16.5	11645	593.5

230 KV WEST END SWITCHYARD: DEVELOPMENT & RECREATION

	A	100	B	25	C	11	D	6	E	4	F	3	TOTAL
SCHEME 1	6	600	41	1025	56	616	19	114	24	96	8	24	2475
SCHEME 2	76	7600	169	4225	82	902	124	749	99	396	24	72	13939
SCHEME 3	57	5700	127	8175	90	990	121	726	136	544	27	81	11216
SCHEME 4	72	7200	186	4650	82	902	115	690	124	512	20	60	14014
SCHEME 5	25	2500	51	1275	33	363	83	498	54	216	3	9	4861
SCHEME 6													

230 KV EAST END SWITCHYARD: DEVELOPMENT & RECREATION

	A	100	B	25	C	11	D	6	E	4	F	3	TOTAL
SCHEME 1	0	0	3	75	2	22	53	318	87	348	14	42	805
SCHEME 2	69	6900	173	4325	91	1001	121	726	95	380	25	75	13407
SCHEME 3	69	6900	125	3105	36	396	112	672	151	604	18	54	11731
SCHEME 4	66	6600	70	1750	29	319	120	720	141	574	21	63	10026
SCHEME 5													
SCHEME 6													

115 KV SWITCHYARD: DEVELOPMENT AND RECREATION ONLY

	A GRID CELL COUNT	A SUB-TOTAL	B GRID CELL COUNT	B SUB-TOTAL	C GRID CELL COUNT	C SUB-TOTAL	D GRID CELL COUNT	D SUB-TOTAL	E GRID CELL COUNT	E SUB-TOTAL	F GRID CELL COUNT	F SUB-TOTAL	TOTAL
SCHEME 1	3	300	51	1255	16	176	6	36	101	404	27	81	2252
SCHEME 2	72	7200	141	3525	48	528	114	684	125	500	18	54	12491
SCHEME 3	75	7500	134	3450	42	462	128	968	151	604	32	96	13080
SCHEME 4	77	7700	154	3850	88	968	128	968	149	596	27	81	14163
SCHEME 5	23	2300	77	1925	41	451	67	402	56	224	9	27	5329
SCHEME 6													

95. Ibid., p. 51.

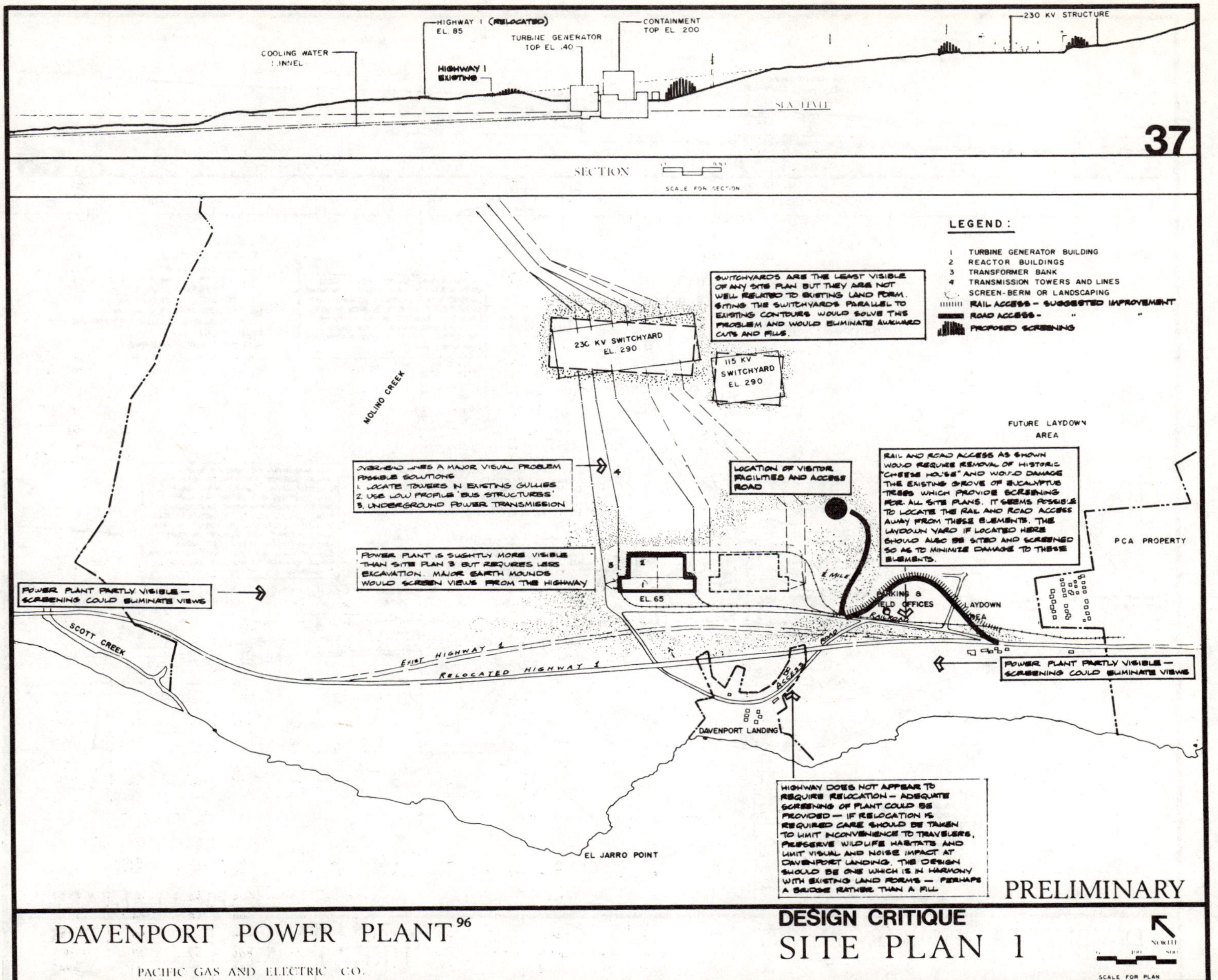

HIGHWAY 1 (RELOCATED)
EL. 85
COOLING WATER TUNNEL
TURBINE GENERATOR
TOP EL. 40
CONTAINMENT
TOP EL. 200
HIGHWAY 1
EXISTING
SEA LEVEL
230 KV STRUCTURE
37
SECTION
SCALE FOR SECTION
LEGEND:
1 TURBINE GENERATOR BUILDING
2 REACTOR BUILDINGS
3 TRANSFORMER BANK
4 TRANSMISSION TOWERS AND LINES
SCREEN-BERM OR LANDSCAPING
RAIL ACCESS – SUGGESTED IMPROVEMENT
ROAD ACCESS – " "
PROPOSED SCREENING
SWITCHYARDS ARE THE LEAST VISIBLE OF ANY SITE PLAN BUT THEY ARE NOT WELL RELATED TO EXISTING LAND FORM. SITING THE SWITCHYARDS PARALLEL TO EXISTING CONTOURS WOULD SOLVE THIS PROBLEM AND WOULD ELIMINATE AWKWARD CUTS AND FILLS.
230 KV SWITCHYARD
EL. 290
115 KV
SWITCHYARD
EL. 290
MOLINO CREEK
FUTURE LAYDOWN
AREA
OVERHEAD LINES A MAJOR VISUAL PROBLEM
POSSIBLE SOLUTIONS
1. LOCATE TOWERS IN EXISTING GULLIES
2. USE LOW PROFILE 'BUS STRUCTURES'
3. UNDERGROUND POWER TRANSMISSION
LOCATION OF VISITOR FACILITIES AND ACCESS ROAD
RAIL AND ROAD ACCESS AS SHOWN WOULD REQUIRE REMOVAL OF HISTORIC "CHEESE HOUSE" AND WOULD DAMAGE THE EXISTING GROVE OF EUCALYPTUS TREES WHICH PROVIDE SCREENING FOR ALL SITE PLANS. IT SEEMS POSSIBLE TO LOCATE THE RAIL AND ROAD ACCESS AWAY FROM THESE ELEMENTS. THE LAYDOWN YARD IF LOCATED HERE SHOULD ALSO BE SITED AND SCREENED SO AS TO MINIMIZE DAMAGE TO THESE ELEMENTS.
POWER PLANT IS SLIGHTLY MORE VISIBLE THAN SITE PLAN 3 BUT REQUIRES LESS EXCAVATION. MAJOR EARTH MOUNDS WOULD SCREEN VIEWS FROM THE HIGHWAY
PCA PROPERTY
POWER PLANT PARTLY VISIBLE – SCREENING COULD ELIMINATE VIEWS
EL. 65
PARKING &
FIELD OFFICES
LAYDOWN
AREA
RAILROAD
SCOTT CREEK
EXIST HIGHWAY 1
RELOCATED HIGHWAY 1
ROAD
ACCESS
POWER PLANT PARTLY VISIBLE – SCREENING COULD ELIMINATE VIEWS
DAVENPORT LANDING
EL JARRO POINT
HIGHWAY DOES NOT APPEAR TO REQUIRE RELOCATION – ADEQUATE SCREENING OF PLANT COULD BE PROVIDED — IF RELOCATION IS REQUIRED CARE SHOULD BE TAKEN TO LIMIT INCONVENIENCE TO TRAVELERS, PRESERVE WILDLIFE HABITATS AND LIMIT VISUAL AND NOISE IMPACT AT DAVENPORT LANDING. THE DESIGN SHOULD BE ONE WHICH IS IN HARMONY WITH EXISTING LAND FORMS — PERHAPS A BRIDGE RATHER THAN A FILL
PRELIMINARY
DAVENPORT POWER PLANT[96]
PACIFIC GAS AND ELECTRIC CO.
DESIGN CRITIQUE
SITE PLAN 1
NORTH
SCALE FOR PLAN

96. Ibid., p. 53.

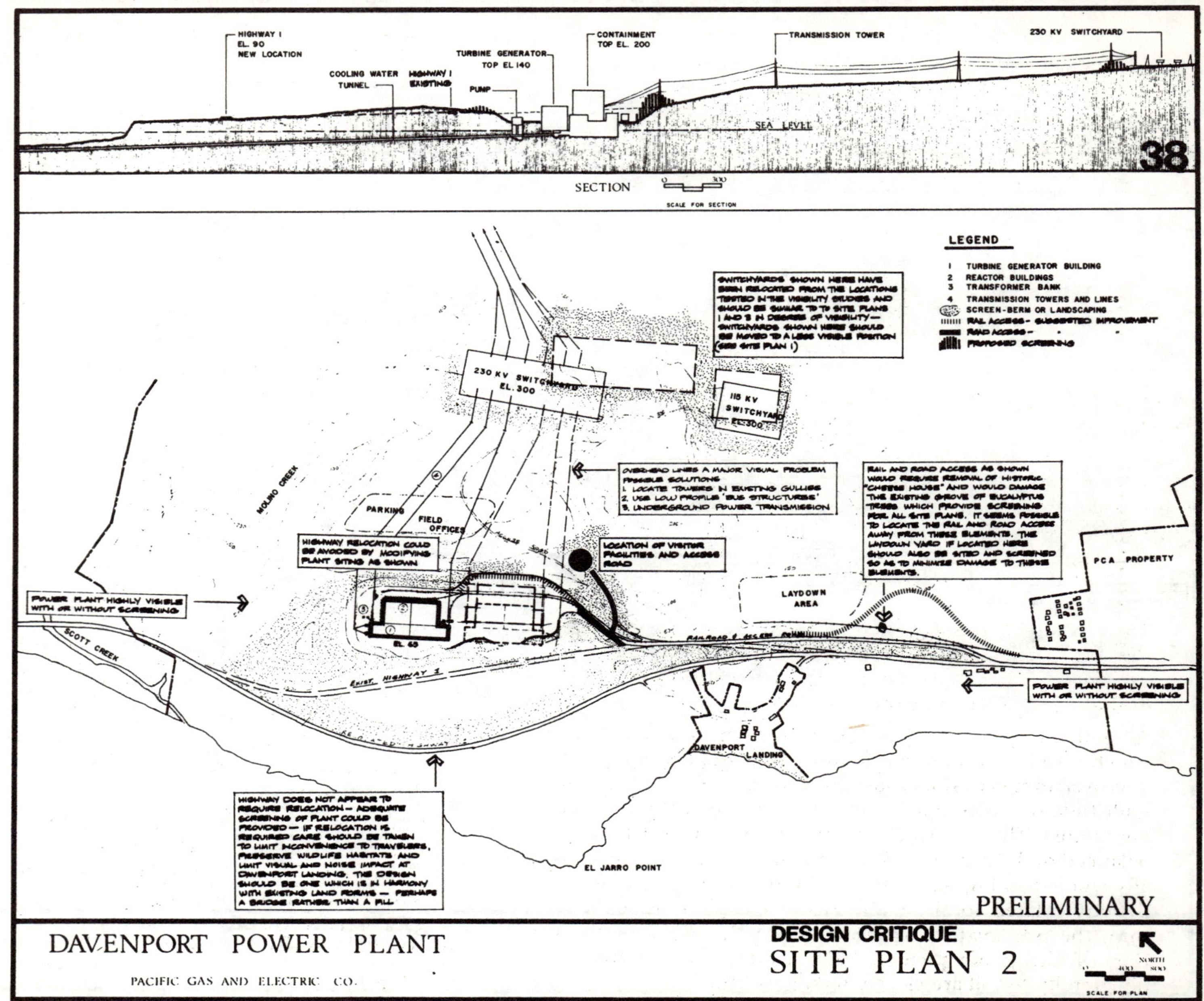

97. Ibid., p. 54.

97. Ibid., p. 54.

- A site is chosen and a power plant of some type is constructed. Then the environmental designer or, in this particular instance, the landscape architect, is able to work with the electric utility company in the actual development of the site itself in order to:
  1. Make it more functional and usable,
  2. Cause the least possible damage to the natural environmental system,
  3. Ameliorate any environmental problems which still exist on the site,
  4. Guide in the further development of the site beyond that which is indicated initially.

One of the outstanding examples of the way in which a landscape architect has worked with an electric utility company in the development of a power plant site was that of the Trojan Nuclear Power Plant. This project was undertaken by the landscape architectural firm of Lawrence Halprin & Associates for the Portland General Electric Company for the Trojan Nuclear Power Plant site at Prescott, Oregon.

Mr. Halprin states in the introduction to this report the following:

> The essence of the problem that confronted us is the same that all of us increasingly face in our daily lives: how to use the new tools and devices of our technology without degrading or desecrating our environment.
>
> A wholistic approach to the problem is the only means by which a valid solution can be found. In such an approach one overriding consideration does not blindfold one to the multitude of other equally important and equally demanding factors. The various diverse factors must be continually balanced and their interrelations and interactions continually studied and weighed; the height of the cooling tower is directly related to the maintenance of the river temperature, the preservation of the Whistling Swans to the location of the group picnic area. No factor can be treated unilaterally for all are closely linked together.
>
> We have shown, we believe, that it is possible to integrate massive forms into the landscape without diminishing either; rather, each supports and augments the other. The import of this transcends the immediate use of the plan, for if this is possible with a large nuclear plant, by extension it is possible with any other large industrial complex.[98]

Mr. Halprin then goes on to say in the conclusion to the introduction to this report:

> It is our hope that this plan will serve to generate similar design approaches to other nuclear plants as well as other large scale industrial complexes. Wholistic, ecologically valid development, well done and beautiful, is the only intelligent way to utilize our finite resources. We can afford no other way.[99]

The Pacific Northwest has traditionally been the prime baliwick of plentiful, inexpensive hydroelectric power. It is not surprising, therefore, that the Trojan Nuclear Power Plant proposed on the Columbia River at Prescott, Oregon, downstream from Portland, will be what is tô be the first nuclear power plant in the Pacific Northwest. The process is described by the consultants as follows:

> The site is of considerable natural beauty and the Portland General Electric Company, recognizing the extent of the impact that the plant would have upon the landscape, commissioned the environmental design firm of Lawrence Halprin & Associates to prepare a master plan that would result in an aesthetically pleasing and ecologically sound development.
>
> Underlying the Halprin approach to the planning of the site is a basic principle: any disturbance of the natural environment must be kept to an absolute minimum. Consequently, the preservation of the integrity of the site was

98. Lawrence Halprin and Associates, *The Trojan Nuclear Power Plant,* (San Francisco, California, A Project for the Portland General Electric Company, at Prescott, Oregon, 1970), p. 1.
99. Ibid., p. 1.

a factor of prime importance. There exists on the Trojan site a unique grouping of biotic communities which, when combined with the topographical aspects of the site, produce an area of charm and beauty. To retain the elements that give these qualities to the site, development must be restricted to as small an area as possible. This results in the intensive use of a relatively small portion of the site.

A closely related concept is the full utilization of all of the resources of the site. By confining the intensive development, the industrial, recreational and natural areas are separated and clearly defined. An equilibrium is established which allows each area to support and harmonize with the others.

During the planning process other design criteria evolved, specific to the project. For the Trojan plant the major components of the generating complex required a design approach that accepted their scale and majesty. The huge cooling tower is on such a scale that it must be treated as a sculptural element of the landscape. Hence, the major components are accentuated and are made extensions of the landscape.

Linked to the scale was the desire to express the essence of nuclear power. The very massiveness of the cooling tower, rising 589 feet above the Columbia, conveys something of the grandeur of man's achievement in harnessing the atom as well as the awesomeness of the power unleashed.

The overall approach to the development of the Trojan site has been ecological, involving the total environment and the interaction of the various components of the landscape, natural and man-made. This requires the understanding of not only the facilities and their functions but of the biological, geological, meteorological and other natural features of the site. It also requires bringing together the specialized knowledge and skills of all of the firms and individuals engaged in the project.

An environmentally sound master plan is based on the consideration of all of the factors, aesthetic, biological, economic, engineering, geographical, and sociological that might affect the development. These factors mesh together, reinforcing one another, and, in effect, dictate the form that the final plan will take.

The site of the Trojan Nuclear Power Plant of the Portland General Electric Company lies 42 miles north of Portland on the Oregon bank of the Columbia River. Here the river, a half mile wide and fast moving, flows north before broadening and making its westward turn for the final run to the sea some 73 miles distant. Longview, known for its aluminum and forest products, 7 miles northeast, is the nearest city of substantial size.

The existing site shown on the accompanying drawing consists of 634 acres, with over a mile and a quarter of frontage on the Columbia River. At its deepest point, the site extends a mile inland from the river. The elevation of the site varies from a few feet above sea level up to 640 feet in height on the bluffs beyond the river. A line of the Burlington Northern Railroad and a section of U.S. Highway 30 converge at the southern end of the property. An abandoned quarry exists on the river's edge. The small town of Prescott is on the northern section of the site on the Columbia River itself.

According to the consultants, ". . . in the analysis of the site, heavy emphasis is placed on the natural ecological communities of the site itself." The Halprin staff characterized these in the following words:

Six major biotic communities are represented on the site: the river, and progressing inland, the Columbia River mixed hardwood forest, the riparian woodland, swamp and marshland, agricultural land, and the northwestern coniferous forest. Each of these communities has its own particular, and distinctive, associations of plants and animals. The terrestrial communities are rich in the variety of wildlife.

In planning the development of the site a prime criterion was to avoid disturbing these plant and animal communities. There will be disturbances in those areas where there is construction and some of these areas will be changed forever. However, upon completion of the plant and recreational facilities those areas that can be re-estab-

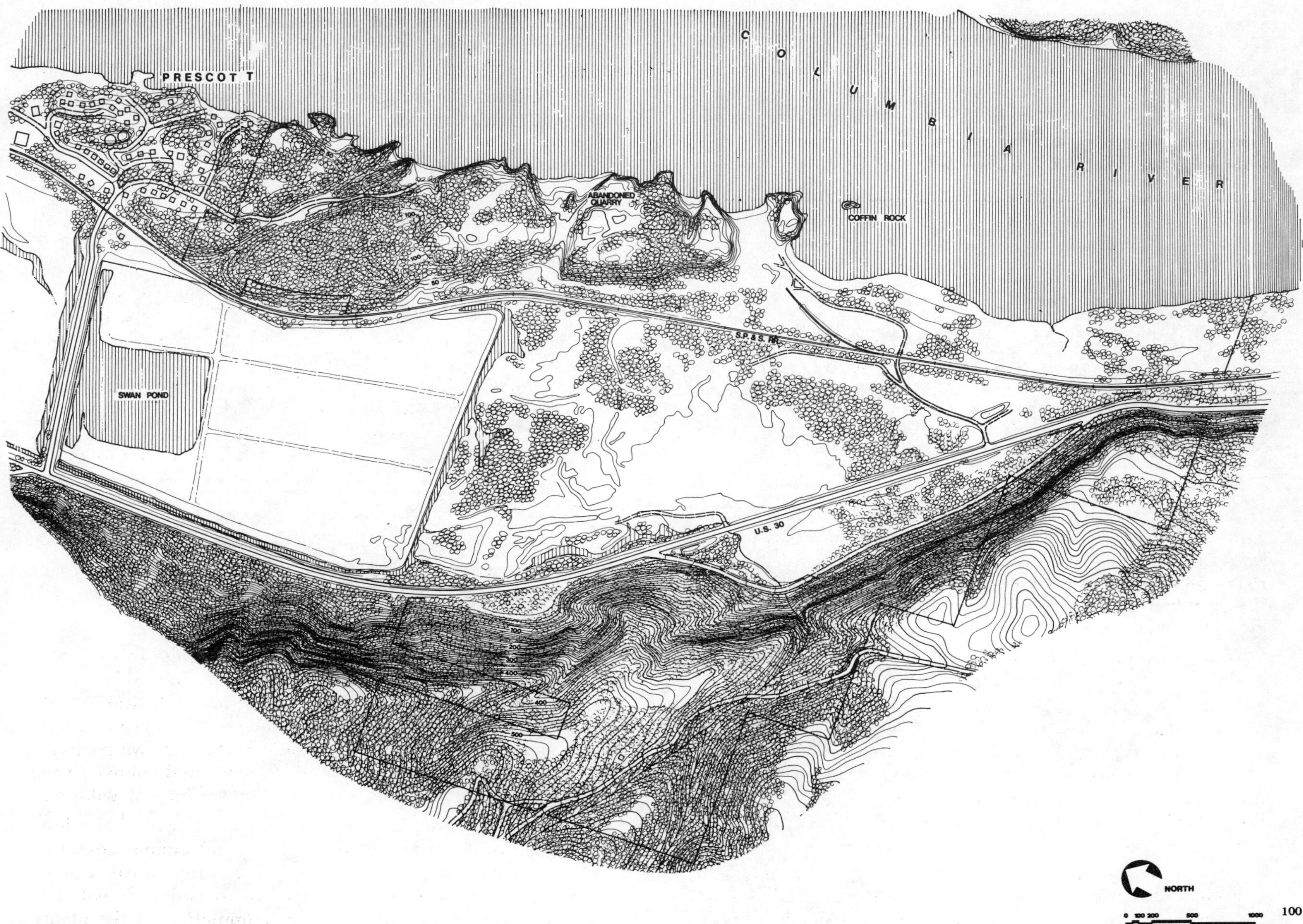

100

100. Ibid., p. 8.

lished will be restored to a close approximation of the communities existing at the present.[101]

Biotic Communities[102]

The following illustration shows the basic biotic communities which appeared on the site before development.

101. Ibid., p. 12.

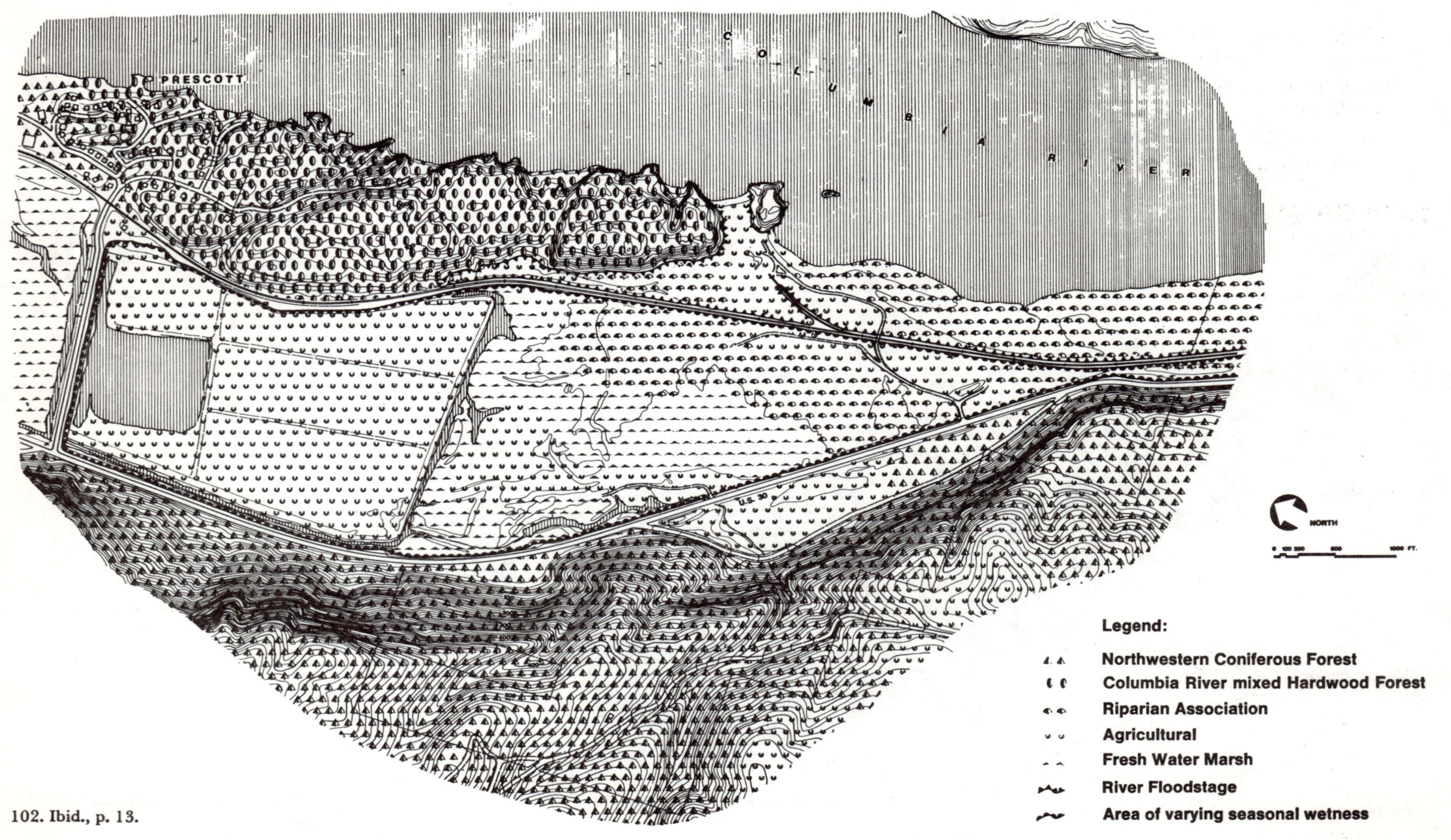

102. Ibid., p. 13.

The following two cross-sections through the site indicate the natural wildlife and plant material distribution on the site before development, and the potential impact of the power plant facilities on the site after development has been completed.

As a result of careful analysis of the site and the requirements of the electric company, the landscape architects developed a diagramatic land use which was characterized the following way.

> Visual, physical and ecological studies of the site and its biotic communities revealed patterns that established the general usage of the land. Thus the extensive damage to the hardwood forest caused by the quarrying makes this the logical location for the reactor. Its use would result in a minimizing of the damage to the landscape. This decision based on aesthetic and ecological criteria was reinforced by the engineering and economic requirements. The location of the reactor would in turn determine the placing of the other components of the generating and distributing systems.
>
> From the land suitability analysis arose the schematic land-use plan. The site divides into three general use areas, the power generating and distributing plant, the area for intensive recreational use, and the natural area. Each of these is a distinct and separate entity with no overlap of function. Natural barriers, such as the river edge, the outcrop, or the hills, and existing man-made barriers, such as the railway or the highway, serve as borders to define each area.[104]

104. Ibid., p. 16.

Environmental Impact[103]

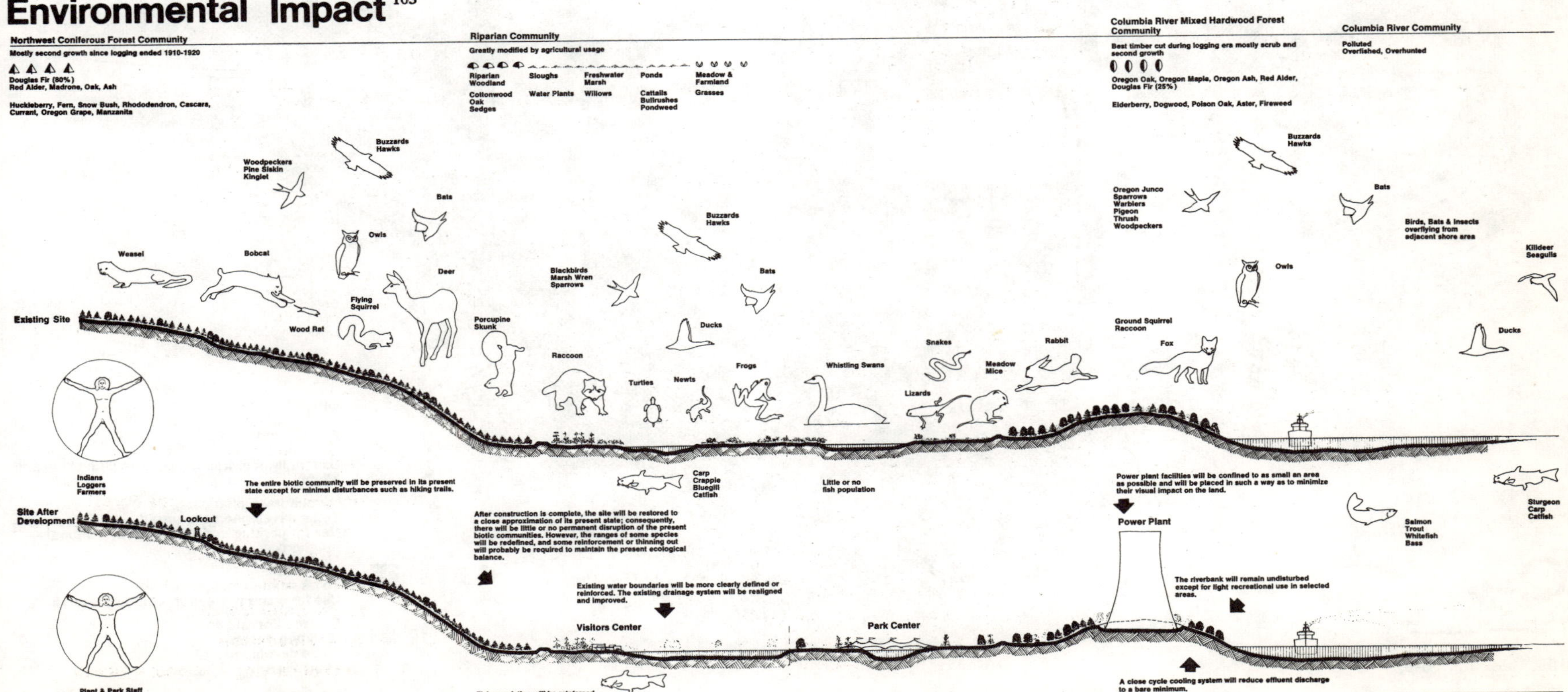

103. Ibid., pp. 14-15.

The following illustrates the diagramatic land use on the site itself.

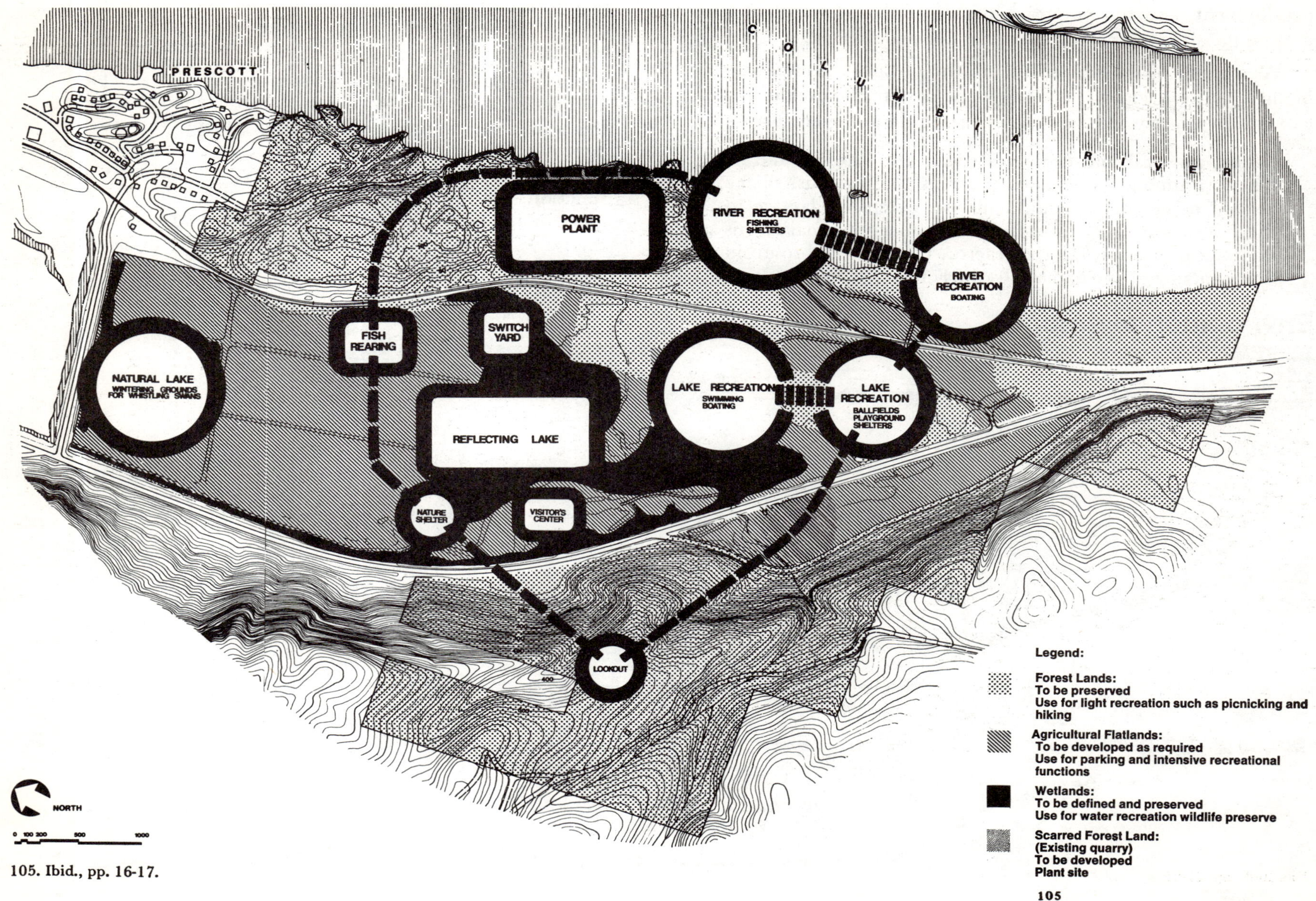

105. Ibid., pp. 16-17.

From this diagramatic land use was developed an overall master plan. The master plan itself is shown on the following illustration.

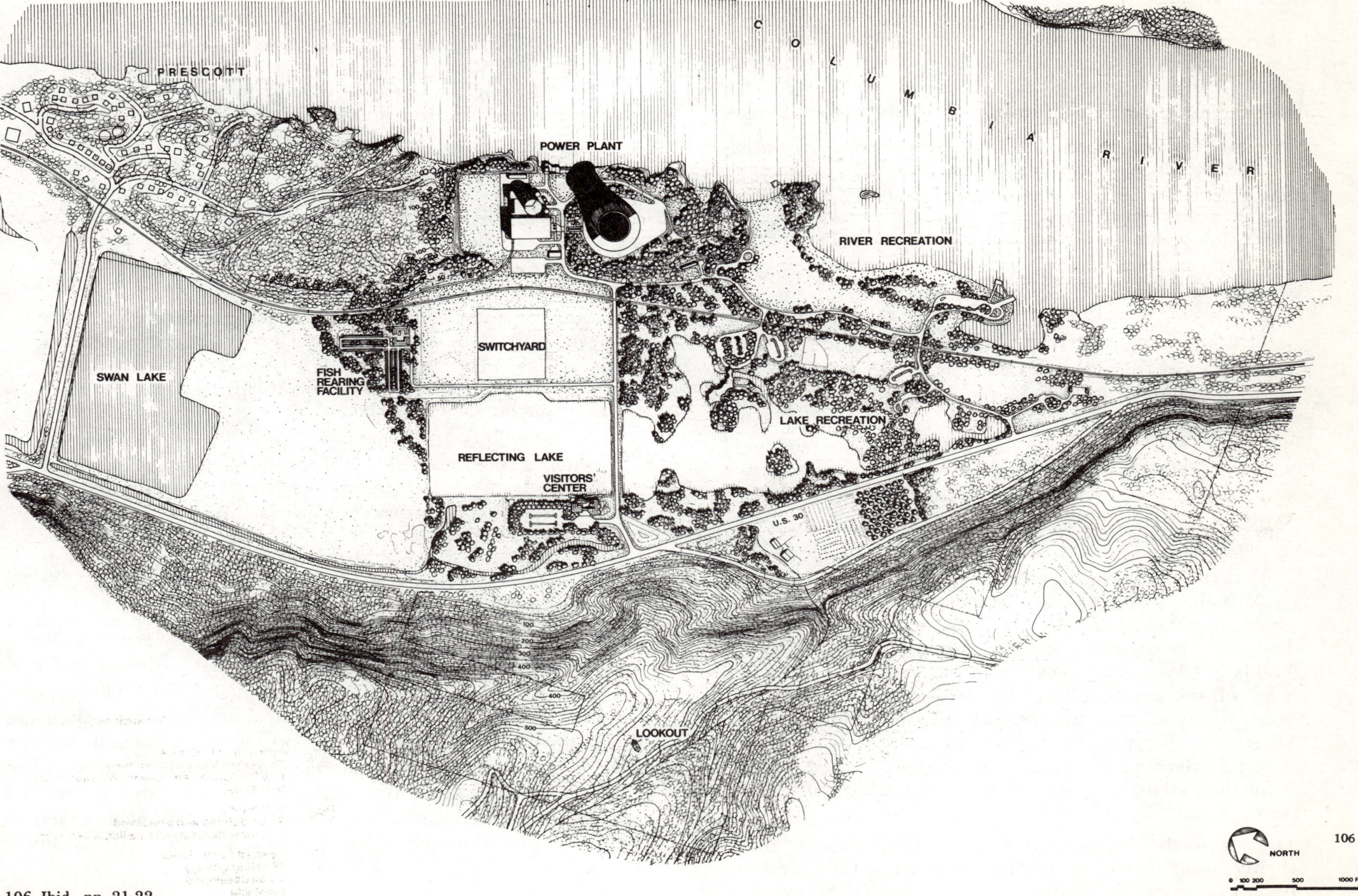

106

106. Ibid., pp. 21-22.

The overall master plan has been described by Lawrence Halprin & Associates in the following way:

> When in full operation the power generating and distributing systems will occupy less than a sixth of the total site area. However, the scale of the components of the plant will make it a dominant feature of the landscape.
>
> West of the reactor complex and across the railroad is the switchyard where the power is routed to the transmission lines. Treated as a twentieth century sculpture garden, the switchyard is situated in the middle of a twenty-five acre level plot planted with grass and low shrubs. Parking for the staff and visitors to the reactor complex is located between the switchyard and the reactor.
>
> Along the embankment at the edge of the switchyard a road leads to the fish rearing facility to the north. Here there is a low, earth-banked building for equipment and a series of shallow rearing ponds.
>
> Separating the plant from the visitors center, near the main entrance, is a thirty acre reflecting lake. This, coupled with the low plantings around the switchyard, provides an unobstructed view of the reactor and tower from the visitors center. The land and the planting also help to set the park-like tone of the development.
>
> The first building encountered upon entering the main drive is the visitors center. The main entrance itself is defined by a low berm and the entry road is so patterned that the visitor is led directly to the center. The center building is low and designed so as not to distract from the view of the plant. Adjacent to the center, by the entry road, is a water garden. This is an outdoor sculpture where slabs of concrete and sheets and pools of water intermingle, offering a visitor a place for quiet contemplation. With the parking area the center occupies three and a half acres.
>
> Leading north from the visitors center is a path to a nature shelter, located near the northwest corner of the reflecting pool. From here the visitor, during the late fall and winter months, can view the Whistling Swans. The shelter has been placed at a sufficient distance from their feeding and resting areas so that the swans will not be disturbed by the viewers. The feeding and resting areas have been shielded from the bustle of the plant and visitors center by a barrier of trees along the northern edge of the reflecting lake and surrounding the fish rearing facility.
>
> From the visitors center the entry road follows a low causeway into the heart of the site. At the foot of the outcrop the road branches, one spur dead-ending at the plant, the other, roughly paralleling the railroad trace, leading to the recreational facilities. There is a secondary access road at the south end of the site near the maintenance and boat launching installations.
>
> South of the plant and visitors center lies the intensive recreation area, with large, irregularly shaped recreation lake. This lake covers over 40 acres and is for water sports such as canoeing, rowing, fishing or bathing; power boating is not permitted on this lake.
>
> The main access to the lake is from the park recreation center, a small, two acre island connected to the mainland by two bridges. Located on the island are the main concessionaires and various recreational facilities. By concentrating these on the island a focal point for the entire recreation area was defined. It also localizes the main trash producing activities. The island and the nearby parking areas occupy four acres.[107]

Detailed master plans were then prepared by the consultants for the various finite parts of the site. The detailed master plan of the south section of the site is shown in the following illustrations.

The overall view of the site in its final development is shown in the following aerial perspectives. Portions of the site in the site development proposals are treated in greater detail in various other parts of the consultants' report. A boat launching area is proposed on a sculptured cove of the Columbia. A boat launching ramp is supposed to be developed for the trailer-borne sail and power boats.

107. Ibid., p. 20.

Detailed Master Plan[108]

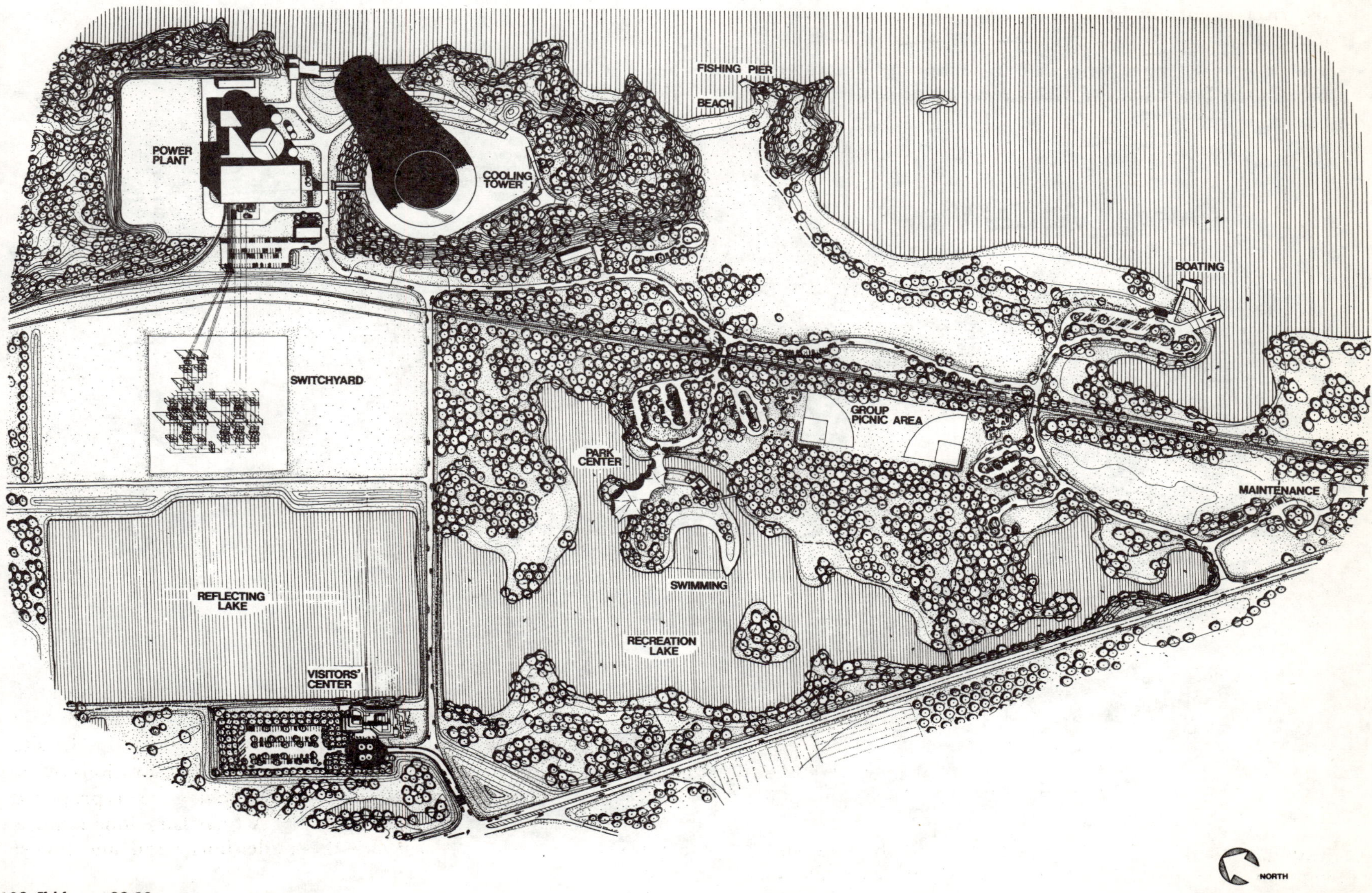

108. Ibid., pp. 22-23.

Connected to the ramp and to a shelter is a floating dock which serves both the small craft passengers and those on the river excursion boats who wish to stop and visit the park and plant.

For those yachtsmen who wish to anchor in calm water or beach their boats and eat, visit the plant, or just relax there is a small lagoon.[109]

109. Ibid., p. 30.

The following illustration shows in plain view this part of the overall development.

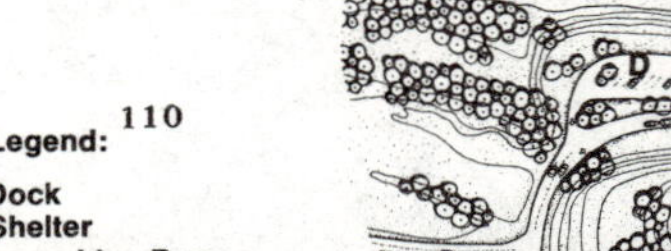

Legend:[110]

A Dock
B Shelter
C Launching Ramp
D Trailer Parking
E Lagoon

110. Ibid., p. 30.

Overall Aerial View[111]

111. Ibid., pp. 24-25.

The following illustration indicates a perspective of the proposed boat launching development on the river side.

Boat Launching Area[112]

112. Ibid., pp. 26-27.

The proposed visitors' center on the site is described in the following way in the Halprin report:

> When the motorist turns off U.S. 30 he is guided imperceptibly to the visitors' center, some 150 yards from the highway. From here he can view to the best advantage the cooling tower and reactor complex.
>
> Carefully blended into the landscape by stepping the walls and terracing the gardens, the center building does not distract the viewer from the main visual elements. These powerful elements, which dominate the landscape, are given further emphasis by the reflecting pool which, in effect, makes them twice as large.
>
> Within the center there is a viewing area, with full glass walls, that gives out on the lake and the view of the plant. Refreshments may be obtained here. Here too is a display area where working models, pictures, and text show the process of generating electricity from nuclear energy. Additionally there is an auditorium for film showing or lectures.[113]

The following illustration shows in greater detail the proposed development of the visitors' center area.

113. Ibid., p. 36.

Visitors' Center[114]

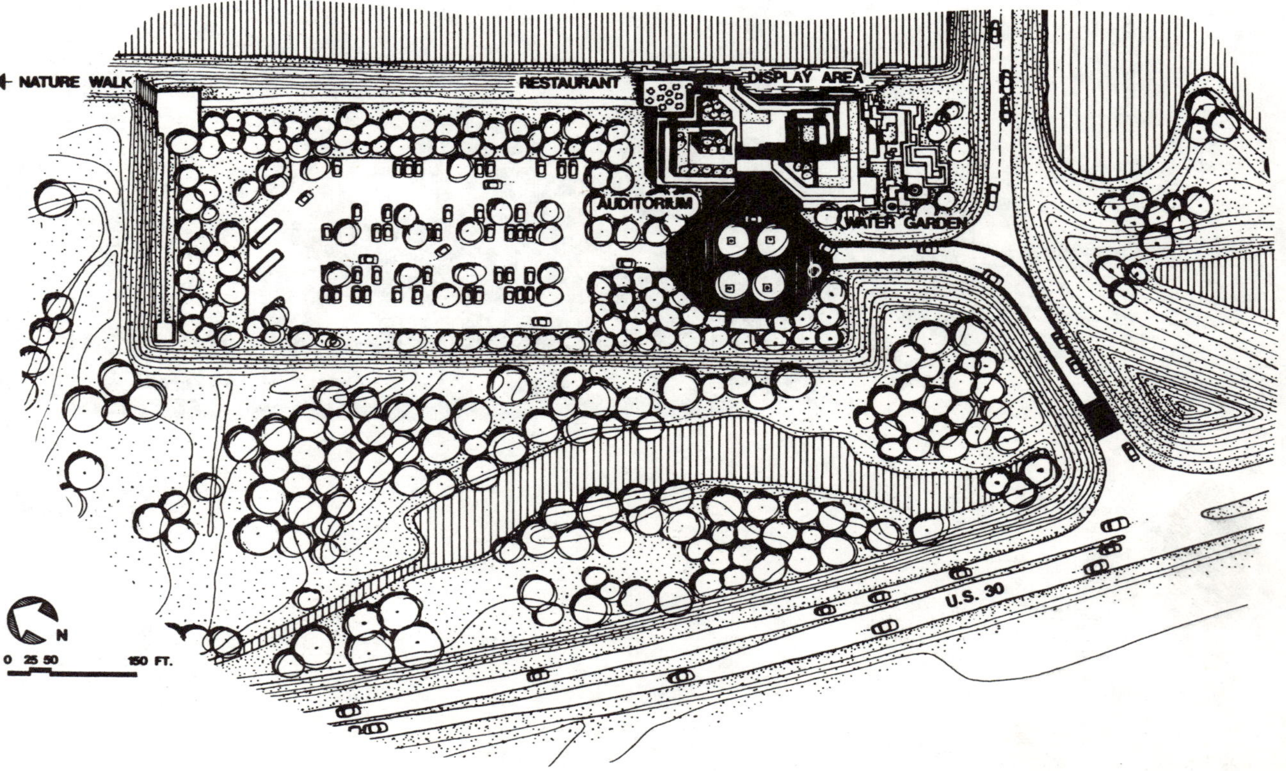

114. Ibid., p. 36.

The proposed fish rearing facility to be developed on the site is described in the following way:

> Situated immediately to the north of the switchyard is the fish rearing facility. Here salmon and steelhead are raised in large outdoor ponds in water slightly warmed by the waste heat from the steam turbine in the plant.
>
> The trees at this facility and along the edge of the reflecting lake serve as a barrier between the high intensity use areas, the plant and the visitors' center, and the undeveloped section to the north. It is in this undeveloped area that the Whistling Swans winter. Visually, the trees serve to break the harsh contrast between the mechanical switchyard and the natural swan area and to engender in both a needed sense of enclosure.[115]

An eye-level perspective showing the fish rearing facility is depicted as follows.

115. Ibid., p. 37.

Fish Rearing Ponds[116]

116. Ibid., p. 37.

The following illustration indicates in plan view the development of the fish rearing facility.

Fish Rearing Facility[117]

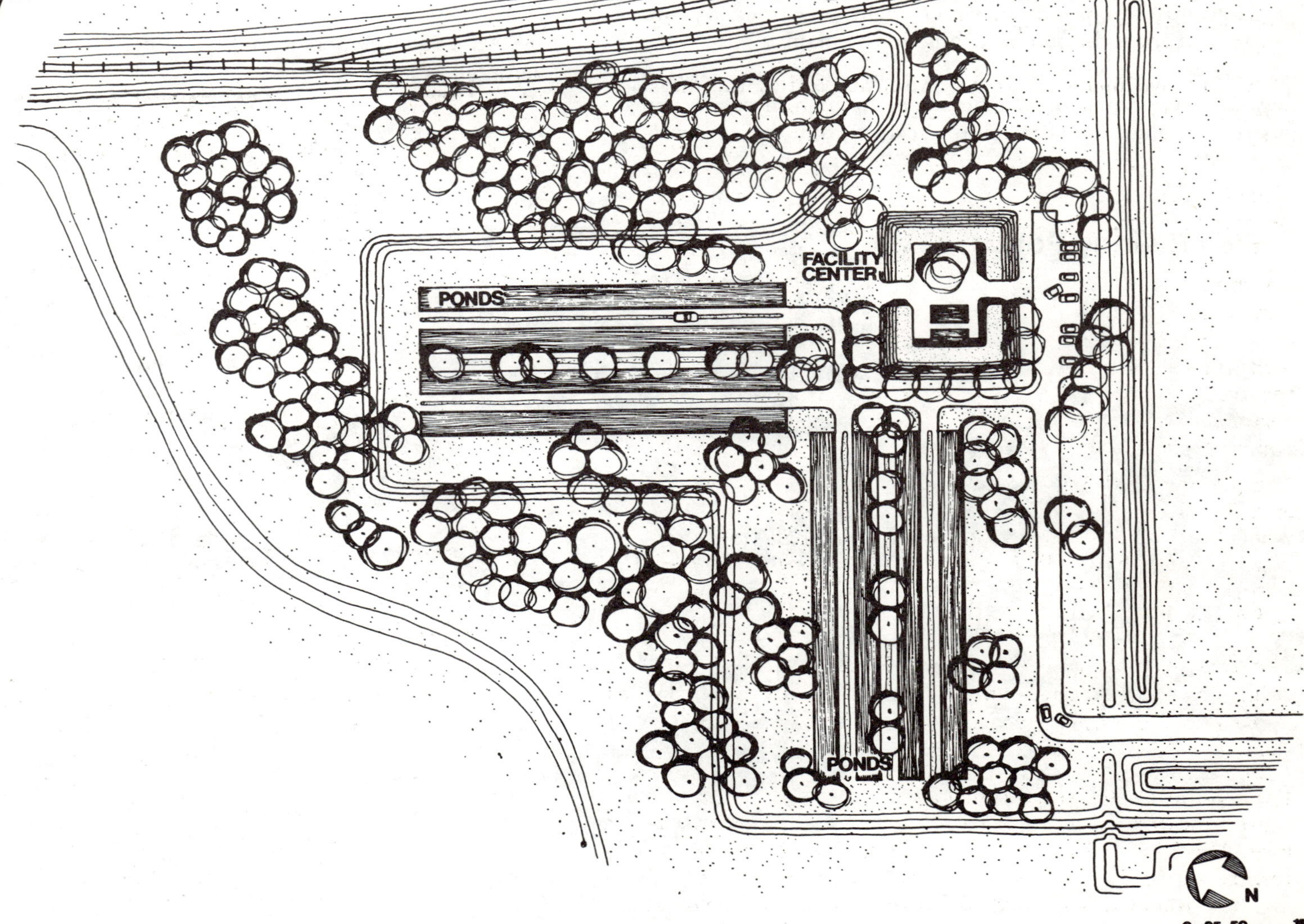

117. Ibid., p. 37.

The group picnic area proposed on this site is shown in the following eye-level perspective in plan view. It is described in the report in the following words:

> A hundred yards from the park center is another focus of intensive use, the group picnic area. It is designed to accommodate 150 families.
>
> There are two picnic areas, one at either end of the main playing field. They are defined by locating the tables, benches, sinks and electric grills on slightly raised ground. For the convenience of busy mothers a sand lot has been placed near, but separated from the cooking areas so that the younger children can be watched without being underfoot.
>
> To shelter the eating areas from the inevitable picnic day rain there is a clear plastic roof made of modules mounted on an unobtrusive light steel frame. The modular arrangement can be adapted to fit around trees without harming the tree or lessening the shelter. The transparent roofing permits an unobstructed view of the surrounding landscape and allows the sun to shine through.
>
> The playing field, which was formed from an existing flat area, is some 600 feet long and 200 feet wide. It is edged with a low berm. Though laid out for baseball (there is a diamond at either end), it can be used for football, soccer, lacrosse and other field games as well as for general picnic and outing events.[118]

The recreation center is described in greater detail with the following aerial perspectives and planned views of the proposed area. The text describing this proposed area for development is as follows:

> Focal point for the recreational areas is the park recreation center located on an island off the eastern shore of the lake. The lake, fed by an all year creek, covers most of the former wetlands and was created with a minimum of diking. The island was formed by cutting a moat-like channel around a spur of land that jutted out into the lake.[119]

118. Ibid., p. 38.
119. Ibid., p. 41.

Group Picnic Area Playing Field[120]

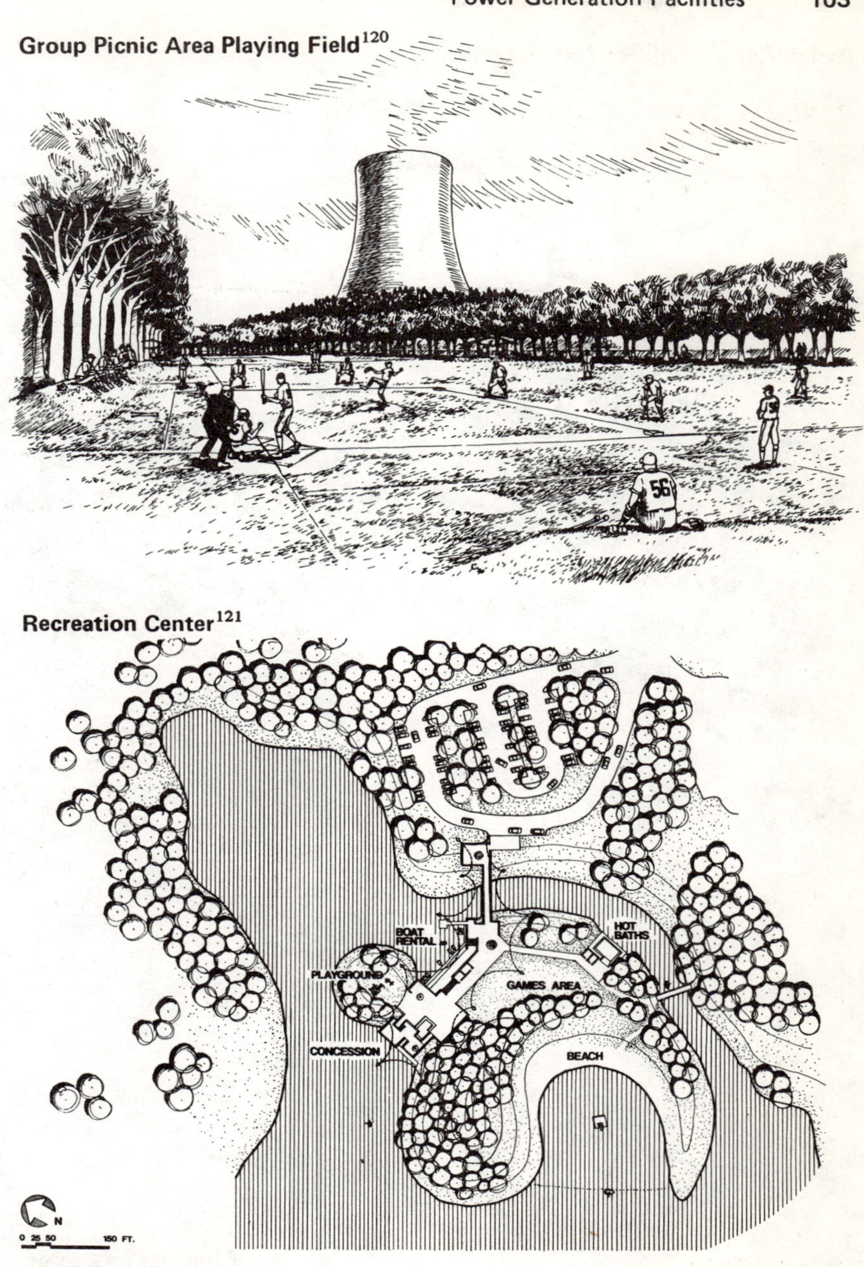

Recreation Center[121]

120. Ibid., p. 38.
121. Ibid., p. 41.

Aerial View of Park Recreation Center[122]

Interior of Recreation Center

122. Ibid., p. 40.

Two bridges connect the island to the mainland, one leads to a forested natural area, the other to the 75 space parking lot. The latter bridge, as well as the main concession area, is covered by a huge tent-like canopy, 400 feet long and 150 feet wide.

All of the recreational facilities whose location was not determined by a natural feature of the land are grouped on the island, creating a structured center for the entire recreation area. Here are found the food, souvenir and fishing supplies concessionaires, the boat rental for bicycle boats, canoes, and rowboats, the tot-lot, the greens for croquet, lawn bowling, and other games, a fishing area, and toilets. Across the island a horseshoe shaped bay defines the swimming area, with its sandy beach. Nearby are the hot baths, veritable spas, that utilize heated water from the plant.

By locating all these facilities in a relatively small area, just over two acres, an almost urban feeling is engendered which contrasts sharply with the feeling of the adjacent natural areas. It also serves the more mundane but functional purpose of concentrating the trash producing activities.[123]

The consultants conclude their report for the site planning of the Trojan Nuclear Power Plant by dealing with some of the natural areas to be left on the site in the following words:

> Fully half of the site, when construction is completed, will remain in the natural state. A broad swatch of development across the central sector of the site divides the natural areas east of the highway in half. To the north all remains natural, to the south the natural areas surround the intensive recreational foci. The natural areas include most of the river's edge, portions of the riparian woodlands and the hardwood forest, the hills and bluffs west of the highway, and the northerly agricultural lands and wetlands.
>
> The only major change will be the strenthening and enlarging of the swan lake at the northern end of the property. This is to provide a more suitable habitat for the Whistling Swans during their winter stay.
>
> There will be nature trails connecting the various points of interest but aside from a few picnic shelters, along the river and elsewhere, and the lookout on the bluffs there will be no construction.[124]

This study and the resultant report is certainly a milestone in the environmental consideration of industrial elements in a contemporary landscape. The report is summarized in the following statement, which actually epitomizes the role of the environmental designer in site planning for electrical generating facilities:

> In an age when men are becoming increasingly aware of the quality of their environment and increasingly concerned over its destruction, every use of the land must be subjected to rigorous analysis and debate. Rather than subjugating the world about him to his own wants and whims, man must learn to live in harmony with this world.[125]

The sixth possible way in which environmental designers are able to deal with electric power generating facilities is through cosmetic treatment to either hide in a superficial way, or to beautify offensive aspects of power generating facilities. There are many examples in reality of this type of approach, though it is extremely dangerous and not in the least productive to point these out. Therefore, none of the actual examples which could illustrate this approach will be pinpointed in this publication.

In keeping with the attempt at a positive approach to show the optimum ways to deal with the environmental and energy problems, the cosmetic way of thinking, however, is characterized by a quote from the 1970 National Power Survey, which alludes to this method or approach.

> Fortunately, there has been a general trend among utility companies to beautify the many facilities that must, of necessity, be accommodated on the landscape. Structure design, architecture, landscaping, and general appearance have been greatly improved. There is no scarcity of models to follow.

123. Ibid., p. 41.
124. Ibid., p. 43.
125. Ibid., p. 47.

- Plants have been carefully sited, thoughtfully designed, and often surrounded by parks and picnic areas.
- Sub-stations have been located where they are unobtrusive.
- Distribution sub-stations have been vastly improved, in some instances being contained within structures indistinguishable from others in the environment.
- It is not unusual for large utilities to have or retain architects and landscape architects, a practice that is gaining, and is doing much to give utilities a leadership role in the field of industrial beautification.
- New types of poles have been developed and transformers have been removed from sight.

Unquestionably, there is much more to be done, but the progress already made would indicate that the industry generally assumes a large measure of responsibility and further programs can be anticipated.[126]

Another major area where environmental designers could have a potential impact in working with utility industry representatives is in the studies for multi-purpose plant sitings. The entire concept of the multi-purpose plant sitings is still in an embryonic stage, but with increased crowding and the utilization of the few remaining optimum sites for single-purpose plant siting, it may become increasingly necessary to plan and utilize to a far greater extent power generating sites for a multiplicity of purposes. When this happens, as it will be coupled to a more crowded environment and a greater sensitivity to the quality of the existing remaining environment, the publication entitled, "Considerations Affecting Power Plant Site Selection" has this to say about multi-purpose plant sitings:

> During the past few years, increased interest has been given to the possibility of combining plant functions to achieve cost saving. There is an almost infinite number of process combinations with power production that one could consider in multi-purpose plants. The idea is not new. Multi-purpose facilities are already in existence throughout the world and in our own country, the multi-purpose nature of TVA hydroelectric power stations is so well known that no further discussion is required here.
>
> The prime examples of such proposed dual-purpose plants are those which produce electricity and desalt water. Other such dual-purpose plant sites would be concerned with effluent disposal, the production of desalted water for irrigation, the production of electricity and the processing of heat for industrial applications, central heating of municipalities, the production of electricity, and the reclamation of waste in which the waste is used as fuel for power production and the utilization of such multi-purpose sites in combination with an agro-industrial complex.
>
> The critical role of land planning and land analysis in such multi-purpose site plant-type planning is evident and will undoubtedly be realized and utilized to a greater extent in years to come.[127]

The seventh basic way in which the environmental designer is able to work with electric utility companies is through the development and utilization of water reservoir areas for a variety of purposes.

Almost throughout the history of reservoir development in the United States, landscape architects have played an important part in the design and development of facilities to utilize more fully the full resources of the water reservoir for the storage of water for hydroelectric facilities. Landscape architectural firms, and individual landscape architects too numerous to mention, have worked extensively throughout the entire United States on the full development of these areas. The work of the landscape architects with the Tennessee Valley Authority, of course, is historic in this regard.

Throughout the 1930's and 1940's a large number of landscape architects worked on the development of these hydroelectric reservoirs in order to insure recreational use, preservation and en-

126. Federal Power Commission, *The 1970 National Power Survey: Guidelines for Growth of the Electric Power Industry*, (Washington, D.C., U.S. Government Printing Office, Superintendent of Documents, 1971), p. I-22-8.

127. The Energy Policy Staff, Office of Science and Technology, *Considerations Affecting Steam Power Plant Selection*, (Washington, D.C., U.S. Government Printing Office, Superintendent of Documents, 1968), p. 69.

hancement of environmental values, and the encouragement of the wildlife potential of the region.

With the advent of the nuclear plants, large masses of water are required to be maintained for cooling purposes for the various parts of the power plant. These water reservoir facilities, once again, can be extensively developed utilizing the services of the landscape architect. The water body required as an adjunct to energy generation is able to be an extremely positive element in the environment of the surrounding area. One of the most innovative new programs in this regard was the land use plan for North Anna Reservoir in Virginia, prepared for the Virginia Commission of Outdoor Recreation by Theodore J. Wirth & Associates in 1971. In the introduction to that study the full relationship of all of the organizations involved was spelled out in the following words:

> A vast new resource is being created in the State of Virginia. That resource is the North Anna Reservoir, built by the Virginia Electric and Power Company as a source of cooling water for their nuclear electric generating station, now under construction. This new resource will have benefits beyond those of simple cooling however. The recreational potential of the reservoir will bring about a vast growth in boating, swimming and other water based recreational activity. This activity will create a demand for summer homes, marinas, boat launching sites and a host of other shoreside developments. If improperly managed, shoreside growth and recreational activity could have an adverse effect on the reservoir environment and on water quality. To help avoid that possibility an environmental study and plans were undertaken to help guide state and local officials as well as residents of the reservoir area.[128]

The location of this reservoir, on the southern edge of the east coast megalopolis provides recreation opportunities in an extremely strategic location. The consultants, in a description of the site itself, said the following:

> Dropping some 267 feet over its 68 mile course, the North Anna River drains almost 600 square miles of Virginia's piedmont. The drainage area is predominantly rural, with gently rolling hills, wooded or farmed, but little urbanized. The North Anna Reservoir drains 343 square miles of the basin and at capacity will impound 305,000 acre feet of water.[129]

The following illustration indicates the character of the North Anna drainage basin.[130] The consultants, as part of their program, prepared a general land-use plan for the more complete utilization of the reservoir for a multiplicity of purposes. That plan is shown in the following illustration:[131]

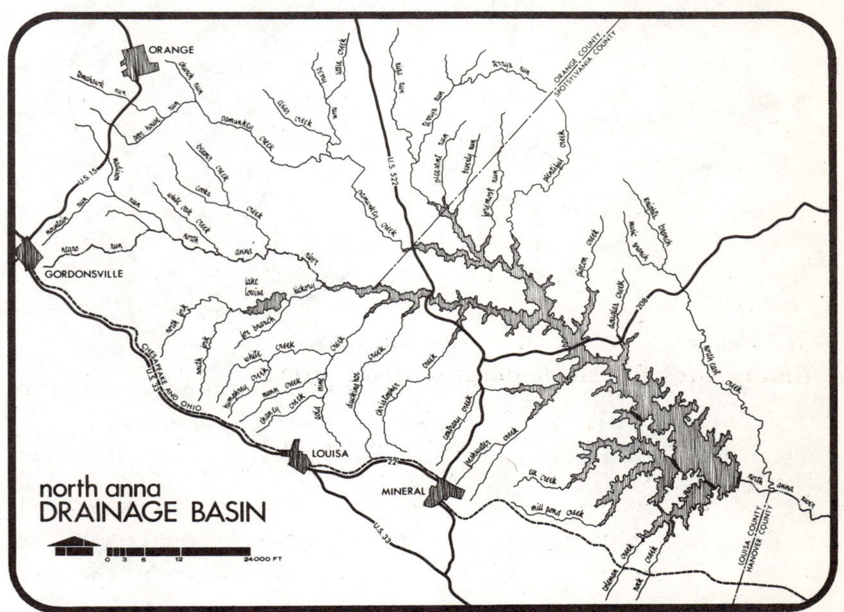

128. Theodore J. Wirth and Associates, *A Land Use Plan for North Anna Reservoir,* (Chevy Chase, Md., prepared for the Virginia Commission of Outdoor Recreation, 1971), p. 1.
129. Ibid., p. 8.
130. Ibid., p. 8.
131. Ibid., p. 17.

Study Area

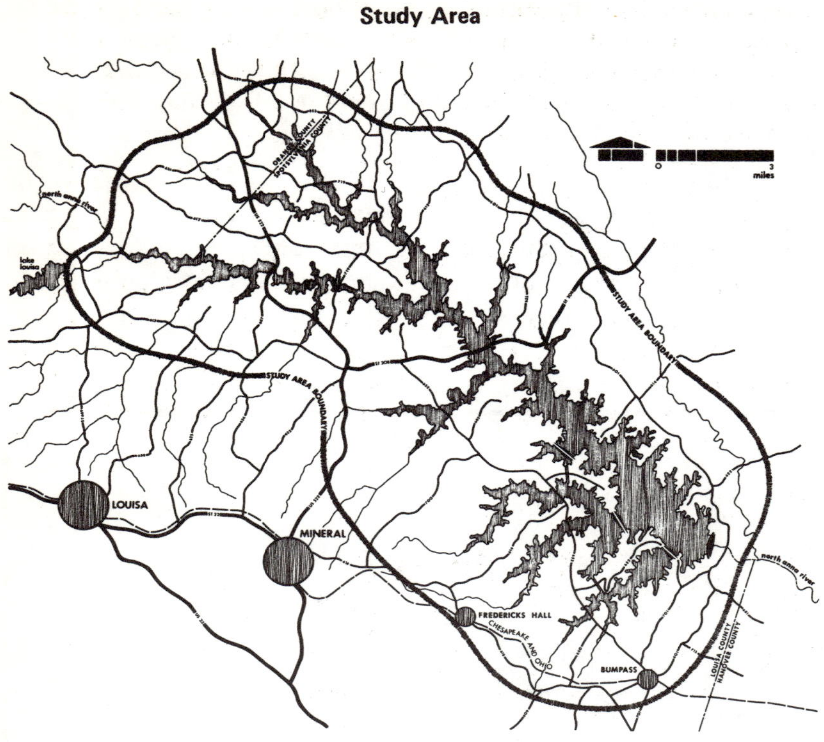

The consultants suggested a series of development concepts on the peninsulas of land extending into the reservoir itself. The following illustration shows a typical state park development concept for the proposed area for the state park.

The consultants also suggested that two additional regional or county park areas be provided to allow public access to the reservoir. The Wirth staff also made suggestions for residential development concepts for the areas around the reservoir. The following illustration indicates "a cluster concept relating to water-oriented facilities within the residential area which is also necessary to the good of the overall reservoir shoreline development reducing or eliminating 'strip' boat dock shoreline development."

state park development concept[132]

residential development concepts[133]

132. Ibid., p. 40.
133. Ibid., p. 41.

This project is an excellent example of the ways in which the landscape architect has worked with a state agency and a power company to develop maximum utilization of the areas immediately surrounding the reservoir itself. This and similar sorts of reservoir projects point the way to sensitive incorporation of functional electrical energy components into a fragile environment.

The 1970 National Power Survey made the following statement about the possible utilization of reservoirs for recreational purposes.

> The warm waters of cooling ponds can provide important recreational areas. Lands adjacent to the 2,600-acre Lake Sangchris are being developed by the State of Illinois for recreational use. In addition to fishing, facilities are to be provided for boating, camping, and picnicking. This lake was created by Commonwealth Edison Company to provide a source of cooling water for its 1,200-megawatt Kincaid generating station. The cooling pond for Virginia Electric and Power Company's 1,140-megawatt Mt. Storm plant is used for boating and water skiing. Kansas City Power & Light Company placed its Montrose Lake under the jurisdiction of the Missouri Conservation Commission which maintains facilities for various types of recreation. The following illustration shows some of the possibilities in the use of cooling ponds for various recreational and residential purposes. However, care must be exercised in selecting the types of uses, to avoid those uses contributing significantly to the eutrophication process of the pond and adversely affecting its usefulness for cooling purposes.

Possible Multiple Uses of a Cooling Lake[134]

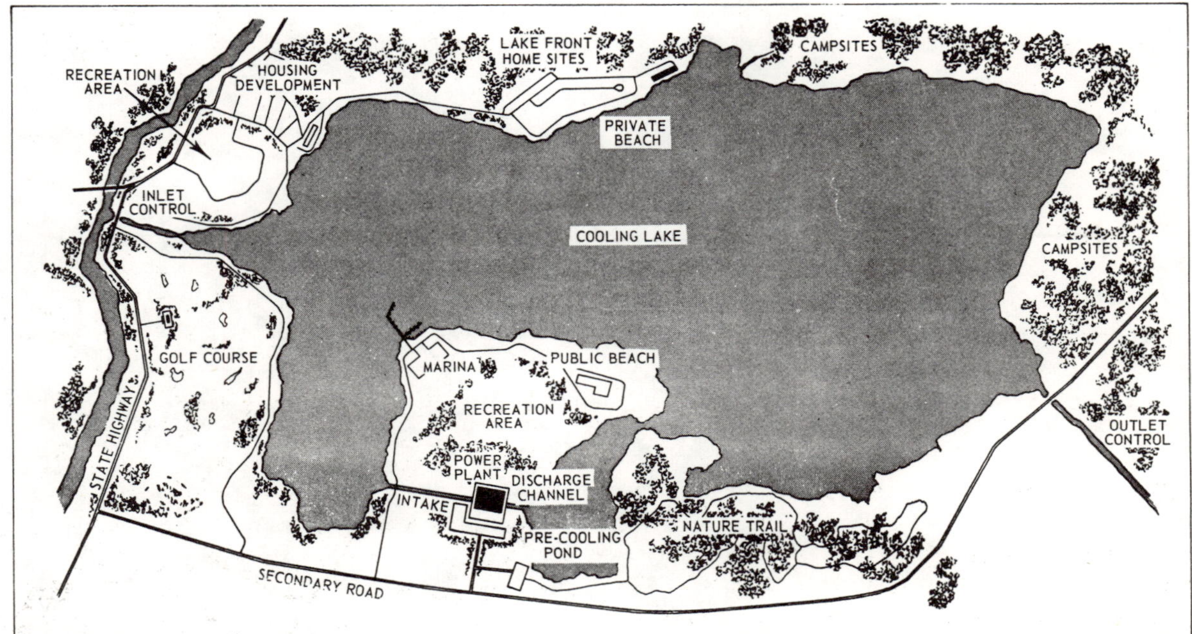

134. Federal Power Commission, *The 1970 National Power Survey: Guidelines for Growth of the Electric Power Industry*, (Washington, D.C., U.S. Government Printing Office, Superintendent of Documents, 1971), p. I-10-12.

Power Transmission Facilities

Introduction

Among the various physical components of the electric utility industry, the transmission lines of various sizes are the most pervasive, obvious and ubiquitous of the infrastructure associated with the industry.

Transmission lines of some type intrude or are at least evident in all types of landscape throughout most civilized nations. The height and size of the structures, the fact that the lines pervade all qualities and levels of landscape, and that in the past little concern or few guidelines existed to guide in the optimal location for the minimum destruction of environmental values by transmission lines, all are reasons for the general feeling that they are destructive of the quality of the environment.

The publication "Electric Power and the Environment" previously mentioned, deals with the transmission lines in the following words:

> Transmission lines are the vital links between generating plants and the distribution systems that deliver electricity to the consumer. A strong transmission grid is vital for reliable economic electric service. And of even greater significance for our purposes here is the fact that a strong regional and interregional system provides additional flexibility in selecting sites for generating plants.
>
> In the future, land may not be available to meet expected demands for rights-of-way by the electric power industry and other industries if present day practices in establishing

single purpose rights-of-way for each industry continue to prevail. Literally millions of acres of rights-of-way land would be required in the next two decades for the growth of the utilities and their tremendous needs for transmission lines along with the expansion of highways, interstate gas lines, water supply lines, etc. It is imperative that land be used more efficiently for all competing demands. This calls for locating utility services whenever possible on the same rights-of-way and planning joint use service or 'utility corridors' in order to minimize impact on the environment.

Local opposition to rights-of-way would be substantially reduced if utility corridors are established with proper public sanction in a long-range planning program. Communities are increasingly concerned with the appearance of their surroundings and with natural beauty. They have shown much interest in the location and appearance of electric power equipment. In response to this public interest, it is necessary that transmission line planning allow sufficient time and procedures for 'public input' in a timely and meaningful manner.

The prospective land requirements of the electric power industry are only a part of a national pattern. A recent study estimates that by the year 2000 satisfaction of all projected demands for land would mean the use of every acre in the 48 contiguous states including deserts, mountain peaks and marshes and still leave a net shortage of 50 million acres. Such a projection of course is merely to illustrate the magnitude of the potential problem. In fact, the land shortage is already with us in many areas. It is imperative that there be an immediate effort to plan land use for optimum overall benefits to the public. An effort of this nature would generally involve developing broad land-use policy which would, among other things, encourage and aid development of utility corridors even though we recognize the many technical and institutional problems that must first be resolved before utility corridors can become a reality.[1]

History

The history of the electric transmission lines throughout the United States is not a long one. It has made extremely rapid strides in the middle years of the 20th century. The trend in transmission voltage in the history of the transmission line as an environmental problem may be traced in the following quote from the publication entitled, "Considerations Affecting Steam Power Plant Site Selection."

> The steady increase in electric power use and the accompanying expansion of electric power networks with their many connecting transmission lines have produced problems of obtaining transmission line rights-of-way to the extent that this sometimes creates considerable impact on power plant site selection.[2]

The following short passage traces, in some detail, the history and the increase in transmission line voltages:

> From the introduction of 110,000-v. alternating current transmission in the United States in 1908 to about 1950 there was a steady increase in the voltage levels. In the past decade progress has accelerated to the point that one system in the central part of the United States is now engaged in the application of 765,000 kv transmission equipment.[3]

The illustration on page 113 shows the rate at which these increases in maximum transmission voltages have occurred.

The maximum alternating current voltage in use from 1883 to 1970 is shown on the following chart on page 113.

"Power Lines and Scenic Values in the Hudson River Valley," published by the Hudson River Valley Commission in 1968 makes the following statement in regard to the history and the development of voltages on various transmission lines.

1. The Energy Policy Staff, Office of Science and Technology, *Electric Power and the Environment,* (Washington, D.C., U.S. Government Printing Office, Superintendent of Documents, 1970), p. 21.
2. The Energy Policy Staff, Office of Science and Technology, *Considerations Affecting Steam Power Plant Selection,* (Washington, D.C., U.S. Government Printing Office, Superintendent of Documents, 1968), p. 66.
3. Ibid., p. 66.

Maximum Transmission Line Voltage[4]

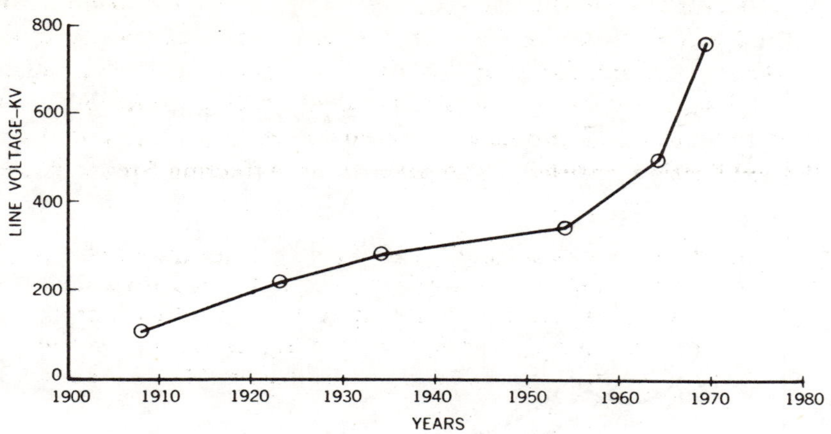

Maximum A-C Voltage in Use 1883-1970[5]

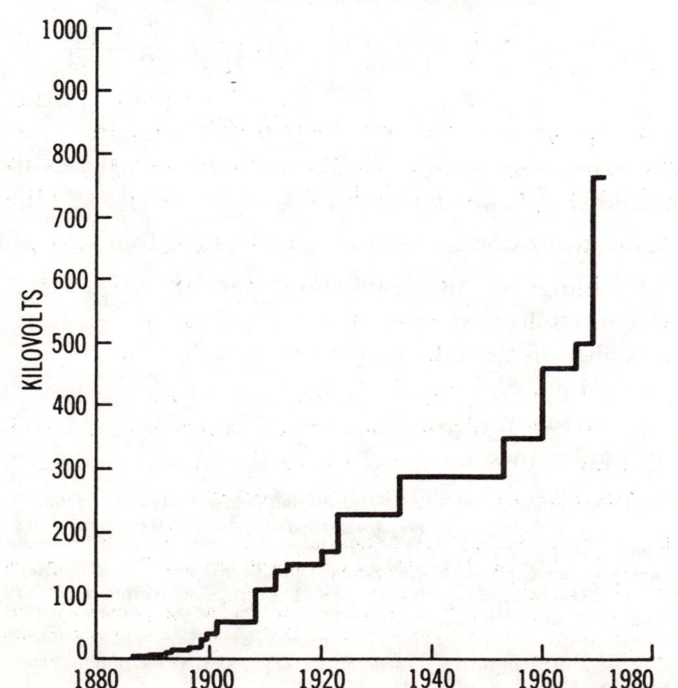

Voltage on transmission lines (as distinguished from local distribution lines) seldom is less than 69,000 volts. Because of the current shift from self-sufficient power systems to a pattern of interconnected networks and of improved transmission technology, transmission lines constantly are being upgraded to carry increasingly larger voltages. Lines with a 345,000-volt capacity first were built in the early 1950's and now are becoming standard. In 1965, the first 500,000-volt transmission lines were introduced in the United States, and future technological changes promise to bring even higher transmission voltages.

Although larger towers usually are needed, extra high transmission voltages mean fewer conductors to carry the same amount of power. A race thus ensues between improvements in transmission capacity and increases in power demands.

The first significant application of direct-current EHV transmission in the United States is expected to go in service late in 1969. This will be an 800-mile, ±400 kv line between the Pacific Northwest and the Pacific Southwest which will be capable of transmitting about 1300 mw. A single dc line of similar characteristics and length is planned for initial operation in 1972.[6]

The approximate power transmission capabilities of long lines are shown on the following chart on page 114.

The relative cost and capabilities of extra-high voltage alternating current lines may be explained in the following quote:

> Variations in design, conductor materials and configuration, basic insulation levels, and related factors involved in transmission line construction render it difficult to present a simple comparison of transmission line costs and capabil-

4. Ibid., p. 66.
5. Federal Power Commission, *The 1970 National Power Survey: Guidelines for Growth of the Electric Power Industry,* (Washington, D.C., U.S. Government Printing Office, Superintendent of Documents, 1971), p. I-13-5.
6. Bruce Howlett, Frederick J. Ehniger with assistance of Anthony Corkill and John Seddon for the Hudson River Valley Commission, State of New York, *Power Lines and Scenic Values,* (Tarrytown, N.Y., Hudson River Valley Commission, 1968), p. 6.

Approximate Power-Carrying Capabilities of Long Lines[7]

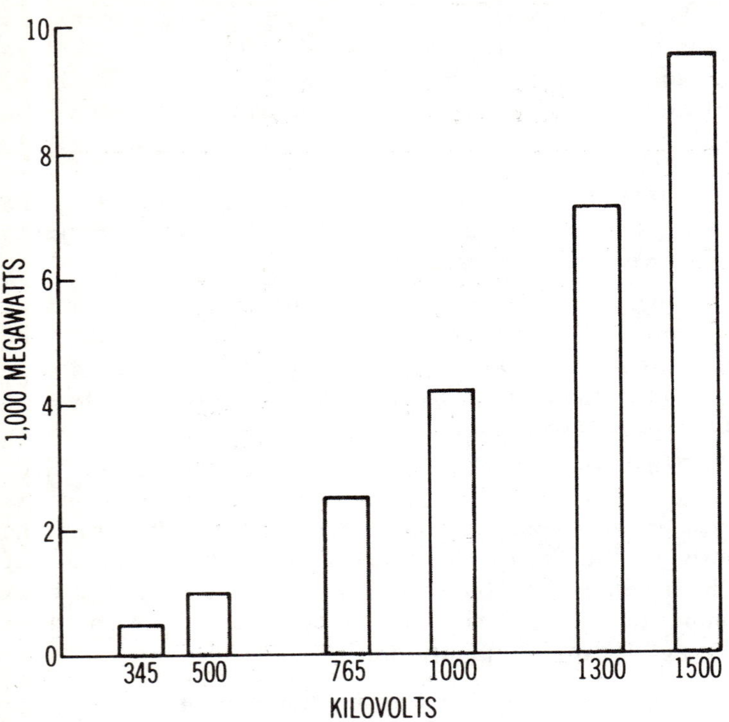

ities. The following table, however, gives an indication of the relative values which apply to several transmission voltage classes now in use or planned for early construction in the United States.[8]

Line voltage (kv)	Range of cost, $1,000 per mile	Approximate capability (mw.)	
		100 miles	300 miles
230	45- 60	275	135
345	60- 80	700	350
500	85-100	2,000	960
750	125-160	4,400	2,150

The Federal Power Survey of 1970 also presents a table showing costs of electric energy transmission for 200 miles.

Cost of Electric Energy Transmission for 200 Miles[9]

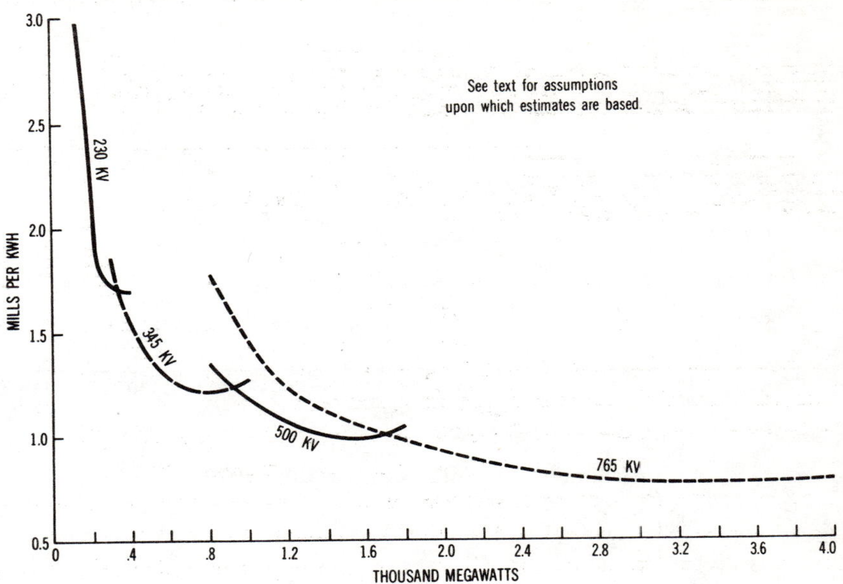

The actual costs of specific alternating current lines are shown on the following table on page 115.

Each transmission line project involves the following stages:

A. Planning and environmental impact analysis
B. Route selection
C. Rights-of-way acquisition
D. Design
E. Construction
F. Maintenance

7. Federal Power Commission, *The 1970 National Power Survey: Guidelines for Growth of the Electric Power Industry*, (Washington, D.C., U.S. Government Printing Office, Superintendent of Documents, 1971), p. I-13-6.
8. The Energy Policy Staff, Office of Science and Technology, *Considerations Affecting Steam Power Plant Selection*, (Washington, D.C., U.S. Government Printing Office, Superintendent of Documents, 1968), p. 68.
9. Federal Power Commission, *The 1970 National Power Survey: Guidelines for Growth of the Electric Power Industry*, (Washington, D.C., U.S. Government Printing Office, Superintendent of Documents, 1971), p. I-13-9.

Actual Costs of Specific AC Lines[10]

Conductors	Cost per Mile *		
	Right of Way and Clearing	Line Construction	Total
EASTERN AREA—500 kV			
2-2037 ACSR	$30,700	$80,800	$111,500
2-2493 ACAR	13,500	128,500	142,000
2-2049 5005	16,700	85,800	102,500
3- 971 ACSR	12,400	65,300	77,400
4- 583 ACSR	10,000	95,500	105,500
2-2032 ACSR	17,000	98,000	115,000
2-2490 ACAR	20,000	142,000	162,000
2-2490 ACAR	59,000	272,000	**331,000
2-2500 ACAR	22,000	118,000	140,000
3- 954 ACSR	12,000	95,000	107,000
CENTRAL AREA—500 kV			
3- 954 ACSR	....	84,200	....
3-1024 ACAR	24,000	95,600	119,600
WESTERN AREA—500 kV			
2-1780 ACSR	7,100	72,200	79,300
2-1852 ACSR	....	82,000	....
2-2156 ACSR	25,000	93,900	118,900
2-2156 ACSR	2,000	124,000	+126,000
ALL REGIONS—735-765 kV			
Average of 735 kV and 765 kV lines	18,700	146,300	++165,000

* 1968–1969 prices (Significant increases in costs have been experienced recently in some areas).
** Line near urban center.
+ Desert construction.
++ Includes line sections built over 4-year span.

10. Ibid., p. I-13-7.

The Problems

Basically, the problems involved in electric transmission lines and structures are manifold. Firstly, our greatest concern probably is the large amounts of rights-of-way accorded the various transmission lines.

Secondly, they are very ubiquitous, in that they pervade the human view throughout the landscape in all sorts of unlikely locations and situations.

Thirdly, the ungainliness of the practical and economical structures required to hold the lines overhead have caused a visual blight in human perception of natural landscape features.

The construction and maintenance of the transmission rights-of-way causes difficulties and alterations in the natural ecology of certain areas through which the lines pass.

In other instances in times gone by, improper or ill-concerned alignment has resulted in insensitive visual infringements of the lines in inopportune locations and situations.

According to the publication entitled, "Power Lines and Scenic Values in the Hudson River Valley" published in December, 1968, by the Hudson River Valley Commission:

> The selection of the route for a new power transmission line is, for the most part, up to the individual electric utility. In the case of hydroelectric generating stations, including pumped storage stations, the Federal Power Commission must review and approve the transmission line routes for the generating station up to the point where the line reaches a distribution substation. The FPC does not pass upon transmission line routes for nuclear or thermal generating stations, however.
>
> Local governments cannot prohibit a utility company from building transmission lines or other facilities needed to provide electric service, but localities can make the construction of such facilities subject to special permit and require them to meet reasonable standards with regard to safety, aesthetics and other factors affecting the community.
>
> In the past, local governments have rarely asserted control over transmission lines and other utility projects.[11]

The extent of the problem can best be epitomized by a review of the basic statistics concerning the state of transmission lines in the United States, and by indicating on maps the pervasiveness of transmission lines of various levels.

Currently, it has been estimated that there are 300,000 miles of overhead electric transmission lines, with an average right-of-way of 110 feet in width. These lines utilize 4,000,000 acres of land for rights-of-way. In the balance of this century 100,000 miles of new transmission line will require one and one-half million additional acres of rights-of-way.

The two maps on the following pages indicate the extent of the electric transmission system in 1970, and the possible pattern of transmission by the year 1990.

The environmental problems created by electric transmission lines have been alluded to in previous sections. It probably would take an entire volume to list completely in some minute detail the environmental problems caused and created by electric transmission lines in the contemporary American environment.

Basically, the problems may be categorized into visual, ecological, and locational. Sylvia Crowe in the British publication entitled, "The Landscape of Power" refers to the contemporary terrain as the "landscape of the grid." Basically, the environmental problems created by transmission lines have to do with the types of structures themselves, with the excessive right-of-way widths, with right-of-way locational problems, and with the insensitive right-of-way configurations or cross sections. Ecological communities are disturbed. Visual perception of the unspoiled landscape is marred by the intrusive qualities of the electric lines, and the standards which hold them.

The construction and maintenance of the transmission lines can cause unsightly scars on the landscape. Natural ecological communities which are untouched by any other man-made appurtenances may be drastically disturbed by the intrusion of the electric power lines.

11. Bruce Howlett, Frederick J. Ehniger with assistance of Anthony Corkill and John Seddon for the Hudson River Valley Commission, State of New York, *Power Lines and Scenic Values,* (Tarrytown, N.Y., Hudson River Valley Commission, 1968), p. 8.

Transmission System 1970[12]

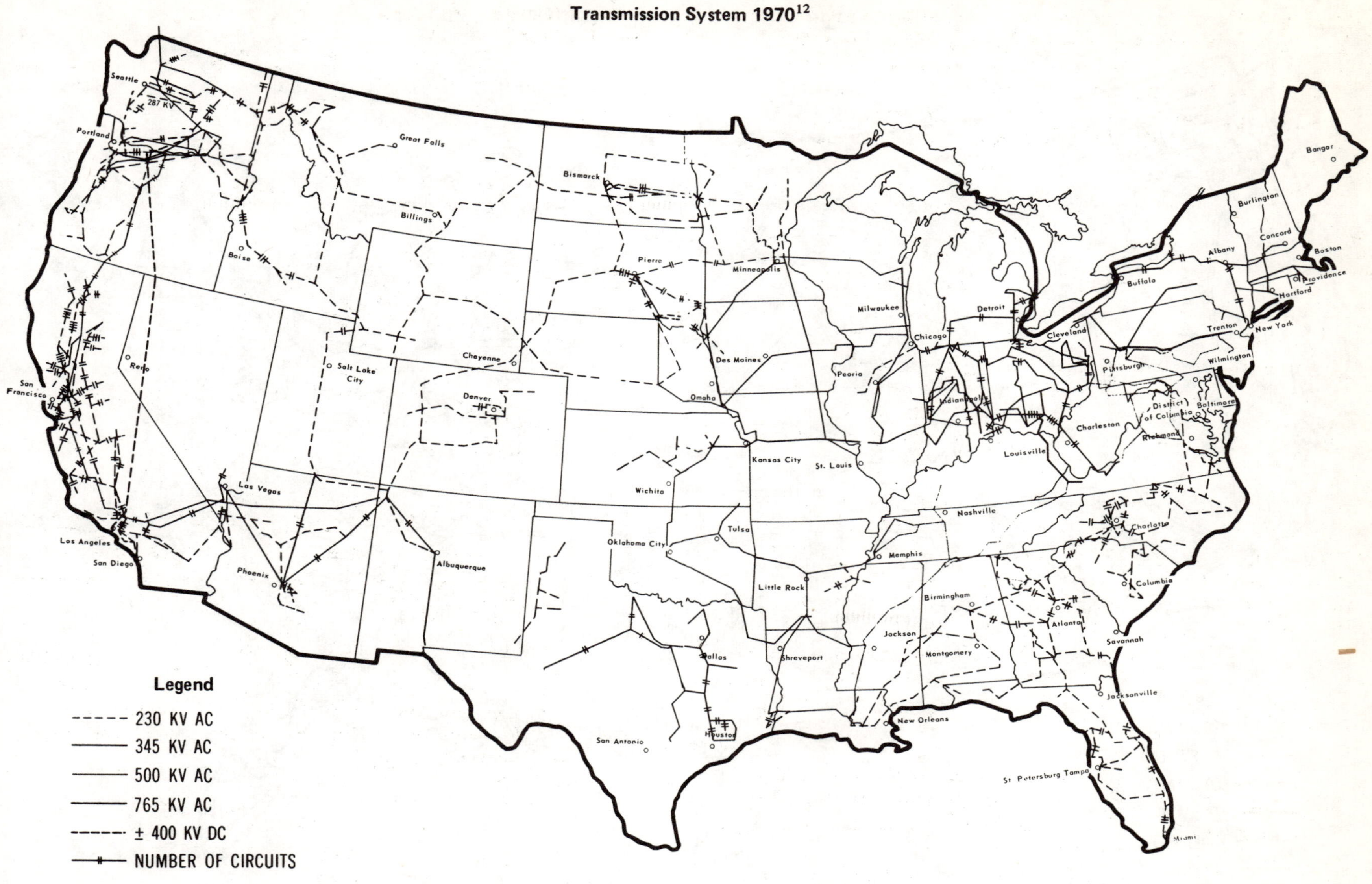

12. Federal Power Commission, *The 1970 National Power Survey: Guidelines for Growth of the Electric Power Industry*, (Washington, D.C., U.S. Government Printing Office, Superintendent of Documents, 1971), p. I-13-2—I-13-3.

Possible Pattern of Transmission 1990[13]

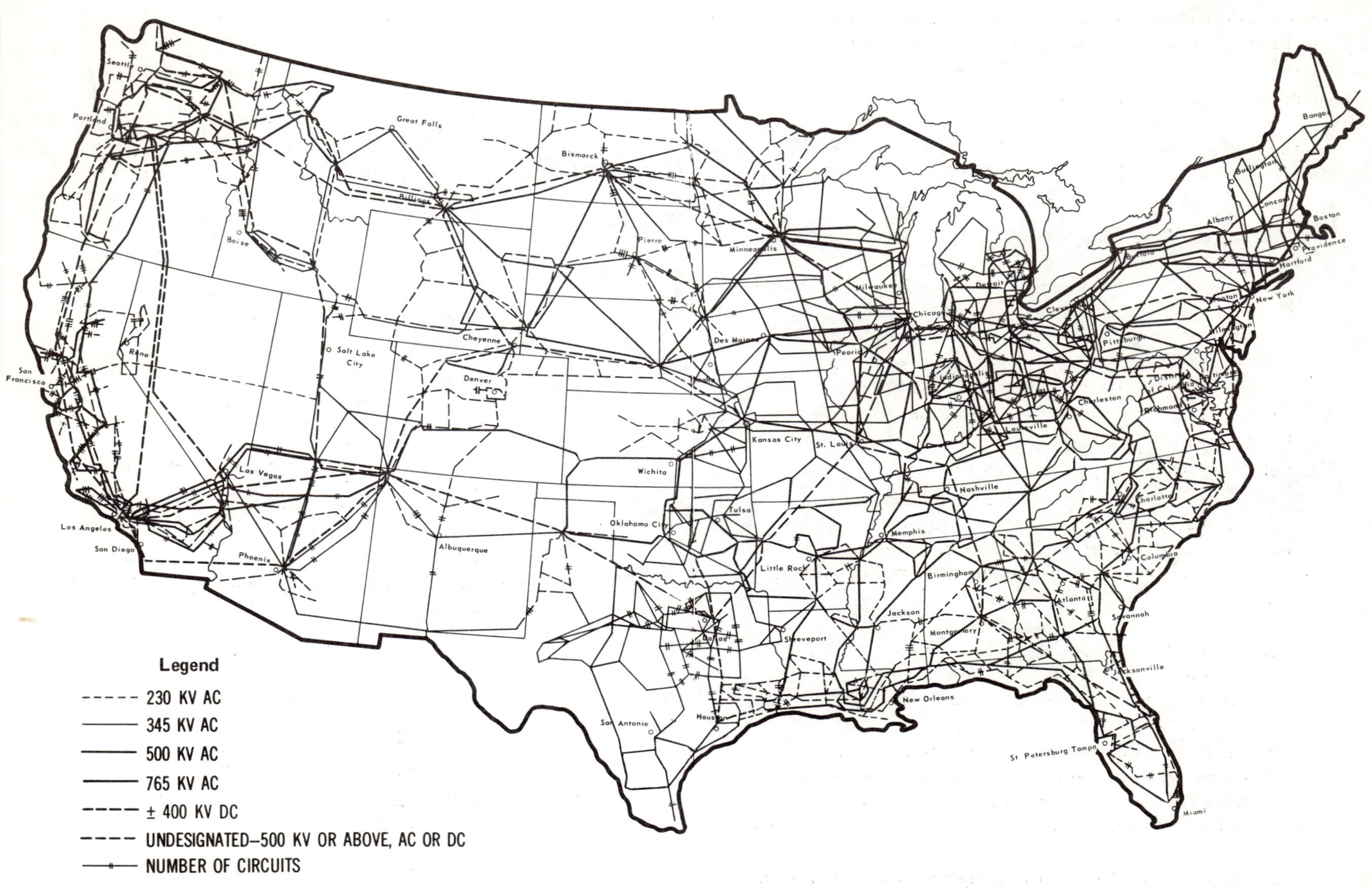

13. Ibid., p. I-18-14–I-18-15.

Nearly all of these environmental problems have been dealt with at some time in the recent past by either study commissions or consultants. One of the most significant publications developed by the Federal government was entitled, "Environmental Criteria for Electric Transmission Systems" and was published by the U.S. Department of Interior and the U.S. Department of Agriculture in 1970.

In the section on transmission lines the following guidelines and statements were made in that publication:

A. General Comments

Past experience has demonstrated that compliance with the following criteria will safeguard aesthetic and environmental values within the constraints imposed by the current state of high-voltage transmission technology. The development of an economically feasible underground transmission technology will require a redefinition of environmental transmission criteria.

These criteria are intended for use in rural areas and may not apply specifically to urban areas.

B. Selection of a Proposed Route or Corridor

Concealment of transmission towers and lines is virtually impossible, but much can be done to make them less obtrusive and more attractive. Adherence to the following criteria will minimize the impact and optimize the compatibility of transmission facilities with the environment.

1. Rights-of-way should be selected to preserve the natural landscape and minimize conflict with present and planned uses of the land on which they are to be located.
2. Where possible, retirement or upgrading of existing lower voltage transmission circuits should be required to allow construction of higher voltage, higher capacity circuits on the existing right-of-way.
3. Properly sited established rights-of-way should be used where warranted for the location of additions to existing transmission facilities.
4. The joint use of electric transmission facilities by two or more utilities should be encouraged, when feasible, to reduce the total number of transmission lines constructed.
5. The joint use of rights-of-way with other types of utilities should be coordinated in a common corridor wherever uses are compatible.
6. The relative advantages and disadvantages of locating a new line either adjacent to or widely separated from existing transmission lines should be considered. Right-of-way boundaries should be so located as to avoid creating usable hiatus areas.
7. Rights-of-way should avoid heavily timbered areas, steep slopes, proximity to main highways, shelter belts and scenic areas.
8. Where possible, transmission line crossings of major roads in the vicinity of intersections or interchanges should be avoided.
9. Long views of transmission lines parallel to existing or proposed highways should generally be avoided. Alternative routes away from highways should be considered. Where ridges or timber areas are adjacent to highways or other areas of public view, overhead lines should be placed beyond the ridges or timber areas.

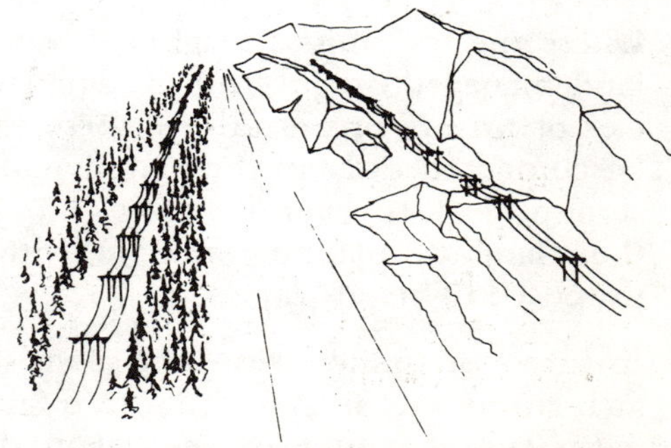

Place beyond ridges or timber

10. Avoid crossing at high points in the road so that the towers cannot be seen from a great distance. Instead, where possible, cross the highway between two high points, at a dip, or on a curve in the road.

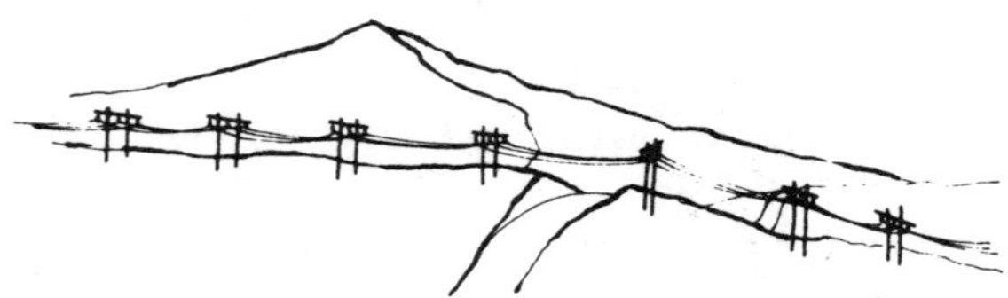

Avoid crossing at high points

Cross at a dip

11. Avoid open expanses of water and marshland and particularly those utilized as flight lanes by migratory waterfowl and as heavily used corridors by other birds. Avoid areas of wildlife concentrations such as nesting and rearing areas.

12. Where the transmission rights-of-way cross areas of land managed by government agencies, state agencies or private organizations, these agencies should be contacted early in the planning of the transmission project to coordinate the line location with their land-use planning and with other existing or proposed rights-of-way.

13. In forest or timber areas long spans should be used at highway crossings in order to retain much of the natural growth or provide a planted screen along the highways.

Use long span towers **To retain natural growth**

14. Long views of transmission lines perpendicular to highways, down canyons and valleys or up ridges and hills should be avoided. The lines should approach these areas diagonally and should cross them at a slight diagonal.

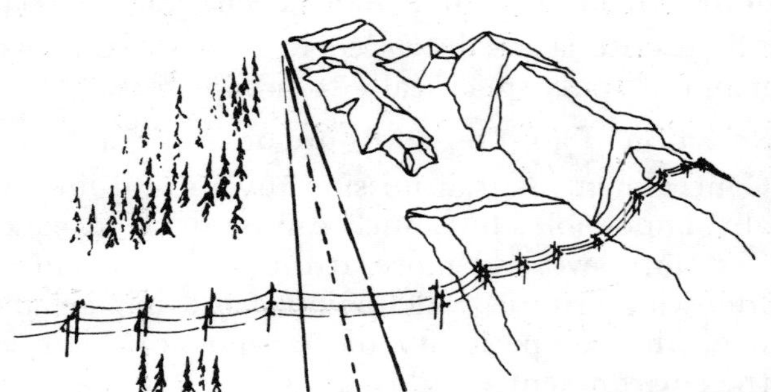

Avoid long views of transmission lines

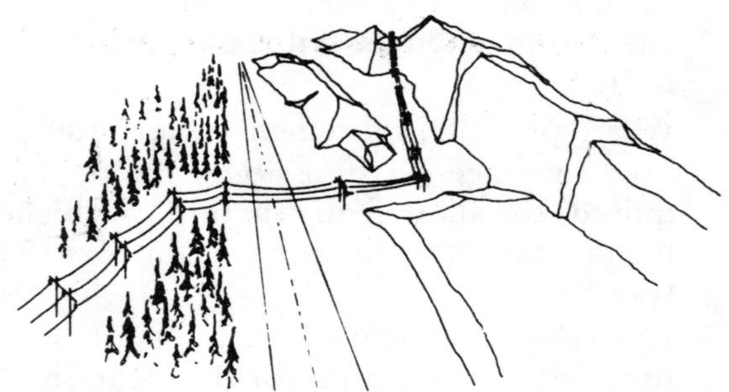

Approach and cross diagonally

15. Transmission lines should cross canyons up-slope from roads which traverse the length of the canyon.

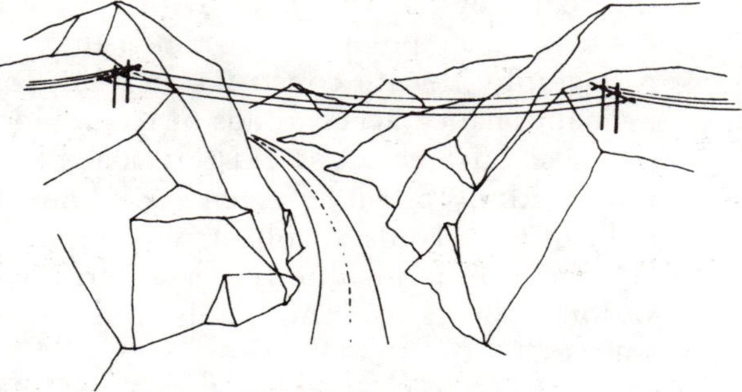

Cross canyons up-slope

16. Transmission facilities should be located part way up slopes to provide a background of topography and/or natural cover where possible. Screen these facilities from highways and other areas of public view to the extent possible with natural vegetation and terrain.

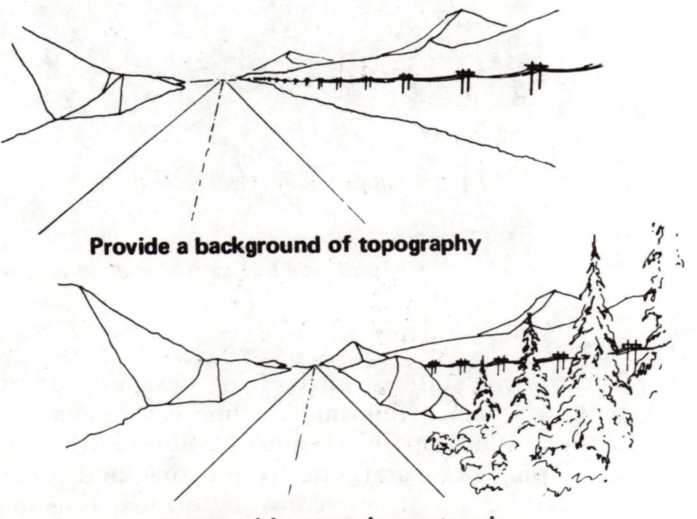

Provide a background of topography

Screen with vegetation or terrain

17. Rights-of-way should not cross hills and other high points at the crests. To avoid placing a transmission tower at the crest of a ridge or hill, space towers below the crest or in a saddle to carry the line over the ridge or hill. The profile of the facilities should not be silhouetted against the sky.

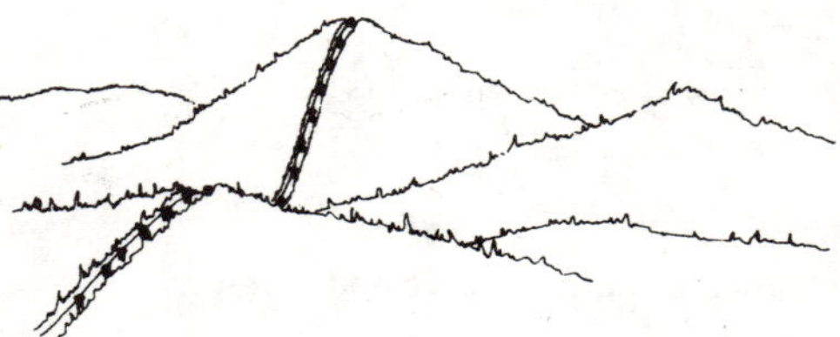

Avoid crossing at crests

Facilities should not be silhouetted

18. Rights-of-way should avoid parks, monuments, scenic, recreation, or historic areas. If a line must be located in or near these areas, the feasibility of placing the line underground should be clearly determined. If the line must be placed overhead, it should be located in a corridor least visible to public view. Other criteria or conditions as necessary to minimize adverse impacts may be imposed by the agency administering the lands involved.
19. When crossing a canyon in a forest, high, long-span towers should be used to keep the conductors above the trees and to minimize the need to clear all vegetation from below the lines. Clearing in the canyon should be limited to that which is necessary to string the conductors.

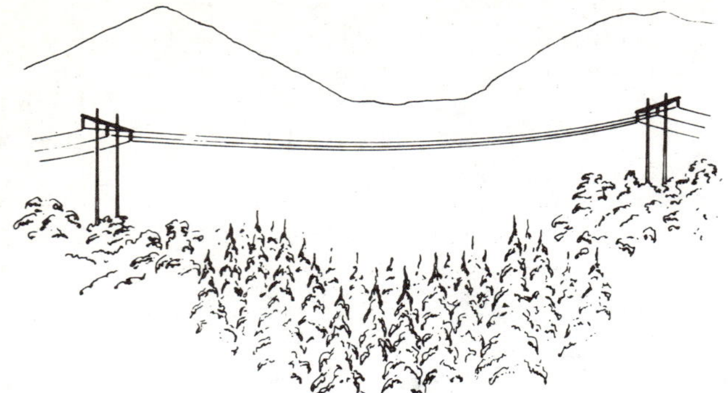

When crossing a forrested canyon

20. It may be desirable to occasionally deflect right-of-way strips through scenic forest or timber areas. The resulting irregular patterns prevent the rights-of-way from appearing as tunnels cut through the timber.

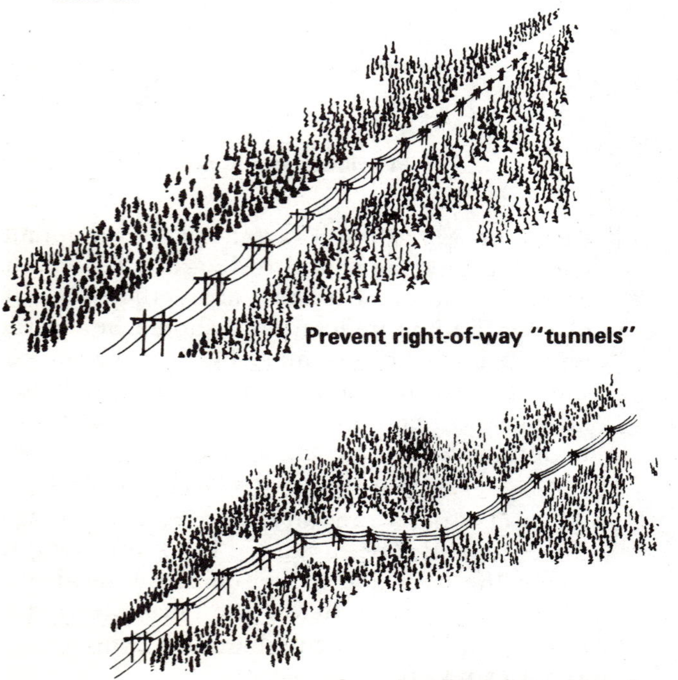

Prevent right-of-way "tunnels"

Occasionally deflect right-of-way

21. Locate access and construction roads in a manner that will preserve natural beauty and minimize erosion. Locate road grades and alignments to follow the contour of the land with smooth, gradual curves when possible. Commensurate with the topography, locate construction roads for later use as maintenance access roads or to provide access to recreational areas. Use existing roads to the maximum extent possible. Agencies administering the right-of-way lands involved may limit access and construction roads due to certain fragile or conservation aspects of the lands and associated resources.
22. Select a route that will maximize the use of natural screens to remove transmission facilities from view.
23. Where lines cross roads, the right-of-way should be left in its natural state as far back from the road as possible.

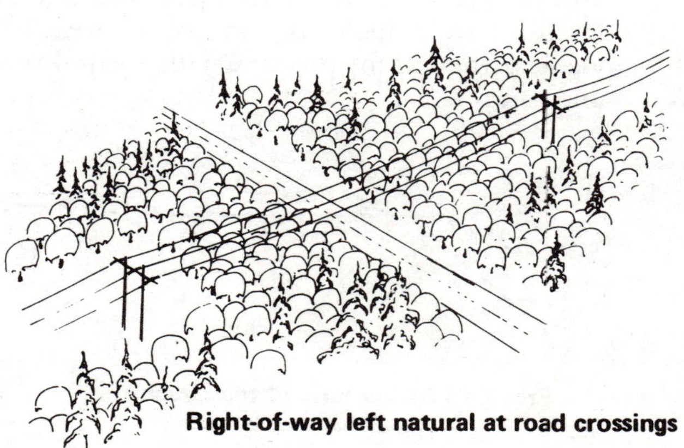

Right-of-way left natural at road crossings

C. Design

The present state of engineering design limits the extent to which a transmission line can be made unobtrusive. Additional work must be done in the areas of developing new, aesthetically pleasing, and functional towers; the use of new construction materials in these towers; and eventual development of economically feasible methods of undergrounding high-voltage, high-capacity transmission circuits. In the interim, design of transmission lines should consider placing of lines underground when crossing roads or traversing aesthetically sensitive areas of limited scope and distance.

The following factors used in the design of surface transmission facilities will minimize the adverse effects:

1. Towers should be strategically located to make maximum use of existing topography, vegetation, etc., for screening.
2. New and more attractive concepts in the design of transmission line towers should be considered.
3. Coloring of transmission line towers to blend with the landscape may be desirable where they must be located in or near areas of high scenic value.
4. The materials used to construct transmission towers should harmonize with the natural surroundings. Self-protecting bare steel is appropriate in many areas. Towers constructed of material other than steel, such as concrete, aluminum, or wood, should be considered.

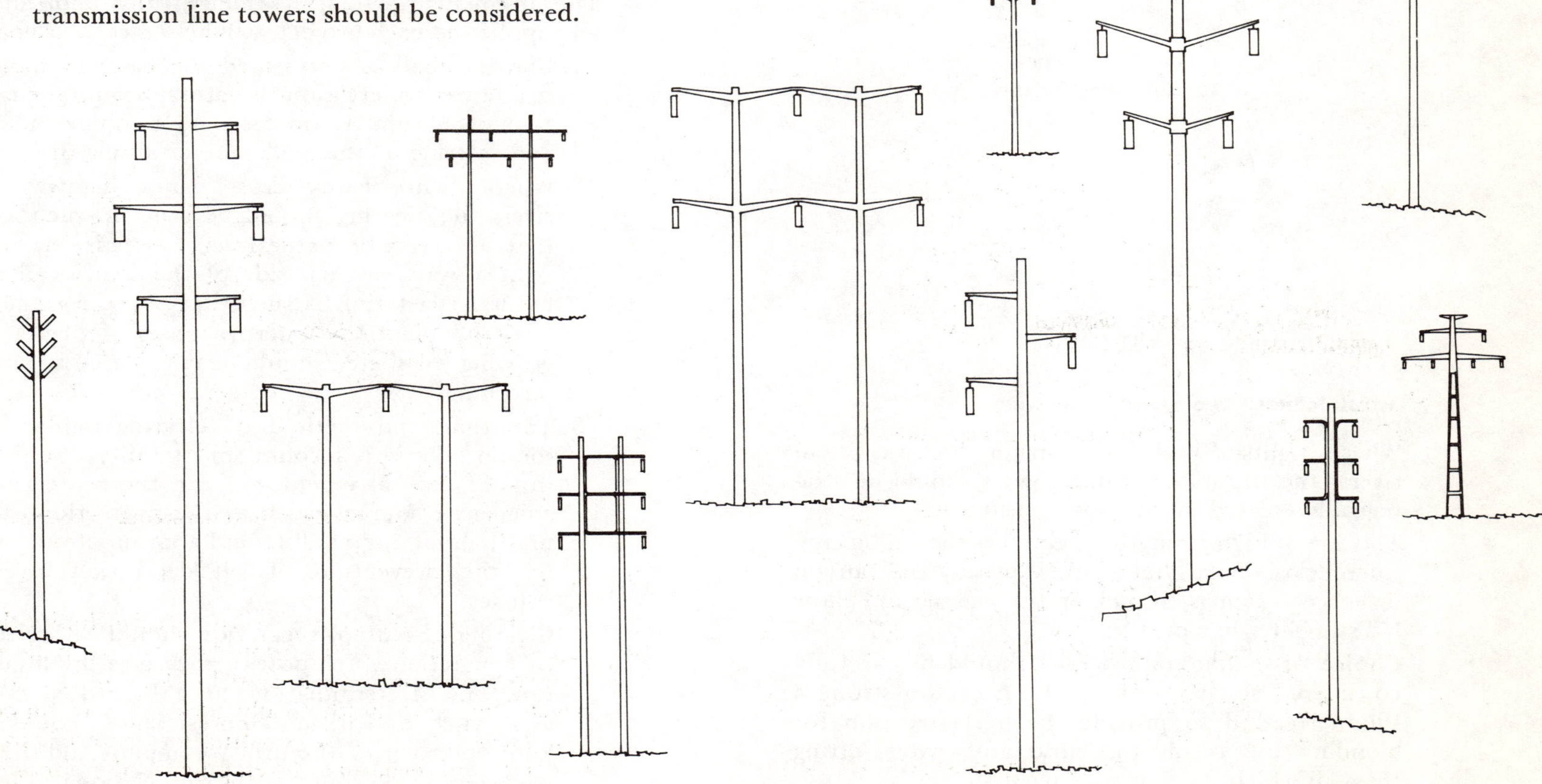

5. At road crossings of two or more circuits, and where only a portion of the line is visible from the highway, the use of multiple circuit towers may be effective in minimizing the impact of the lines at that point.

Consolidating is desired

Two structures + two circuits = too wide

6. Where rights-of-way cross major highways and rivers, the transmission line towers should be strategically located for minimum visibility.
7. The use of high strength conductors should be considered, particularly at road, waterway and canyon or gorge crossings to pick up the line sag and allow for straighter line profiles.
8. Choice of conductor material should be carefully considered so as to avoid sheen or too strong a silhouette and to provide the best selection for blending the conductors into any given setting through which the line must pass.
9. When lines are adjacent to highways, guyed towers should be avoided whenever possible.
10. In situations where the minimum visibility sought in items 3, 4, 6 and 8 conflicts with safety regulations, the safety regulations shall govern. These requirements are outlined in Part 77, F.A.A. Regulations 'Objects Affecting Navigable Air Space.'

D. Clearing

Clearing plans, methods and practices are extremely important for success in any program designed to minimize the adverse effects of electric transmission lines on natural environment.

The following factors, if thoughtfully implemented and applied to each project, will help meet this goal:

1. Clearing shall be performed in a manner which will maximize preservation of natural beauty, conservation of natural resources, and minimize marring and scarring of the landscape or silting of streams.
2. Where rights-of-way cross major highways and rivers the clearing should be done in such a way that a screen of natural vegetation is left in the right-of-way on each side of the road or river. If natural vegetation is such that a screen cannot be left, the planting of native types of plants, low-growing trees, etc., should be considered to provide screening.
3. The time and method of clearing rights-of-way should take into account soil stability, the protection of natural vegetation, and the protection of adjacent resources, such as the protection of natural habitat for wildlife and appropriate measures for the prevention of silt deposition in water courses.
4. Clearing of natural vegetation should be limited to that material which poses a hazard to the transmission line. Determination of a hazard in critical areas such as Park and Forest lands should be a joint endeavor of the utility company and the area manager in keeping with the National Electric

Safety code, State or other electric safety and reliability requirements.

5. The use of 'brush blades' instead of dirt blades on bulldozers is recommended in clearing operations where such use will preserve the cover crop of grass, low growing brush, etc.
6. Where possible, right-of-way strips through sensitive forest and timber areas should be cleared with curved, undulating boundaries. The notched effect of a right-of-way cross section should be avoided.

"Feather-back" the right-of-way

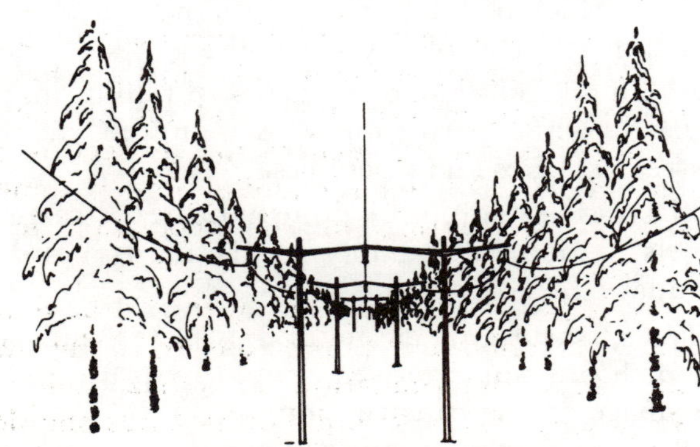

Avoid this notched effect

Careful topping and pruning of trees can contribute to this. Also, small trees and plants should be used to feather back the rights-of-way from grass and shubbery to larger trees. If the rights-of-way are through dense areas of timber where trees are of equal height, and the right-of-way must be cleared to a straight line, the maintenance plan should provide for ultimately reaching the above desired results through seeding, planting and selective cutting of native material. Consideration should be given to the establishment of native vegetation of value as food and cover for wildlife.

7. Special care should be used when clearing through shelter belts, orchards, natural stands of trees, or other special areas with high exposure to public view which cannot be avoided.
8. Where rights-of-way enter dense timber from a meadow or where they cross major roadways, streams or rivers in forested areas, a screen of natural vegetation should be retained along the right-of-way.

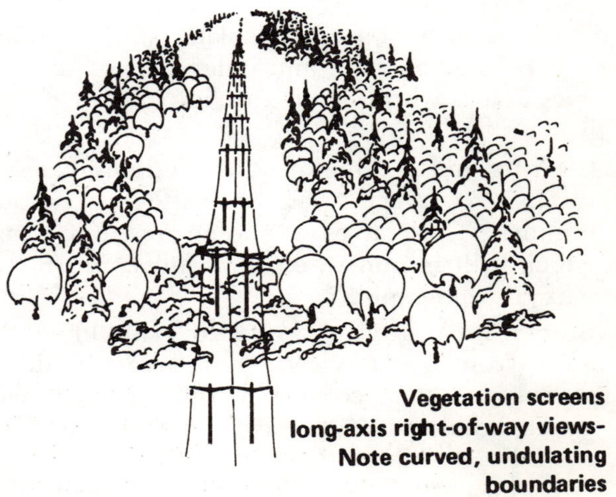

Vegetation screens long-axis right-of-way views- Note curved, undulating boundaries

9. Brush or small trees cleared and not otherwise disposed of may be piled in a way to provide cover habitat for small game animals and birds. Such brush piles should be screened from public view.
10. Trees and brush should be cleared only when necessary to provide electrical clearance, line reliability or suitable access and construction roads for operation, maintenance, and construction. Straight swath appearance should be avoided.
11. In any timber region where particular forest species can be grown for Christmas tree market, the right-of-way should be used for Christmas tree growing purposes if economically feasible. This will help to preserve the scenic beauty.
12. In other areas, consideration should be given to other uses of the right-of-way.

E. Construction

The best environmental planning can be reversed or defeated by uncontrolled construction activities. The entire force employed by contractors should be advised that all aspects of their construction activities are to be geared to the preservation and enhancement of natural beauty and the conservation of natural resources. The following criteria will help to attain this goal. (These criteria are particularly subject to adjustment befitting the rules and judgments of the various agencies across whose lands the line may be permitted to go.)

1. Clearing and grading of construction areas such as campsites, storage areas, setup sites, etc., should be minimal. These areas should be graded in a manner which will minimize erosion and conform to the natural topography.
2. Borrow areas, rock quarries, etc., should be located away from public view. These areas should be restored to such condition that erosion will be avoided and appearance is acceptable. Borrow pits can be tastefully and aesthetically created to blend into the natural landscape and in some instances used for active resource purposes; for example, fish and wildlife management purposes.
3. Soil which has been excavated during construction and not used should be evenly back filled onto the cleared area or removed from the site. The soil should be graded to conform with the terrain and the adjacent land. Top soil should be replaced and appropriate vegetation planted and fertilized.
4. Terraces and other erosion control devices should be constructed where necessary to prevent soil erosion along the right-of-way.
5. The tops of soil piles in arid locations should be shaped in a concave manner to retain moisture. Steep slopes on these piles should be avoided to prevent erosion.
6. Roads should be provided with side drainage ditches and culverts across the roads to prevent soil or road erosion.
7. Roads should not be constructed on unstable slopes. Where feasible, service and access roads should be used jointly.
8. As a general rule, machine clearing (bulldozing) should not be done on slopes which exceed 35%.
9. Clearing and construction activities in the vicinity of stream beds should be performed in a manner to minimize damage to the natural condition of the area. Machine clearing should not be permitted within 100 feet of any stream bed.
10. Avoid oil spills and other types of pollution, particularly while performing work in the vicinity of streams, lakes and reservoirs.
11. Blasting should not be done in or near stream channels without adequately protecting fish and other aquatic life.
12. Trees, shrubs, grass natural features and topsoil which are not removed should be protected from damage during construction.
13. Water used for construction purposes and taken

from streams or other bodies of water should be limited to volumes which will not cause harm to the ecology or aesthetics of the area.

14. Every precaution should be taken to prevent the possibility of accidentally starting range or forest fires. Construction programs should include fire prevention planning, training of personnel in fire fighting and a fire inspection program. Full compliance with fire laws and regulations is a necessity.
15. The use of helicopters for construction on rights-of-way should be considered in mountainous or otherwise inaccessible terrain or areas of scenic and historic significance.
16. Undergrounding must be considered for lower voltage subtransmission lines when alignments parallel or cross major highways, natural, scenic and historic sites, recreation areas, wildlife refuges, national and state monuments, etc. Consideration of underground installations must take into account the relative ecological damage, success of landscape restoration and the overall environmental quality retention as independent aspects of feasibility evaluation in comparison to aerial installations.
17. Tension stringing of conductors should be employed wherever possible so as to reduce the amount of vegetation clearing before the final conductor location is established. Tension stringing is also associated with helicopter construction where 'Feed' or 'Stringer' lines are aerial borne and where high-strength conductors are used.
18. When possible, construction should be performed during seasons of low-wildlife occurrence, such as between periods of water-fowl migrations.
19. Construction roads should be located and designed to prevent erosion and sedimentation, and to serve permanent service access requirements.
20. Soil disturbance during construction should be kept to a minimum, and restorative measures should be taken promptly.

F. Cleanup and Restoration

The following criteria provide for the cleanup of construction debris and the restoration of the area's natural setting. Further requirements may be imposed by land management agencies across whose lands a line may be permitted to go.

1. Scars, cuts, fills, or other aesthetically degraded areas should be seeded or reseeded as soon as possible to reduce erosion and restore a natural appearance, and to provide food and cover for wildlife.
2. Temporary roads should be obliterated by restoring original slopes and planting natural ground cover.
3. Approach roads and existing low-standard roads which are to be used for access roads should be improved to provide proper drainage and control erosion.
4. Fertilize and seed restored areas as required to encourage growth of grass and other vegetation which is ecologically desirable.
5. Construction campsites, storage areas, etc., should be restored to their original or natural condition.
6. Dismantle and remove all abandoned or useless buildings, equipment, supplies, and personal property.
7. Material to be burned should be piled in a manner and in such locations as will cause the least fire risk. Care must be taken to prevent fire or heat damage to desirable trees and shrubs within and adjacent to the right-of-way, to conform with local fire regulations, and to minimize air pollution.
8. If the natural vegetation cannot be effectively saved to provide an adequate screen, trees and shrubs native to that area should be planted to ultimately provide the necessary screening. Considerations should be given to the establishment of native vegetation of value as food and cover for wildlife.

9. If it is necessary to clear down to the mineral soil, the topsoil should be saved, replaced and stabilized without undue delay by the appropriate seeding of grass, shrubs, and other native vegetation compatible with the surrounding ground cover.
10. Waste, cleared and trimmed material shall be ground up, burned, removed, concealed, or scattered as required prior to completion of the contract.
11. Replacement of earth adjacent to water crossings for access roads should be at slopes less than the normal angle of repose for the soil type involved. Sodding or seeding should be accomplished without undue delay.
12. Brush, timber and other wood products can be disposed of by chipping or shredding. After reduction in this manner the materials can be dispersed to serve as mulch, rather than burned.
13. Where site factors make it unusually difficult to establish a protective vegetative cover, other restoration procedures may be advisable, such as the use of gravel, rocks, concrete, etc.

G. Maintenance

Preservation of both the environmental and natural resource conservation factors designed and built into a transmission system will require a thoughtful, comprehensive program for maintaining the facility. The following factors should be incorporated into such a program.

1. Native vegetation, particularly that of value to fish and wildlife, which has been saved through the construction process and which does not pose a hazard to the transmission line should be nurtured and allowed to grow on the right-of-way.
2. Native grass cover should be maintained if ecologically appropriate in the areas immediately adjacent to transmission towers.
3. Native trees, shrubs, herbs and grass should be allowed to grow and where ecologically appropriate in critical areas vegetation of this type should be planted at acceptable distances from transmission facilities.
4. Once a cover of vegetation has been established on a right-of-way, it should be properly maintained.
5. Access roads and service roads, if and where permitted, should be maintained with native grass cover, water bars and the proper slope in order to prevent soil erosion.
6. Chemicals, when used, should be carefully selected to have a minimum effect on desirable indigenous plant life and selective application should be used wherever appropriate to preserve the natural environment. In scenic areas, the impact of temporary discoloration of foliage should be considered; and where this factor is critical, either mechanical means of vegetative control should be used, or the work should be scheduled in the early spring or late fall. It is essential that chemicals be applied in a manner fully consistent with the protection of the entire environment, particularly of the health of humans and wildlife.
7. Maintenance inspection intervals should be established so that routine maintenance occurs when access roads are firm, dry or frozen. Maintenance vegetative clearing in particularly critical areas should be done on a short cycle to satisfy minimal requirements and avoid heavy, long-term cutbacks.
8. Aerial and ground maintenance inspection activities of the transmission line facility shall include observations of soil erosion problems, fallen timber and conditions of the vegetation which require attention. The use of aircraft to inspect and maintain transmission facilities should be encouraged.
9. Public acceptance of rights-of-way is generally broadened when compatible multiple use of a right-of-way is allowed or encouraged. Transmission line rights-of-way can be made available for

appropriate types of multiple use concepts, such as:

Game food plots	Christmas tree nursery
Recreation areas	Other nursery stock
Access to recreation areas	Wildlife sanctuaries
Parks	Wildlife refuges
Golf courses	Wildlife management areas
Equestrian or bicycle paths	Hiking trail routes
Orchards	General agriculture
Picnic areas	Athletic facilities
Storage facilities	Game cover[14]

Types

Allusions have been made in previous sections to the various types of extra-high voltage transmission lines. There is basically a hierarchy of utility lines. As mentioned in earlier references, voltage on transmission lines is seldom less than 69,000 volts.

Voltage on distribution lines is almost always below that figure. Transmission lines of 135 kv and above have the greatest impact on the environment. Transmission lines of 345 kv and above are extra-high voltage lines, and are referred to as "electrical superhighways." Basically, the levels of transmission lines are:

230 kv
345 kv
500 kv
375 kv
765 kv

The distribution of these lines in the transmission system in 1970 has been shown in an earlier map.

The Federal Power Survey in 1970 made the following statement in regard to transmission and interconnection:

> The need to transport electrical energy efficiently and economically from one location to another in ever-increasing amounts has led to continual development of higher and higher transmission voltages. Virtually all transmission in the United States is by means of alternating current and the past decade has seen a rapid expansion in the use of 345 and 500-kilovolts as primary transmission levels. The first 765-kilovolts equipment in the United States was energized in the spring of 1969. The first extra-high voltage (EHV) dc installation in the United States, a ±400-kilovolts system reaching from northern Oregon to southern California, was placed in commercial service in May, 1970. The trend to higher voltages, accelerated in the past primarily by the economic advantages of transmitting large blocks of power, is now influenced significantly by the need to make maximum use of rights-of-way and thereby to minimize the environmental intrusion.
>
> A fundamental approach to achieving reliable power supply requires that extensive transmission systems operate as integral parts of a strongly interconnected network. Three principal objectives in providing adequate interconnection transmission capacity are:
>
> 1. To support immediately any load area suddenly faced with a serious and unexpected deficiency in its normal generators in the surrounding interconnected network.
> 2. To transfer, without serious restriction, capacity and energy within regions and between regions to meet power shortages. Emergencies can arise from innumerable causes, such as delays in commercial operation of new generation, problems with new equipment, the failure of major generating units or other elements of the system, and unexpected peak demands caused by weather extremes.
> 3. To exchange power and energy on a regional and interregional scale, and to achieve important reductions in generating capacity investment and in cost of energy production.
>
> The latter objective is often a major factor in utility transmission and generation agreements. Frequently, individual

14. U.S. Department of the Interior, U.S. Department of Agriculture, *Environmental Criteria for Electrical Transmission Systems,* (Washington, D.C., U.S. Government Printing Office, Superintendent of Documents, 1970), pp. 3-27.

systems are not able to finance or economically use the large steam-electric generating units now technically feasible, but by appropriate planning and mutual agreement and the strengthening of transmission interties, the economies inherent in the construction and operation of large units can be jointly shared. Other benefits include:

1. Scheduling of bulk energy transfers over tie lines to take advantage of energy-cost differentials between the areas.
2. Diversity exchanges made possible by differing load characteristics due to seasonal patterns, time zones, and weather.
3. Sharing of operating reserve to take advantage of different types of generation and to maximize efficiency in unit commitment and scheduling.
4. Improvement in transient stability.

Utilities recognized the advantages of interconnections at least as early as 1914, when it was observed that appreciable economic benefits could accrue from a transmission tie between two systems in New England—one with only hydroelectric and the other with only steam-electric sources of generation. Elsewhere throughout the country, the rapidly improving transmission and generation technologies provided strong economic and reliability incentives to interconnect small systems. By 1920, many small generating plants were either abandoned or interconnected into systems established to serve the needs of relatively large geographical areas. With the exception of the Texas Interconnected Systems group, virtually all of the large utilities in the United States now operate synchronously, although the power transfer capabilities of some tie-lines are lower than those needed at times to meet peak demands in power shortage areas. In recent years the eastern two-thirds of the contiguous United States was interconnected with the western interconnected systems through what have become known as the 'east-west ties' in the Rocky Mountain area. These ties have very limited capacity and are vulnerable to frequent tripping on power swings.[15]

The following chart shows the various transmission line mileages in the United States of 230 kv and above for the period from 1940 through 1990.

15. Federal Power Commission, *The 1970 National Power Survey: Guidelines for Growth of the Electric Power Industry,* (Washington, D.C., U.S. Government Printing Office, Superintendent of Documents, 1971), p. I-13-1.

Transmission Line Mileages in U.S., 230 kV and Above[16]

	230 kV	287 kV	345 kV	500 kV	765 kV	±400 kV(dc)	Total
1940	2,327	647					2,974
1950	7,383	791					8,174
1960	18,701	1,024	2,641	13			22,379
1970	40,600	1,020	15,180	7,220	500	850	65,370
1980	59,560	870	32,670	20,180	3,540	1,670	118,490
1990*	67,180	560	47,450	33,400	8,940	1,670	159,200

* By 1990 there may be significant applications of ac voltages higher than 765 kV and more extensive use of HVDC than that shown in the table.

16. Ibid., p. I-13-4.

Location

Basically, transmission lines of various capacities are utilized to transport, transfer or transmit generated electricity from generating plants to interconnecting grids, or to lesser levels of transmission lines. Interconnecting networks or power grids call for even more transmission lines. Therefore, as transmission technology becomes more sophisticated, undoubtedly, it will call for increased transmission lines.

The publication, "Power Lines and Scenic Values in the Hudson River Valley" indicates and suggests the following concerning the location of transmission lines:

> The alignment and construction of transmission lines can be divided into six aspects:
> 1. Choice of route.
> 2. Overhead vs. underground placement of lines.
> 3. Use of bridges for transmission-line crossings.
> 4. Tower design.
> 5. Multi-purpose use of rights-of-way.
> 6. Landscaping and maintenance of rights-of-way.[17]

Mr. Frank Burggraf in a paper entitled, "Power to the People," which appeared in the New York State Planning News, indicated the following aspects of right-of-way for transmission facilities:

> Ideally, a right-of-way for transmission facilities should:
> 1. Utilize existing rights-of-way and avoid making new intrusions on the landscape.
> 2. Respect existing land uses.
> 3. Avoid above-grade changes.
> 4. Avoid areas of environmental sensitivity.
> 5. Respect property boundaries and minimize severances.
> 6. Minimize the required clearing of vegetation.
> 7. Utilize marginal and inactive land.
> 8. Avoid airport approach zones.
> 9. Avoid undisturbed or wild lands.
> 10. Respect natural features, such as lakes, streams, scenic areas and geological provinces.
> 11. Avoid areas where geological conditions would create construction problems.
> 12. Avoid compromising recreational resources.
> 13. Avoid areas of population concentration, including urban areas and residential development.
> 14. Avoid interference with the orderly development of the area traversed.[18]

Mr. Burggraf then goes on to give the following warning:

> These criteria cannot be employed like a mathematical formula to determine an appropriate route, nor is it sufficient to create a route that merely avoids problem areas or links segments of modest environmental impact. All the criteria represent a range of values. Therefore, it is likely that there will be many conflicts in the application of the criteria. The resolution of such a conflict is a matter of professional judgment giving the proper weight to all relevant factors. The most serious problem associated with transmission lines siting is 'skylining': The silhouetting of structures against the sky. Any place a transmission line crosses a ridge or follows the top of a geological promise, a visual impact of the structure is maximized.
>
> When a transmission line traverses deep slopes or ridges, the area cleared of vegetation becomes all the more obvious, again maximizing the visual impact of the facility. This is without regard to any additional visual impact on the structures sited on slopes or ridges. The adverse environmental effects of clearing vegetation in protected areas is also felt. The soils in such locations tend to be thinner than soils in the valleys and flats. Finally, the thin soils on steep slopes cannot readily retain rain water or organic material. Hence, the environmental impact of any intrusion on steep slopes, mountain tops, or ridge lines is amplified because of these conditions, and the time required for the recovery from any intrusion is lengthened.

17. Bruce Howlett, Frederick J. Ehniger with assistance of Anthony Corkill and John Seddon for the Hudson River Valley Commission, State of New York, *Power Lines and Scenic Values,* (Tarrytown, N.Y., Hudson River Valley Commission, 1968), p. 7.
18. Frank B. Burgraff, Jr., "Power to the People," *Planning News* (Albany, New York, New York State Planning Federation, Vol. 35, No. 4, July-August 1971), p. 6.

> Transmission facilities are less of a visual intrusion in areas of industrial land use, or where the land has been dedicated to junk yards, land fills, or extractive industries. Thus, the adopted plans of a community in established land use are particularly important in judging the appropriateness of a route. We have argued that it is more reasonable to expect a utility to underground the line where the community has an established program for the control of other land use abuses, billboard controls, etc., than where the community has not implemented or even considered planning action. The presence of junk, abandoned auto hulks and refuse is indicative of a low regard for the environment. It was difficult to argue that a transmission line strung through such an area is a significant intrusion.
>
> Assuming that we must have them, the incremental aesthetic damage is least in an area already committed to unattractive uses, or programmed for such uses, and is most damaging in unspoiled areas or those programmed for preservation.[19]

One of the areas of greatest concern, and hence has generated the greatest number of studies in recent years has been in regard to the location of the transmission lines, and the development of a more attractive right-of-way.

The publication previously referred to entitled, "Power Lines and Scenic Values in the Hudson River Valley" contains some useful sections on the location of transmission lines and the development of the areas either under or immediately adjacent to such transmission corridors. It does so with the following listing of guidelines:

> Even though the alignments of transmission lines are under no agency's direct control, there are principles that can be useful in locating a route to do minimum damage to the scenic, natural or other resources of an area:
>
> 1. Avoid sites of great scenic value, including prominent ridge lines, lakes, barren sides of mountains or hills.
> 2. Keep alignments along the bottoms of lower slopes and valleys between hills.

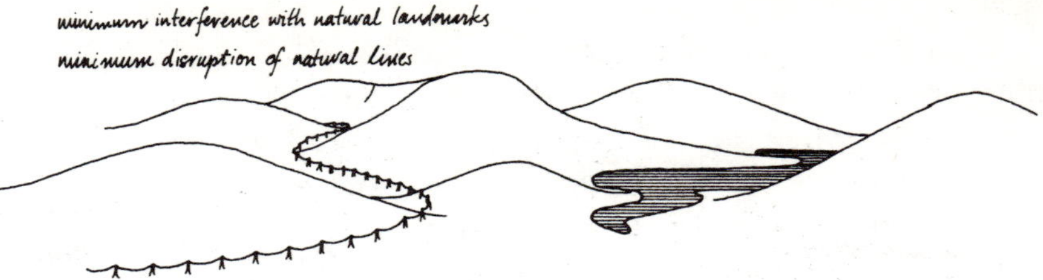

> 3. Avoid crossing hill contours at right angles; avoid steep grades which expose the right-of-way to view.

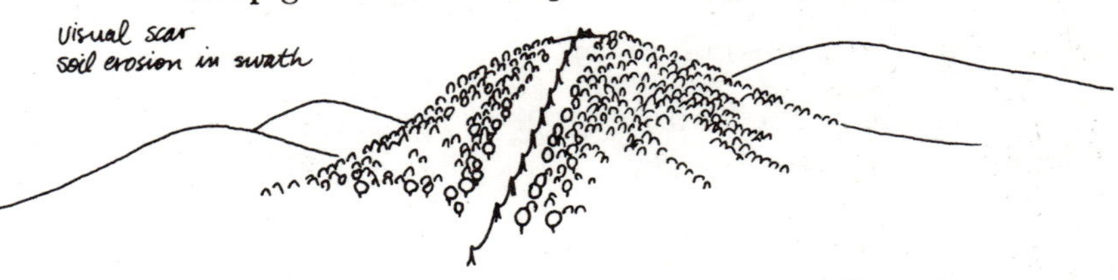

> 4. When crossing principal roads, jog alignments to avoid extended tangents along the right-of-way on either side of the road.

> 5. In rough or very hilly country, change the alignment continuously in keeping with the scale of topographic change.

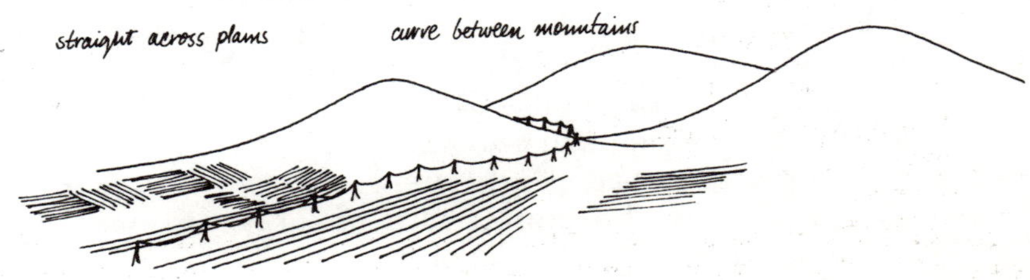

19. Ibid., p. 6.

6. Where feasible and in keeping with the prior criteria, run lines along the edges of differing types of land use, with special emphasis on preservation of forest growth.

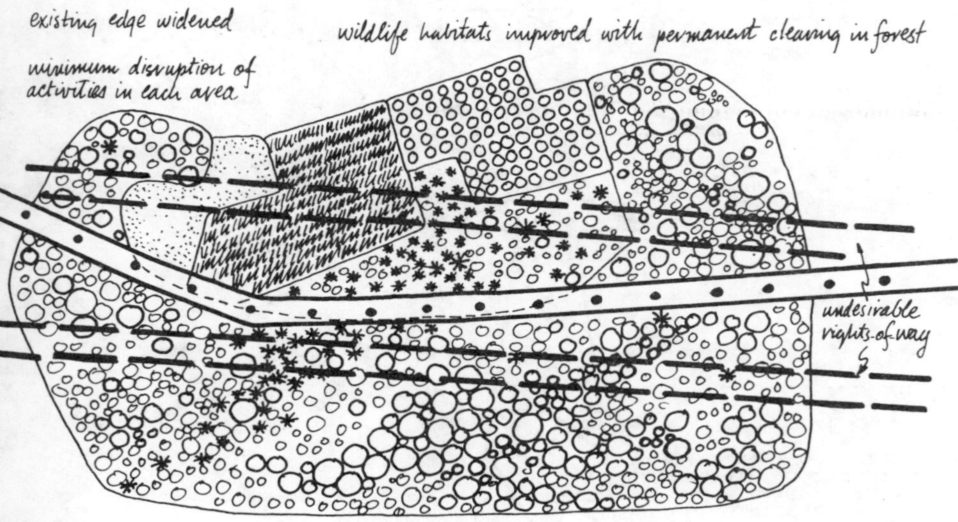

7. For lower voltage sub-transmission lines, undergrounding is desired when alignments parallel major highways or scenic areas. In these instances, undergrounding should be considered when construction is undertaken.
8. If a proposed route realistically can be shifted from a scenic area to one already industrially developed, the route should be placed through the latter.

The general principle involved in the location of transmission lines, then, is the reduction of their visibility. Assuming these suggestions are followed, further efforts should be made to select alignments least visible from highways, scenic lookouts and other locations where many people pass or congregate. Simple tests of visibility can be carried out by touring roads and points close to a proposed right-of-way to see what the line's visual impact might be prior to construction.[20]

More will be said concerning approaches to development of optimum alignment in the following section dealing with approaches, processes and solutions.

Standards

In a hierarchy of problems caused by transmission lines, the location of the line would be closely followed by the standards or the uprights on which the transmission lines are carried which probably causes as much comment and as much visual pollution as any other element in the utility industry.

Many times the inherent design of the existing utility structure is offensive, more often it is insensitive in its design and placement. Basically the poles or standards are symbolic, in that they are oriented toward ease and economy of construction and maintenance. Incidentally, then, they just turn out to be unpleasing aesthetically because of the engineering aspects of carrying the weight of transmission wires, insulators, cables and lines. Poles and other structures are intrusive, marching off across the landscape and presenting an obvious manifestation of the electric utility industry presence. They are usually tall, and they present strong verticals. Since in the past little concern was exhibited in regard to the standards themselves and their alignment, they have caused more problems at the present time than they probably will in the future.

A good deal of work has been undertaken in recent years studying alternative standards for carrying electric transmission lines. A study conducted for Consumers Power Company in Michigan by the Landscape Architecture firm of Johnson, Johnson and Roy entitled, "Transmission and Distribution of Rights-of-Way, Selection and Development" indicates in diagramatic form the standards for the electric transmission and distribution system.

In this publication Johnson, Johnson and Roy, as landscape architects, deal to some extent with the design of the structure with the following quotes and illustrations:

20. Bruce Howlett, Frederick J. Ehniger with assistance of Anthony Corkill and John Seddon for the Hudson River Valley Commission, State of New York, *Power Lines and Scenic Values*, (Tarrytown, N.Y., Hudson River Valley Commission, 1968), p. 0.

Of the several types of electric power lines, transmission lines (138 kV and above) are considered to have the greatest impact upon the environment. Due to the high voltages utilized, it is not now technologically nor economically feasible to consider the extensive placement of transmission systems below ground entirely out of view. Exceptions include unique situations of significant scenic and/or historical value, densely populated areas and intensively developed commercial districts. Underground placement of future distribution systems, however, is considered realistic.

The major portion of the criteria developed in this report, therefore, is applied to transmission and sub-transmission lines. Although it is not the intent of these guidelines to determine designs for transmission structures, it is important in arriving at criteria for their placement to understand the various types of structures employed and their characteristics. In the case of transmission lines, the extra-high voltage conductors (345kV and above) are carried on steel towers of several configurations; lower voltages are carried either on steel or wood structures. Heights range from fifty feet to one hundred fifty feet and occasionally greater depending upon voltage, tower type and situation. Wood poles vary in height in five foot increments, steel structures in ten foot increments. Generally, the higher the tower or pole, the greater the spacing between them. As the span increases, the resulting increase in sag requires taller towers to provide required clearances from the ground and obstructions.

The impact of a tower on the environment generally relates to its size, complexity, design and color. The dimensions of a structure are fairly dependent upon the line voltage, the number of circuits, and the ratio of span to sag. The size of a tower significantly affects its cost normally influences the number and spacing of towers. Size is further determined according to the line angle imposed upon a tower with the weight and complexity increasing with the degree of line angle.[21]

The following illustration indicates the transmission structure characteristics:

Transmission Structure Characteristics[22]

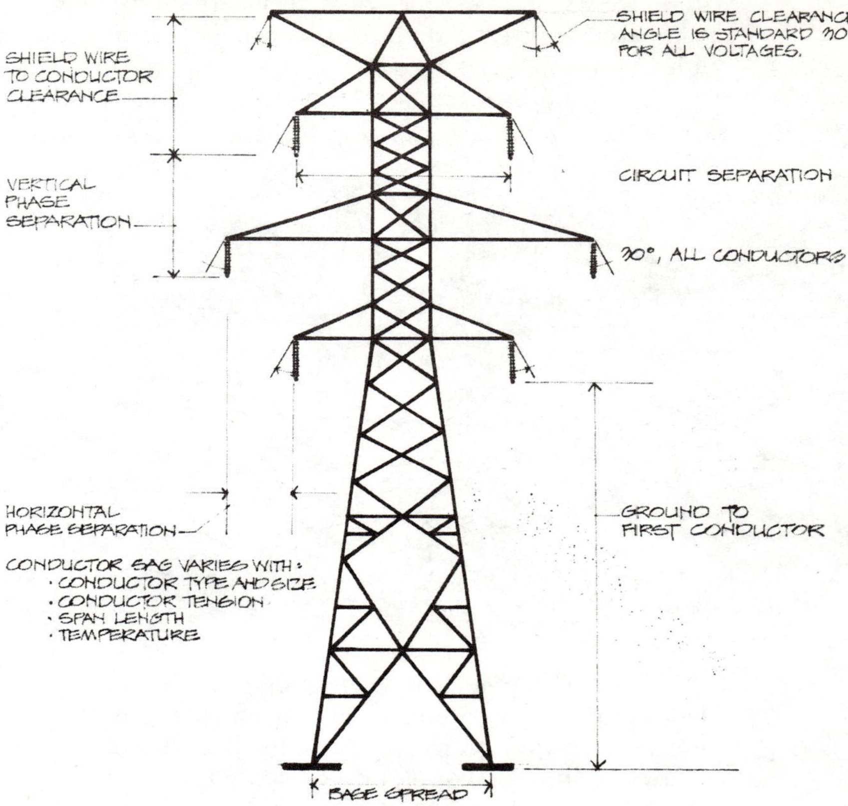

SPAN LENGTHS BETWEEN TRANSMISSION STRUCTURES CAN VARY BY 30%:

NOMINAL SAG AT 120° F: 30' ON 345 KV WITH 1000' SPAN
20' ON 138 KV WITH 710' SPAN
6' ON 46 KV WITH 350' SPAN

The Johnson, Johnson and Roy report for Consumers Power Company goes on to make the following comment in regard to the structures themselves:

> Structure designs that are most visually acceptable in the environment are those of the simplest form. Technical re-

21. William J. Johnson, Carl Johnson and Clarence Roy, *Transmission and Distribution Rights-of-Way Selection and Development,* (Ann Arbor, Michigan for Consumer's Power Company, Jackson, Michigan, 1970), p. 4.
22. Ibid., p. 5.

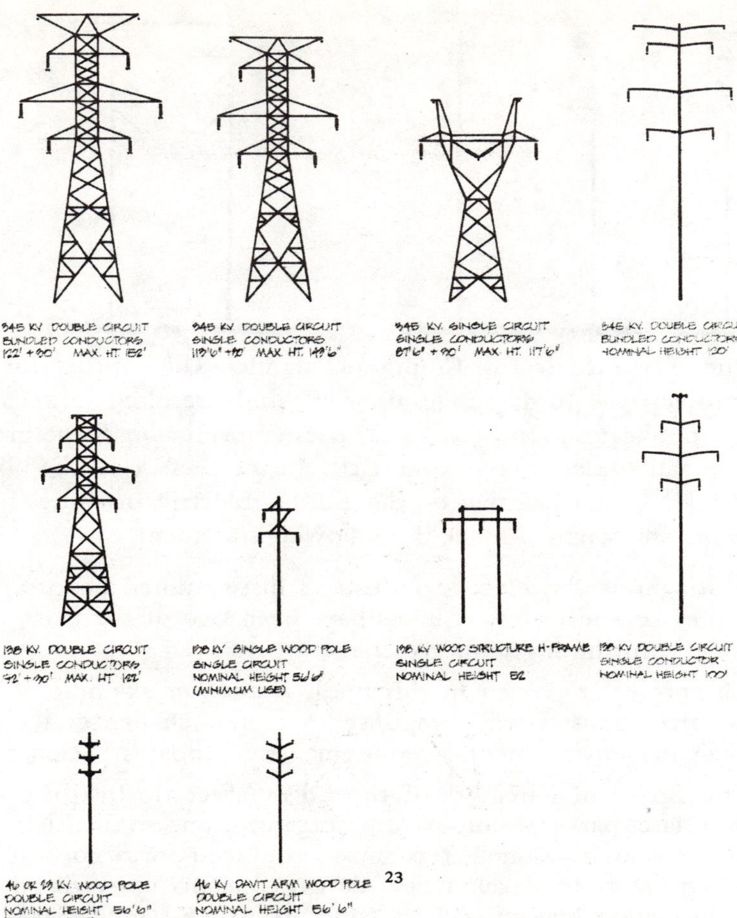

[23]

quirements, however, necessitate that transmission structures be composed of more than a simple vertical element. Simplicity is overall outlines, a balanced or symmetrical arrangement of elements relative to the vertical support, vertical support, parallel and orderly placement of elements within the overall outline and the minimization of pole-top elements are design criteria upon which to evaluate structure designs. Attempts at increasing the attractiveness of towers have sometimes led to exotic designs that only attract more attention than those of less complex silhouette. A more appropriate design purpose would be to achieve designs that are so simple as to be inoffensive and unnoticeable.

Careful siting of rights-of-way for minimum visibility and least physical and visual disruption of the landscape offers the greatest opportunity for success and is the focus of this study. Further investigation of new tower designs should not be discouraged, however, and attempts should be made to achieve simple, clean-lined structures upon which all the essential elements are held close to or within the overall outline of the structure.

The use of new materials such as laminated wood, concrete and Fiberglass or reinforced plastic should be continually explored in addition to steel and aluminum and the selection of color to blend with backgrounds should be a constant goal.[24]

While special situations will need to receive special color considerations, this report recommends criteria for color selections as follows:

1. The basis for color selection assumes the premise that it is impossible to screen or camouflage towers through the use of color.
2. Tower color (hue) is meaningful to the viewer up to a maximum distance of one thousand feet. Beyond that point, the hue becomes indistinguishable and only the value of the color can be expected to have any appreciable effect.
3. When viewed from the shaded side, a tower presents itself as a dark silhouette and generally its color is indistinguishable.
4. A great majority of towers are viewed against a backdrop of sky. Because the condition of the sky changes greatly from day to day and from season to season, color cannot be relied upon to diminish the visual impact of the tower.
5. Colors should be selected on the basis of their ability

23. Ibid., p. 4.
24. Ibid., p. 6.

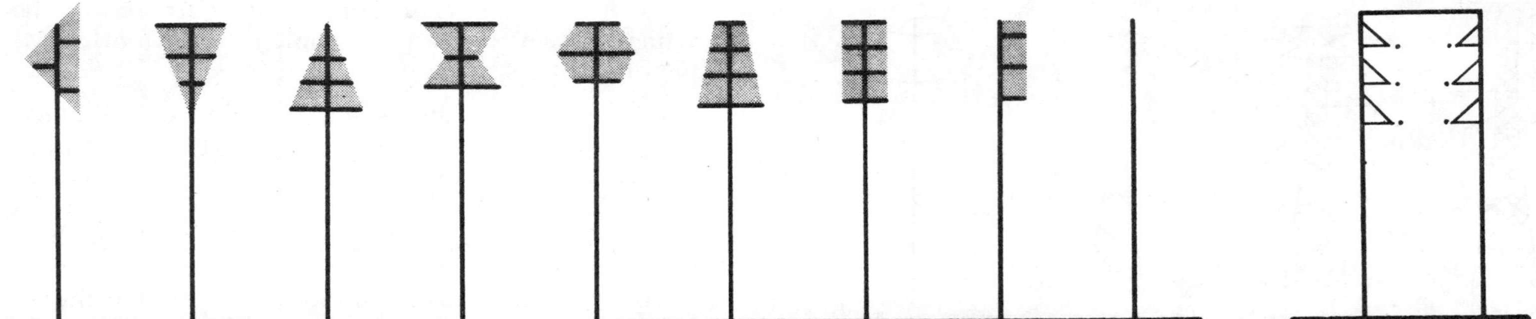

to render the towers more compatible with their immediate environment. The warmest tones are more appropriate for both natural and man-made settings.

6. Except for special cases, all colors should be judged on the basis of their ability to blend with both the sky and the environment in which they are being used.
7. Colors that reflect colors adjacent to it by including them within it are most successful in adapting to their environment.
8. In the interest of simplicity as well as uniformity of company image, a standard color that will blend well with a wide variety of environmental factors (sky, foilage, residential zones, industrial zones, etc.) should be selected for general use.
9. Other colors should be chosen for use in special instances; that is, towers viewed against a heavy evergreen backdrop could be a deep olive green, towers adjacent to sand dunes and beaches could be a sand color, etc.
10. Where natural wood poles are appropriate the color range should be as limited as feasible to present a unified series of poles. Special situations may require that wood poles be painted.[25]

Undoubtedly, the most significant study in the history of electric transmission structures was a design research program sponsored by the Electric Research Council.

This study was conducted by the industrial design firm of Henry Dreyfuss & Associates, with engineering consultation by Severud, Perrone, Sturm, Conlin and Bandel. The coordinator of this project was Jordan Lummis. This study resulted in a hardbound publication entitled, "Electric Transmission Structures" and a small folder of the same title. In the preface to the book Robert N. Coe, Chairman of the Edison Electric Institute Task Force on Environment made the following statement:

> Throughout the electric industry's distinguished history, resilience and resourcefulness have been two of the many traits contributing to its greatness.
>
> The project reported in this book is another example of positive industry-wide response to a new challenge: the changing environmental requirements of today's society.
>
> The Board of Directors of the Edison Electric Institute, with later participation by other organizations on the Electric Research Council, recognized the need for a coordinated effort to develop new design concepts in overhead transmission line structures—concepts which would make these structures more esthetically pleasing to Americans.
>
> Importance of the project dictated that it should have the composite expertise of an eminent industrial designer, a prominent structural engineer and a recognized and experienced power engineer. A careful study of qualified professionals in these and other fields led to retaining the services of Henry Dreyfuss and Associates, Severud Associates and Mr. Jordan Lummis—all experts with impressive credentials.

25. Ibid., p. 7.

Responsibility for overall direction of the project was assigned to the EEI's Task Force on Environment. During the months that the project has been under way, many interested manufacturing organizations have assisted with additional engineering advice.

The cooperation of all of these individuals and organizations has been invaluable in completing this book of designs.[26]

Henry Dreyfuss in his introduction dated January, 1968, to this book made the following succinct statement:

The first man to place a log across a stream was probably criticized roundly by his fellow tribesmen for defacing the countryside. Today we proudly show visitors our bridges as scenic wonders—from the stately George Washington in New York to the harp-stringed Golden Gate in San Francisco. When transmission towers are given the same purity of expression given great bridges, they, too, may be acclaimed as a Twentieth Century art form.

Of course, there is a difference. We encounter bridges in a personal way: we cross them. On the other hand, our contact with transmission towers is remote. The only thing crossing them is electricity. And so we are conditioned, psychologically, to accept a bridge more easily than a power pole.

The towers shown in this book are both simple and functional in form. Particular care was given to fitting them into their environment. Thus in preparing structures for urban areas, we studied urban surroundings and expressed them in the structures. We devoted similar attention to rural structures.

All of us would prefer to place electric power underground or to transmit the power on an invisible beam. But until such techniques prove practical, we hope that the work shown here will serve usefully the dedicated designers and engineers who seek to preserve the integrity of our surroundings.[27]

The challenge met by this book is summarized by the following words:

This is a reference book of designs for high-voltage electric transmission structures. The primary goal of the program which produced these designs was to demonstrate that above-ground utility structures could be aesthetically pleasing. It was to demonstrate as well that such designs could be structurally sound and practical.

It is the goal of this book to encourage and guide the development of high-voltage transmission towers that will enhance the environment of the future.

The designs that follow number more than one hundred variations, each variation having resulted from a consideration of different materials, of utilitarian and technical needs, and of design alternatives. These factors were assembled primarily from a broad study of the conditions of transmission structures in use today.

While the stated challenge is that of developing structures of pleasing appearance, aesthetics must be an outgrowth of the natural characteristics of the function of the structures. For this reason, all designs have been developed with structural analysis and fabrication techniques vitally in mind.[28]

The following pages illustrate the basic designs developed as a result of this study.

The following photographs are of certain of these power structure concepts. These designs and proposals were given wide publicity, and have begun to be adopted in some instances by the electric utility industry. This is one of the most ambitious undertakings by the industry for correcting one of the most obvious environmental problems. The utilization of an industrial design firm to study the possible methodology, and final result the redesign of transmission standards, however, is salutary and without precedent. It certainly is a high goal for other studies concerning transmission standards which has not been paralleled since that time.

26. Electric Research Council, *Electric Transmission Structures, A Design Research Program*, (New York, Edison Electric Institute, EEI Publication, No. 67-61, 1967), p. 4.
27. Ibid., p. 9.
28. Ibid., p. 10.

PICTORIAL INDEX[29,30]

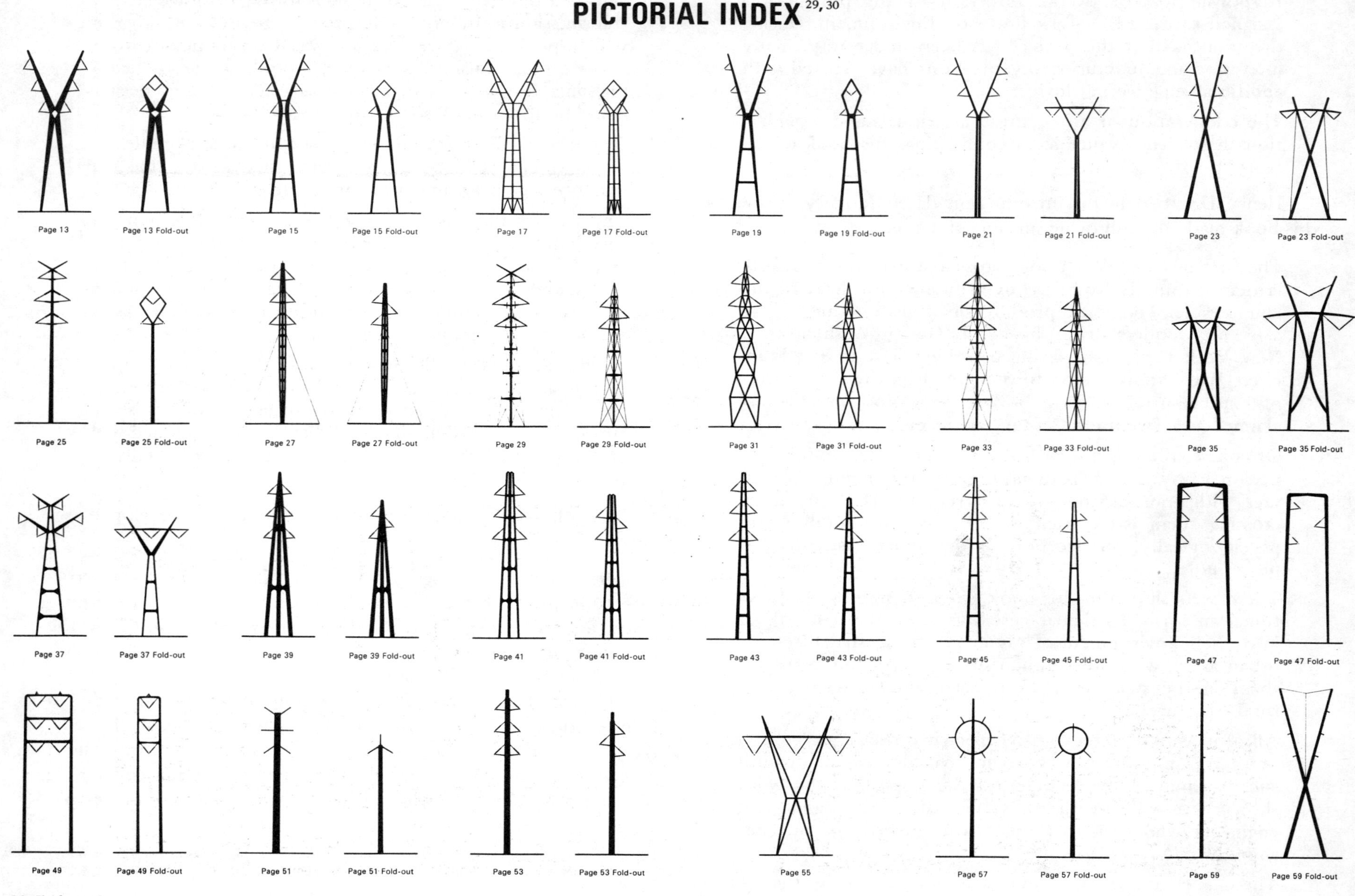

29. Ibid., p. 6.

30. Ibid., p. 7.

31. Ibid., pp. 14, 18, 24, 32.

Right-of-Way Cross Sections

The actual configurations of the right-of-way itself have been the study of much dissention, discussion and study. The landscape architectural firm of Johnson, Johnson and Roy in their previously mentioned study for Consumers Power Company have done probably the best job currently in existence in the United States concerning right-of-way requirements and optimum and altered right-of-way cross sections to enhance and preserve the quality of the environment. The following are quotes and illustrations from their report:

> As the size and configuration of the tower is a factor of the voltage of the transmission lines, so are the dimensional characteristics of the rights-of-way. Ranging from seventy two feet to one hundred and fifty feet of fee right-of-way with tree clearing rights for 'danger trees' ranging from one hundred and ninety feet to two hundred and thirty feet in width, these dimensions increase by sixty feet for each additional 138 kV tower line and one hundred feet for each 345 kV tower line. Rights-of-way for the lower voltage lines are generally in the form of easements, perpetual grants of the right to use portions of private land for special purposes. Fee land is generally reserved for the 138 kV and higher voltage lines. In general, the type of arrangement for use of the land has less impact upon the environment than clearing and maintenance procedures applied to the rights-of-way.[32]

The following are charts from the Johnson, Johnson and Roy study for Consumers Power Company:

32. William J. Johnson, Carl Johnson and Clarence Roy, *Transmission and Distribution Rights-of-Way Selection and Development,* (Ann Arbor, Michigan for Consumer's Power Company, Jackson, Michigan, 1970), p. 8.

MINIMUM GROUND CLEARANCE = 30'
MINIMUM BRUSH CLEARANCE = 20'[33]

33. Ibid., p. 8.

138 KV TRANSMISSION LINE R.O.W.

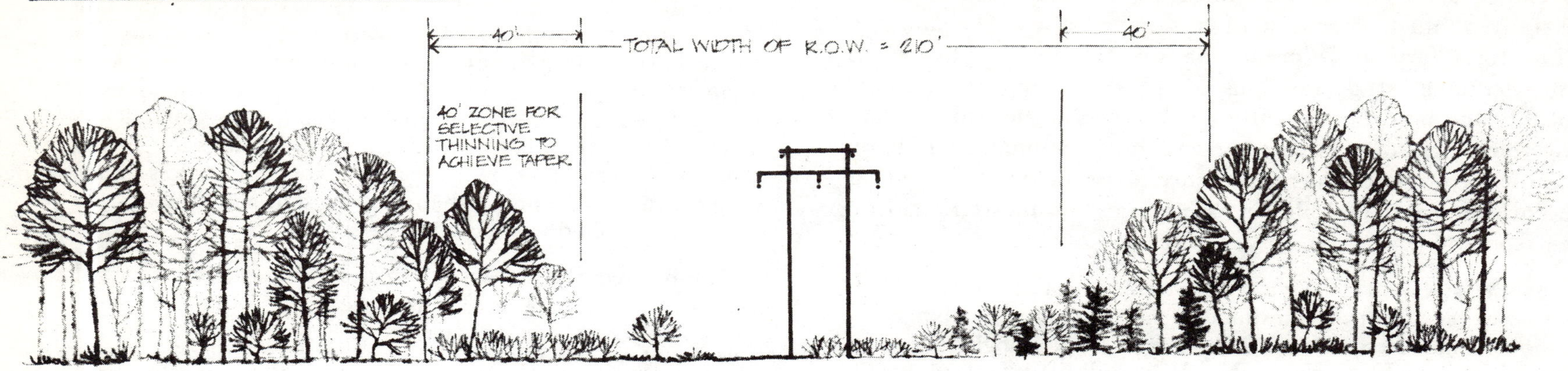

MINIMUM GROUND CLEARANCE = 25'
MINIMUM BRUSH CLEARANCE = 12'

46 KV TRANSMISSION LINE R.O.W.[34]

MINIMUM GROUND CLEARANCE = 22'
MINIMUM BRUSH CLEARANCE = 10'

34. Ibid., p. 9.

MULTIPLE - LINE RIGHTS - OF - WAY
345 KV TRANSMISSION LINES
TOTAL WIDTH OF R.O.W. = 330'
115'
115'
40' ZONE FOR SELECTIVE THINNING TO ACHIEVE TAPER
40' ZONE FOR SELECTIVE THINNING TO ACHIEVE TAPER
100 ℄ TO ℄

MULTIPLE LINE RIGHTS - OF - WAY [35]
138 KV TRANSMISSION LINES
TOTAL WIDTH OF R.O.W. = 270'
105
105'
40' ZONE FOR SELECTIVE THINNING TO ACHIEVE TAPER
40' ZONE FOR SELECTIVE THINNING TO ACHIEVE TAPER
60' ℄ to ℄

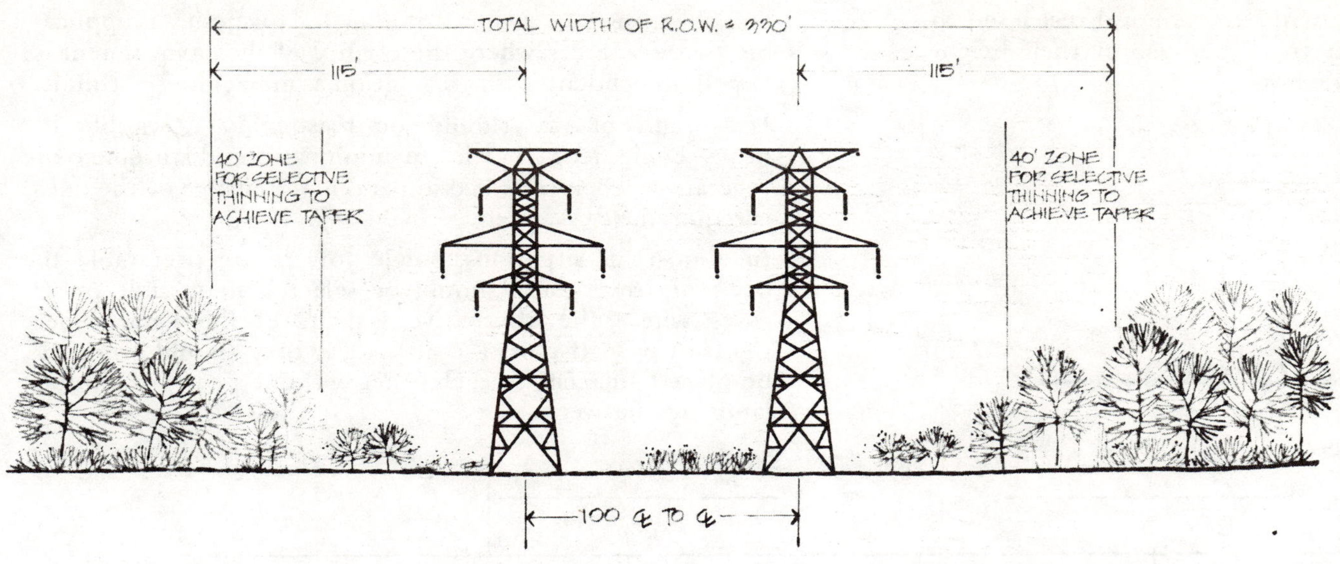

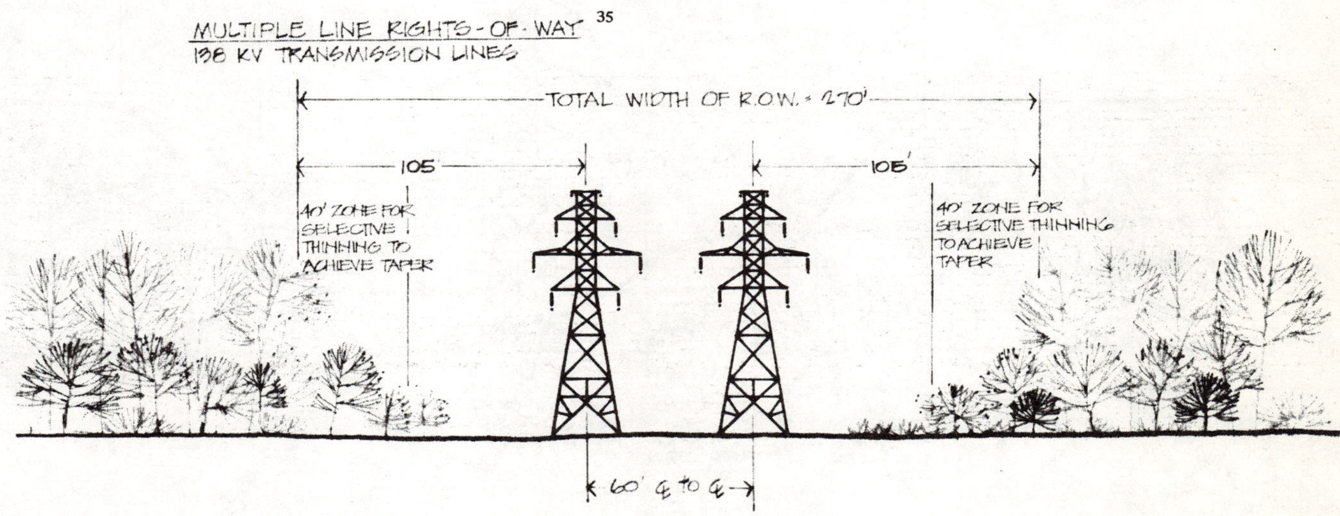

35. Ibid., p. 10.

A growing concern for the natural environment has modified earlier practices of total clearing in some instances to allow the thinning and tapering of the vegetation along edges of the rights-of-way with tree removal restricted to those identified as 'danger trees' because of their height and proximity to the conductors.

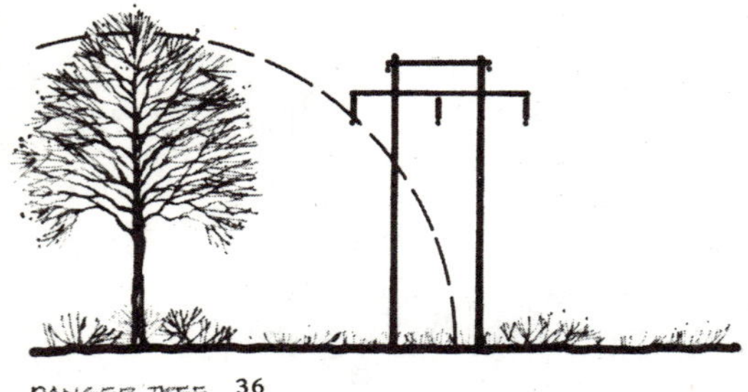

DANGER TREE [36]

Wider application of this newer policy is recommended to allow more vegetation in the rights-of-way, particularly in the area of the structures where the conductors are higher and their angle of swing is minimal. This policy is applicable to those areas where the quality of the environment is largely dependent upon vegetation as an element within it.

Line rights-of-way should be reasonably accessible to heavy equipment in order to minimize the disturbance of vegetation when construction and maintenance of the lines is required.

In almost all situations, single towers are preferable to pairs of towers and should be selected if possible in all cases where the alternative is pairs of lines and wider rights-of-way. If pairs are necessary, they should generally be placed adjacent to each other without a zone of natural landscape between.

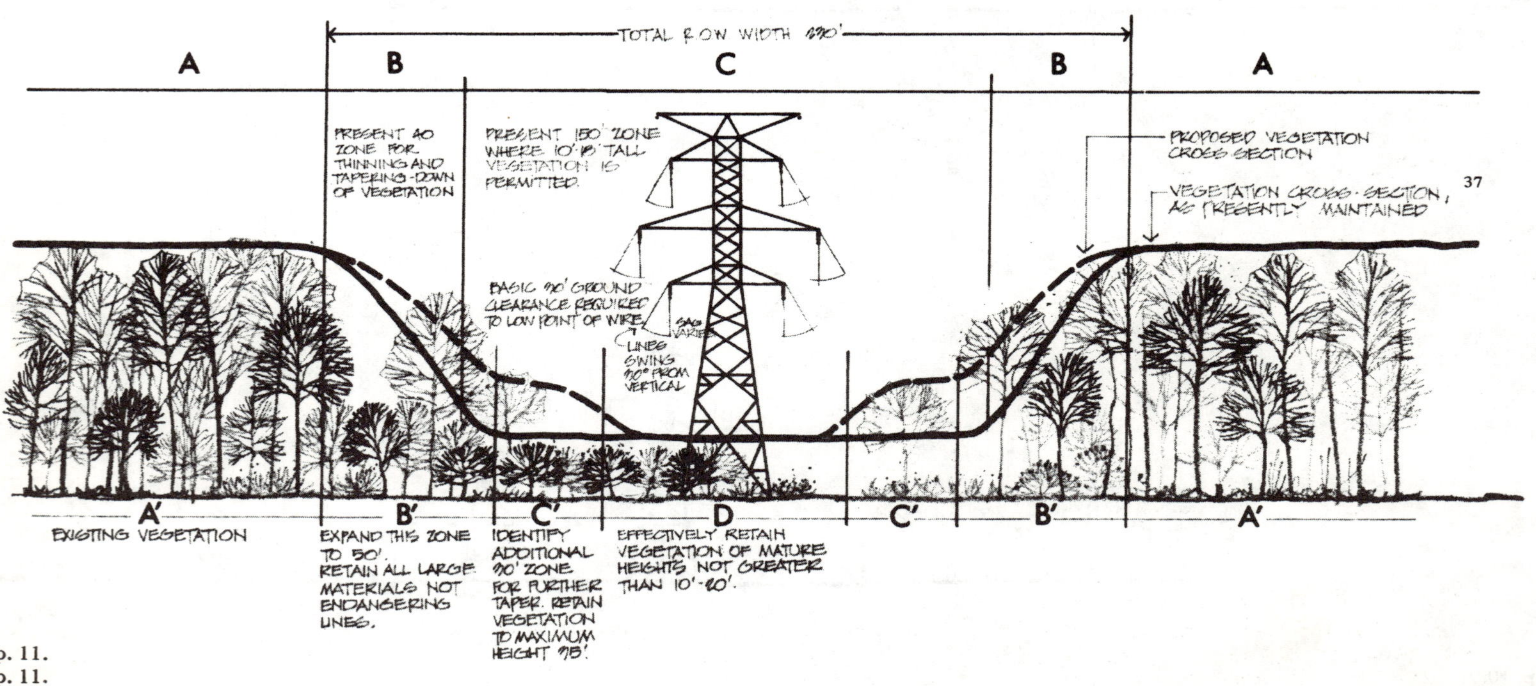

[37]

36. Ibid., p. 11.
37. Ibid., p. 11.

Alignment Analysis and Study

Probably more work has been done by landscape architects in regard to alignment analysis study, design and directives, than in any other area of the electric utility industry. Probably the most significant study in this area is a part of that same Consumers Power Company report prepared by Johnson, Johnson and Roy entitled, "Transmission and Distribution Rights-of-Way Selection and Development."

The following is a major reproduction or major portion from that report dealing with placement of transmission lines in different types of landscape situations:

TRANSMISSION LINES

Land Use

Continuing the concept of locating the alignments of transmission lines along the edges of integral land units, rights-of-way must be studied and selected with strong consideration given to existing and projected land uses. Natural land units should be identified in regard to both use and physical features. A careful effort must be made to avoid the disruption or division of these land units through location of transmission line rights-of-way across them. Transmission lines generally can be most approximately located along the periphery of the land unit or between land uses. This location policy not only respects the integrity of each land unit but also facilitates the efficient provision of service to them. This guideline applies to other types of corridors and facilitates joint use of rights-of-way for various utilities as well as transportation systems. When expansion of a transmission system is necessary, additions to existing rights-of-way are normally preferable to the acquisition of land for new rights-of-way unless an excessive number of structures in one location results.

−

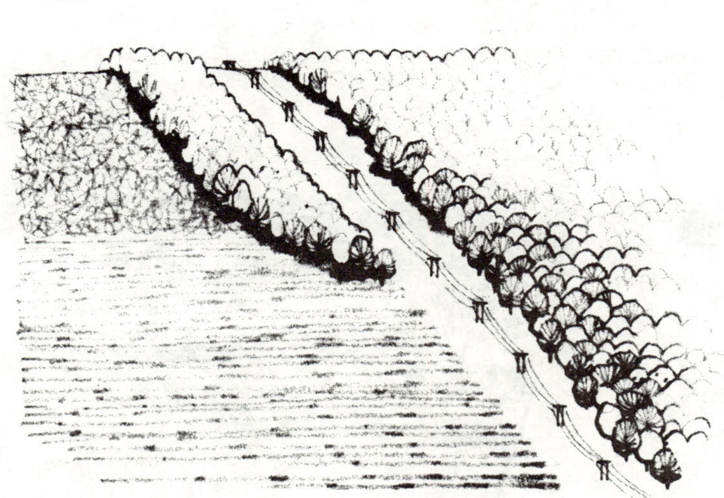

+

LOCATE RIGHTS OF WAY ALONG EDGES OF LAND UNITS TO AVOID DIVIDING LAND USES. HERE THE RIGHT OF WAY HAS BEEN LOCATED BETWEEN AGRICULTURAL LAND AND FOREST, THUS PRESERVING FOREST, USING IT AS BACKGROUND SCREEN FOR THE LINES, AND AVOIDING NECESSITY FOR A WIDE SWATH OF TREE REMOVAL FOR THE RIGHT OF WAY

IN RURAL AGRICULTURAL AREAS FARMING IS CARRIED ON WITHIN THE TRANSMISSION LINE RIGHT-OF-WAY. LOCATE LINES IN CONFORMANCE WITH EXISTING AGRICULTURAL PATTERNS AND BETWEEN LAND USES, IF POSSIBLE. PARALLEL EXISTING FENCE LINES, WINDBREAKS, WOOD LOTS, ROADS, ETC.

—

IN LOCATING TRANSMISSION LINES THROUGH OPEN AGRICULTURAL AREAS, AVOID LOCATING THE RIGHT OF WAY CLOSE TO IMPORTANT LOCAL FEATURES SUCH AS FARM BUILDINGS AND VERTICAL ELEMENTS LIKE WINDMILLS AND SILOS. THE COMPARISON BETWEEN RELATIVE SIZES INCREASES THE VISUAL IMPACT OF THE TRANSMISSION STRUCTURES.

+

Topography

Because topography contributes greatly to the basic character of the landscape, it is critical that it be understood in order that it might help determine appropriate placement of features upon it. Transmission lines should generally be located parallel to the contours of the land and always in conformance with the prevalent direction or patterns of topographic features to reduce their detrimental impact to the minimum. In rough or hilly country, alignments should change in direction in keeping with the scale of the topographic change. Rights-of-way aligned perpendicular to the basic topography serve to call attention to the lines and structures as intrusive elements in the landscape. Alignments should avoid crossing hills at right angles to the contours, especially where the right-of-way is centered on a hillcrest and results in a symmetry that focuses attention upon wooded hills can be especially disturbing to views of landscape. It is important that as much vegetation as possible be retained in order to reduce the impact of the break in forest cover showing against the sky. Where the crossings are made at an oblique angle to the contours, the apparent disruption is reduced to a minimum. Wherever possible, towers should be placed below the crest or horizon from the principal observation zone so as to reduce or obviate their silhouette against the sky. Transmission facilities provided with a background of topography and/or natural vegetation are most successfully accommodated in the landscape.

LOCATE TRANSMISSION LINES IN CONFORMANCE WITH PREVALENT DIRECTION OR PATTERNS OF TOPOGRAPHIC FEATURES. ALIGNING RIGHT-OF-WAY COUNTER TO BASIC LAND ORIENTATION CALLS ATTENTION TO THE TRANSMISSION LINES AND STRUCTURES AS ALIEN ELEMENTS

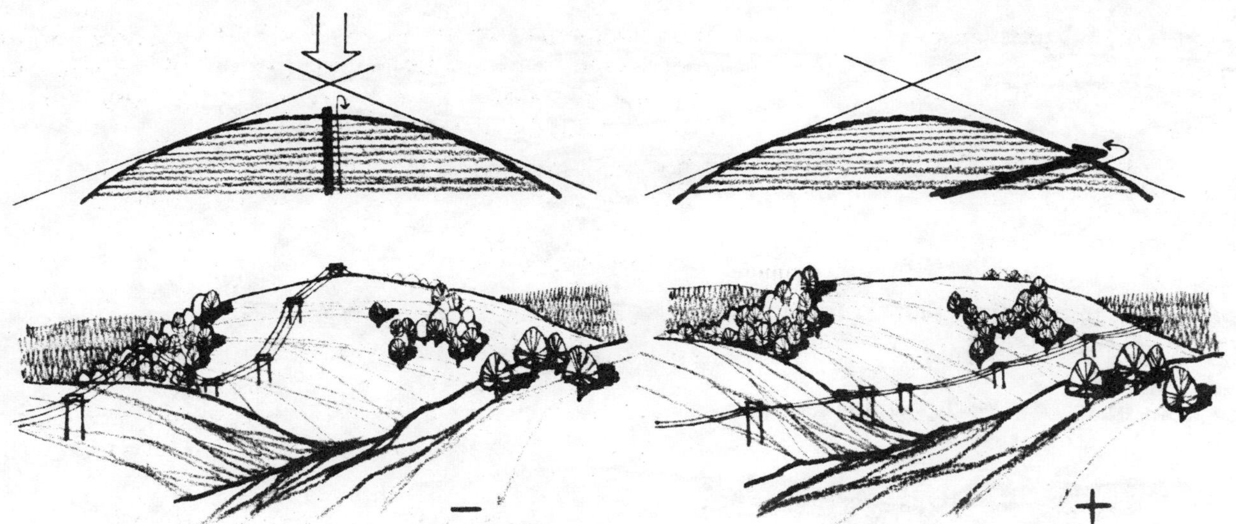

AVOID CROSSING HILLS AT RIGHT ANGLES TO THE CONTOURS, OR WITH THE RIGHT-OF-WAY CENTERED ON THE HILLCREST. THE CENTERING OF LINES ON A FAIRLY SYMMETRICAL FORM SERVES ONLY TO FOCUS ATTENTION ON THEM.

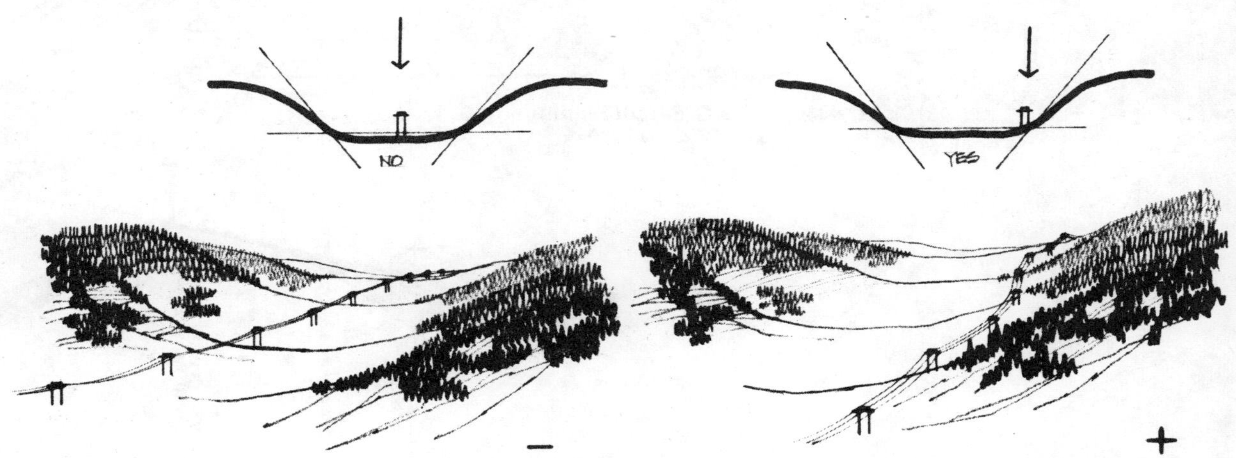

LOCATE TRANSMISSION ALIGNMENTS ALONG THE EDGES OF VALLEYS, INSTEAD OF CENTERING DOWN THE MIDDLE. A CENTER ALIGNMENT FOCUSES ATTENTION ON THE ALIEN ELEMENT, WHILE THERE IS MINIMUM VISUAL DISTURBANCE IF THE ALIGNMENT FOLLOWS THE ZONE OF LAND FORM CHANGE.

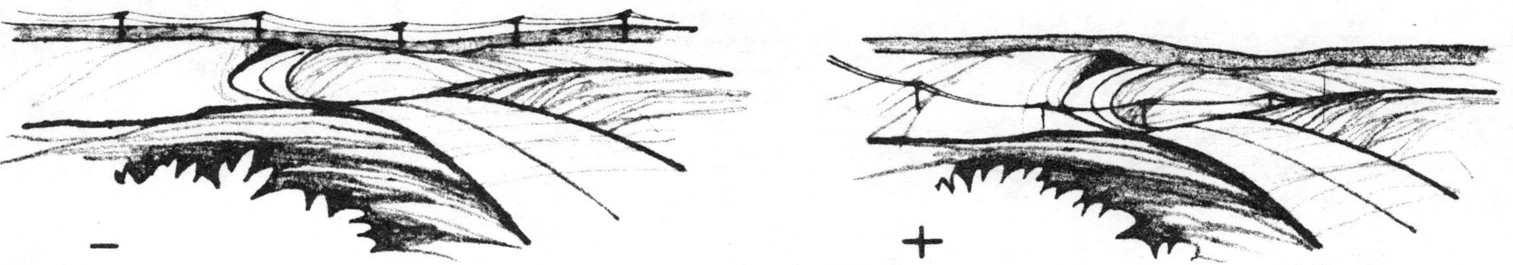

CROSS HIGHWAYS SO THAT TRANSMISSION LINES ARE NOT OUTLINED AGAINST THE SKY. AVOID, HOWEVER, CENTERING THE ALIGNMENT DOWN A VALLEY FLOOR.

ALIGNMENT IS CURVED TO FIT TOPOGRAPHY IN HILLS AND MOUNTAIN AREAS.

ALIGNMENT IS STRAIGHT ACROSS FLAT OPEN AREAS

HILL FORMS DARKER BACKDROP THAN SKY DOES, THUS EFFECTING A GREATER REDUCTION IN IMPACT OF TRANSMISSION LINES. BACKGROUND TREE MASSES ARE VERY EFFECTIVE, SINCE LINES AND POLES BLEND AGAINST THEIR TEXTURE. FOREGROUND TREES SCREEN COMPLETELY.

P.O.Z.

1. VISUAL IMPACT OF POLE IS REDUCED BY BACKGROUND SCREENING OF HILLSIDE.

2. SAME GENERAL SITUATION AS 4.

3. ENTIRE STRUCTURE IS VISIBLE AGAINST SKY FROM BOTH VIEWING ZONES.

4. SLOPE FORMS BACKGROUND FOR LOWER PART OF POLE, BUT MOST UNSIGHTLY UPPER PORTION IS HIGHLIGHTED AGAINST SKY.

5. BACKGROUND SCREENING BY TREES FROM P.O.Z. VIEW.

6. TOTAL SCREENING FROM P.O.Z. VIEW
BACKGROUND SCREENING FROM S.O.Z.

P.O.Z. = PRIMARY OBSERVATION ZONE

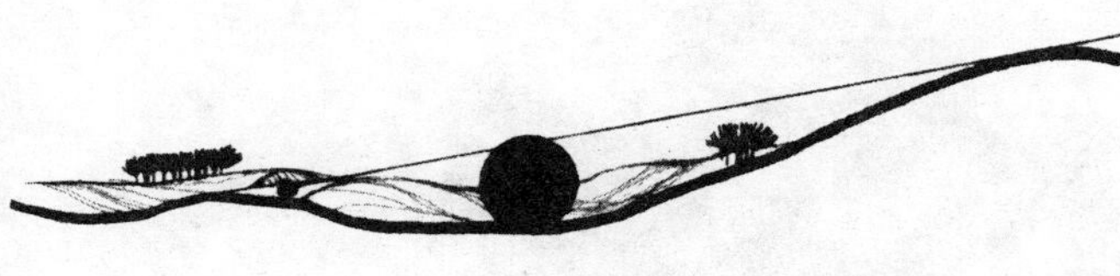

Vegetation

As with topography, vegetation can often be effectively utilized either as a background for transmission lines or to screen views to them from points of public orientation. More importantly, forests and areas of natural vegetation have great landscape and recreational value and must be preserved wherever possible. Where rights-of-way pass through forested areas, efforts should be made when clearing is done to retain within them as much vegetation as possible. The forest edge should be sensitively adjusted during the right-of-way clearance procedures so as to appear natural. The aim in both preservation of existing vegetation as well as the introduction of new plant types should be to achieve a convincingly natural asymmetry that will reduce the disturbing unnatural rhythm and symmetry of the lines and towers. Because the lines are higher and more fixed in position at the tower locations than along their span between towers, the edge of wooded areas can be made more natural by moving it in and out, bringing it closer to the towers than at the mid-points. This should not be done on a regular basis, however, or the constant repetition will cause the view to be as unattractive as the constant-width clearing normally associated with power line rights-of-way.

Where entrances to rights-of-way from open land into wooded areas are necessary, as much vegetation as possible should be retained in the area of entry in order to render it inconspicuous. Changes in right-of-way alignment at relatively short distances from a viewing position, such as along a highway, limits the depth of views along the transmission line and reduces its impact substantially. Structures in parallel alignments should be as close together as possible to minimize the total combined width of the right-of-way. Spacing them apart to permit planting between alignments only increases the apparent dimension and disturbance to the environment.

Retention of existing plant materials is generally preferable to replanting. Where the introduction of new materials is appropriate for either background screening or right-of-way enhancement, the choice of plant material should always be made with reference to presently existing plant types within the area. Road and highway crossings, recreational and residential areas and hill crossings are situations where new planting is especially important if sufficient existing vegetation cannot be retained. In each instance, an inventory of existing plant material within or along the right-of-way should be made prior to clearance to determine materials to be removed, appropriate additions to be made and probable maintenance procedures to be employed. The species listed in the Addendum are all native to Michigan and can be expected to be encountered at various ecological situations throughout the state. The presence of any of these materials in a right-of-way location should be used as the basis for selection of additive plantings within the right-of-way.

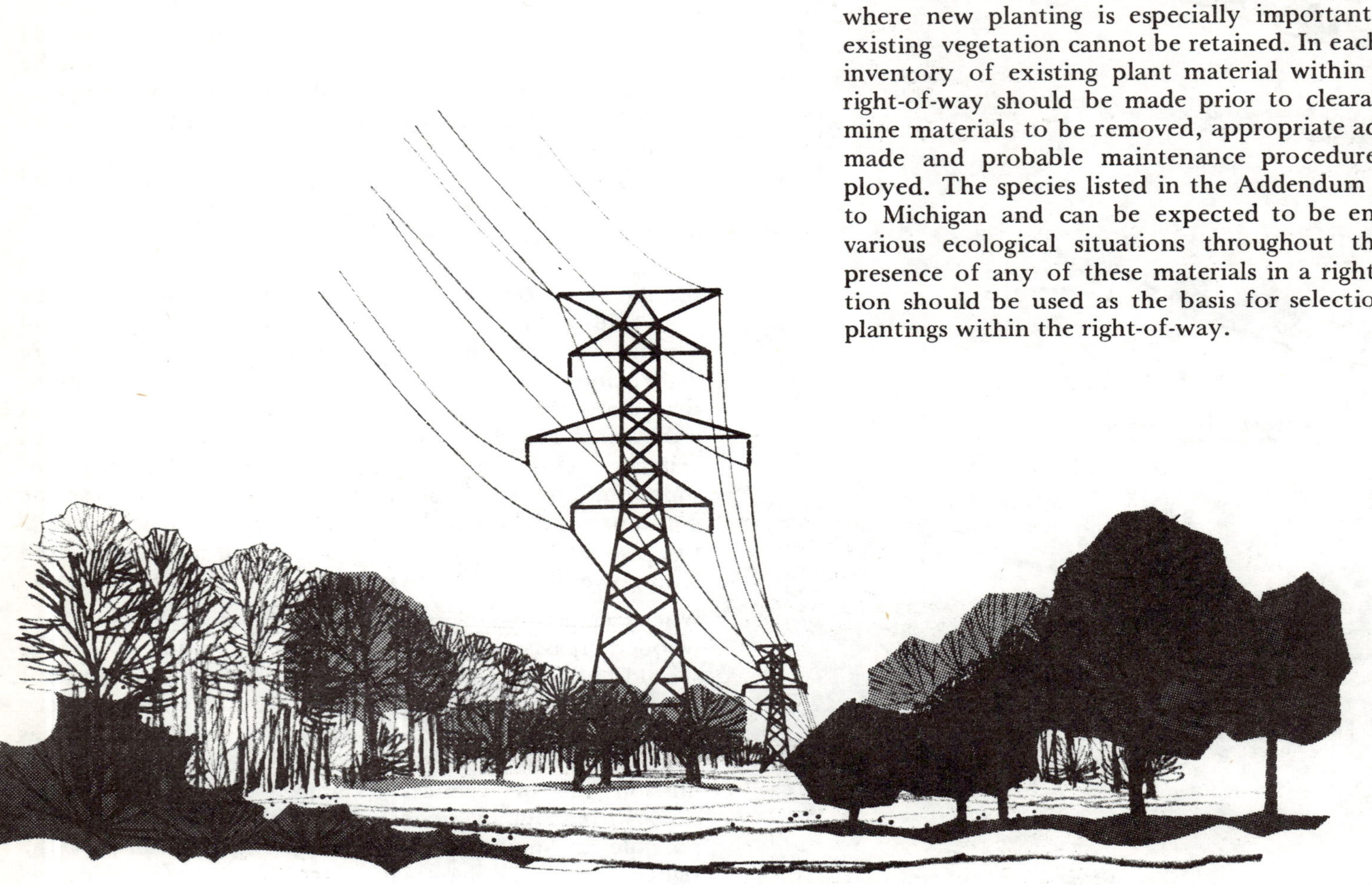

HEIGHT OF CONDUCTORS AT POLES AND TOWERS ALLOWS PRESERVATION OR REPLACEMENT OF LARGE PLANT MATERIALS AT STRUCTURE INTERVALS, INTERRUPTING THE VIEW DOWN THE RIGHT-OF-WAY. AVOID, HOWEVER, ESTABLISHING A RHYTHM OF TALL VEGETATION AT EACH TOWER AND LOW PLANTINGS IN THE CONDUCTOR SAG ZONES BETWEEN STRUCTURES.

APPARENT DIMENSION
−
+

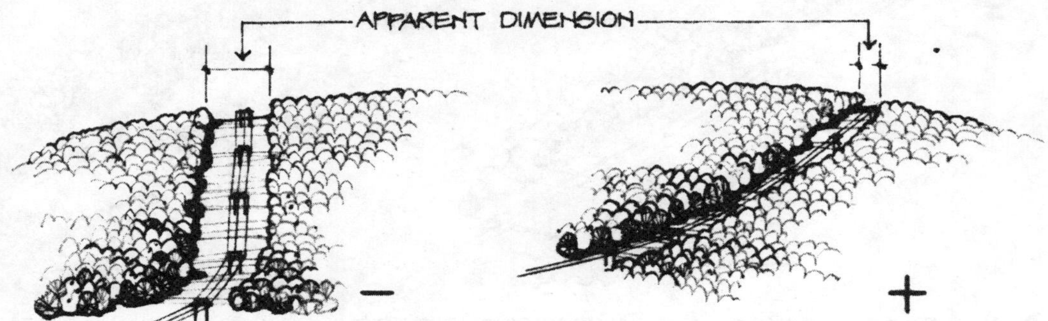

RIGHTS-OF-WAY SHOULD CROSS WOODED HILLS AND MOUNTAINS OBLIQUELY RATHER THAN AT RIGHT ANGLES TO THE CONTOURS.

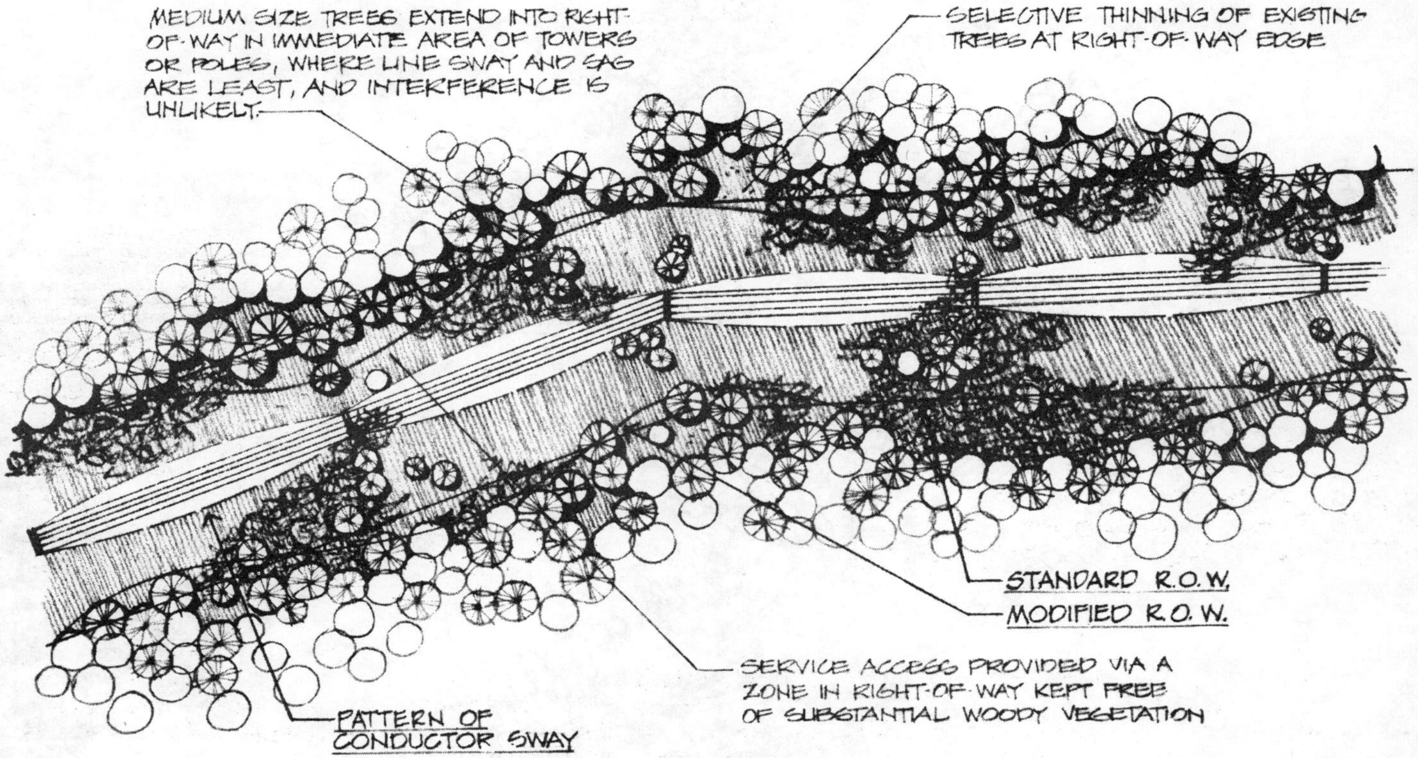
MEDIUM SIZE TREES EXTEND INTO RIGHT-OF-WAY IN IMMEDIATE AREA OF TOWERS OR POLES, WHERE LINE SWAY AND SAG ARE LEAST, AND INTERFERENCE IS UNLIKELY.
SELECTIVE THINNING OF EXISTING TREES AT RIGHT-OF-WAY EDGE
STANDARD R.O.W.
MODIFIED R.O.W.
SERVICE ACCESS PROVIDED VIA A ZONE IN RIGHT-OF-WAY KEPT FREE OF SUBSTANTIAL WOODY VEGETATION
PATTERN OF CONDUCTOR SWAY

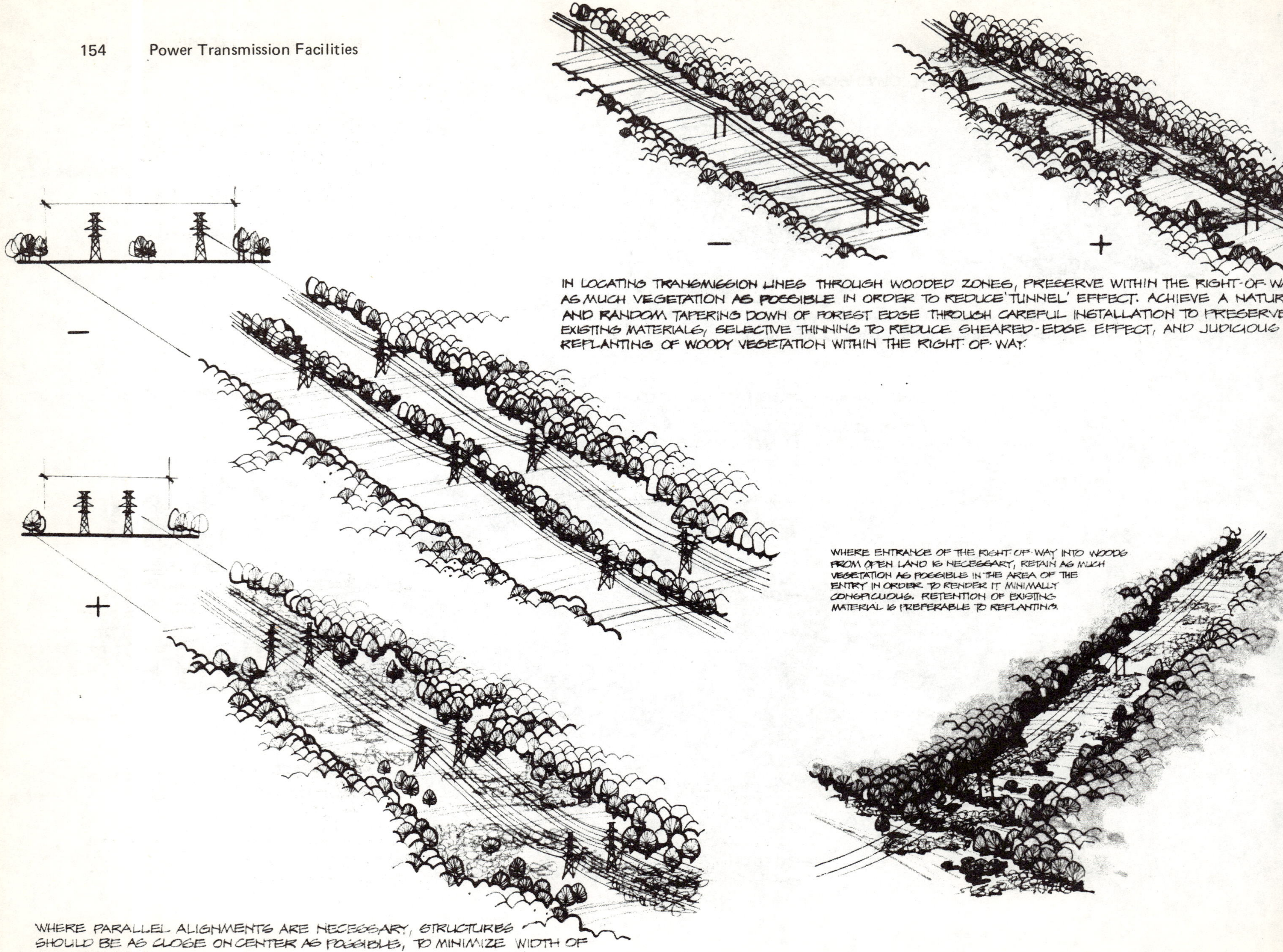
–
+
IN LOCATING TRANSMISSION LINES THROUGH WOODED ZONES, PRESERVE WITHIN THE RIGHT-OF-WAY AS MUCH VEGETATION AS POSSIBLE IN ORDER TO REDUCE 'TUNNEL' EFFECT. ACHIEVE A NATURAL AND RANDOM TAPERING DOWN OF FOREST EDGE THROUGH CAREFUL INSTALLATION TO PRESERVE EXISTING MATERIALS, SELECTIVE THINNING TO REDUCE SHEARED-EDGE EFFECT, AND JUDICIOUS REPLANTING OF WOODY VEGETATION WITHIN THE RIGHT-OF-WAY.
–
+
WHERE ENTRANCE OF THE RIGHT-OF-WAY INTO WOODS FROM OPEN LAND IS NECESSARY, RETAIN AS MUCH VEGETATION AS POSSIBLE IN THE AREA OF THE ENTRY IN ORDER TO RENDER IT MINIMALLY CONSPICUOUS. RETENTION OF EXISTING MATERIAL IS PREFERABLE TO REPLANTING.
WHERE PARALLEL ALIGNMENTS ARE NECESSARY, STRUCTURES SHOULD BE AS CLOSE ON CENTER AS POSSIBLE, TO MINIMIZE WIDTH OF RIGHT-OF-WAY.

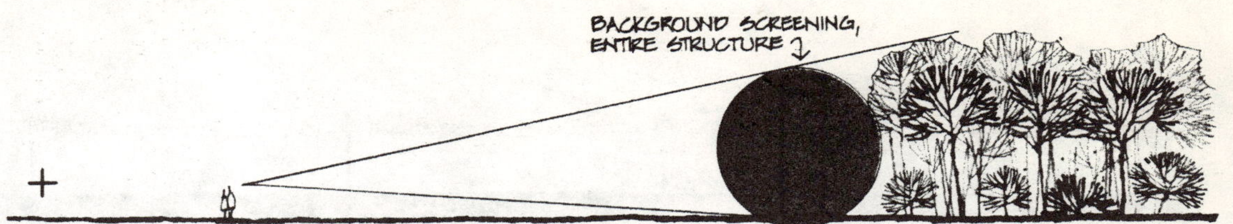

TREES AND WOODLANDS (ESPECIALLY DECIDUOUS TREES IN WINTER) TEND TO BE EFFECTIVE BACKGROUND SCREENING FOR DISTRIBUTION LINES. THE WEBWORK OF POLES, LINES AND WIRES IS MUTED AGAINST THE INTRICATE BACKGROUND OF TRUNKS AND BRANCHES.

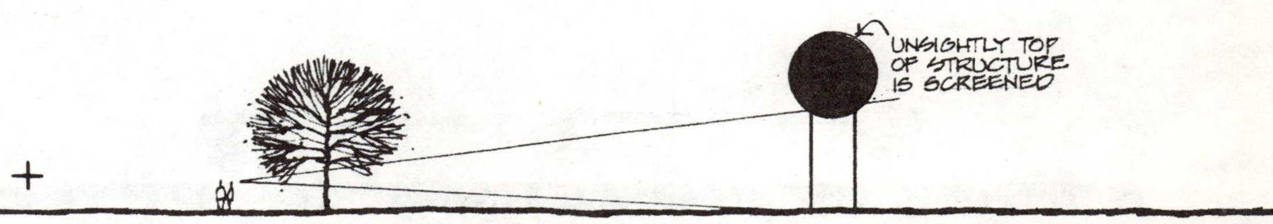

VIEWS FROM BEHIND OR UNDER TALL MATERIALS MAY BLOCK VIEW OF UPPER (AND MOST UNSIGHTLY) PORTION OF STRUCTURE

ATTEMPTS AT FOREGROUND SCREENING OF LINES ARE PARTLY INEFFECTIVE BECAUSE, THOUGH THE LOWER PART OF THE STRUCTURE IS HIDDEN, THE MOST UNSIGHTLY UPPER PORTION IS LEFT IN VIEW AGAINST THE SKY

ONLY FOR VERY SHORT RANGE VIEWS WHERE THE NORMAL ANGLE OF VIEW IS UNLIKELY TO INCLUDE THE ENTIRE POLE OR TOWER ARE LOW MATERIALS GOOD SCREENS.

STRUCTURES AND LINES IN OPEN LAND ARE VERY CONSPICUOUS, ESPECIALLY WHEN SEEN AGAINST THE SKY.

ATTEMPTS TO SCREEN STRUCTURES WITH PLANTING ONLY CALLS ATTENTION TO THEIR RHYTHMIC SPACING

PLANT GROUPINGS OF VARYING SIZES, SPACED IRREGULARLY, PARTIALLY SCREEN STRUCTURES AND LINES, AND INTERRUPT THE REGULAR RHYTHM OF STRUCTURE LOCATION.

NEARLY SOLID PLANTINGS CAUSE WIRES TO BE 'LOST' AGAINST BACKGROUND BRANCHING TRACERY, AS STRUCTURES ARE SCREENED BY BACKGROUND TREES.

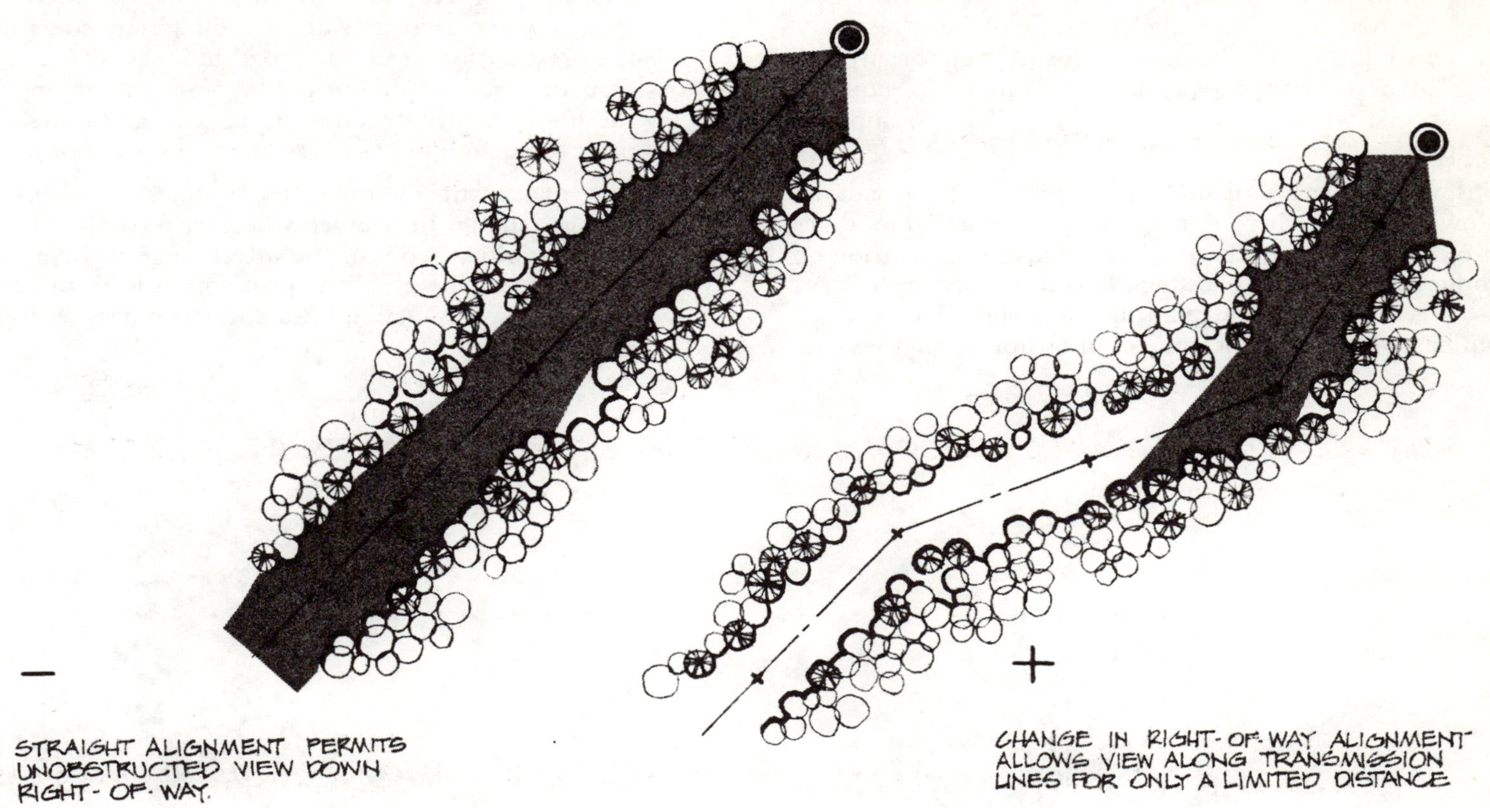

Roads

One of the major recreational activities in the United States is driving and riding, emphasizing the importance of proper visual relationships between transmission rights-of-way and roadways. It is in relation to this recreational pastime that perhaps most of the negative impact of transmission lines is experienced. In general, rights-of-way should not parallel roadways unless separated from them by effective topographic or vegetative buffers. Alignments which result in long views of transmission lines parallel to highways should be avoided; rather they should be located at a sufficient distance from the highway so that vegetation or topography interrupt the view to the transmission line. Where buffering is not possible, the lines should be placed sufficiently back away from the roadway, preferably against a backdrop of trees or topography, so as to have little influence upon the roadscape.

The angle at which transmission lines cross major roadways should be as near to perpendicular as possible to allow for maximum setback of line structures and minimum visibility from roadway into the right-of-way on each side. Special attention should be given to screening the right-of-way by vegetation and topography. In wooded areas, if right-of-way approaches to the roadway cannot be effectively screened, a job in the alignment helps to keep the site distance down the right-of-way to a minimum. In exposed areas, special structure designs can be utilized to improve the appearance.

In instances where existing rights-of-way cross roads at shallow angles or where rights-of-way run parallel to a section of highway, the preservation of existing vegetation or introduction of new plantings becomes very necessary. These need not be solid plantings but should consist of groupings of trees sufficient to interrupt open views of lines of transmission structures. Roadway crossings can be accomplished very successfully by the introduction of long span towers on either side of the highway so as to permit the preservation of existing roadside vegetation.

Structure setbacks from public roads in conjunction with the use of higher structures to provide additional clearance offer excellent opportunities for natural screening.

The placement of transmission lines at highway interchanges should be avoided wherever possible. The need for good visibility requires the interchange to be a very open landscape so there is little or no opportunity to soften the impact of transmission lines and structures with plantings or topography.

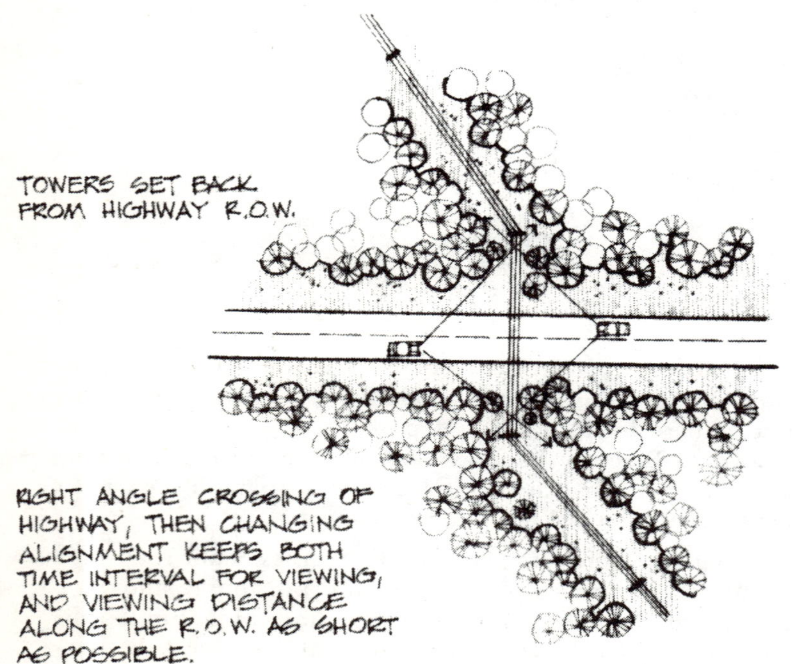

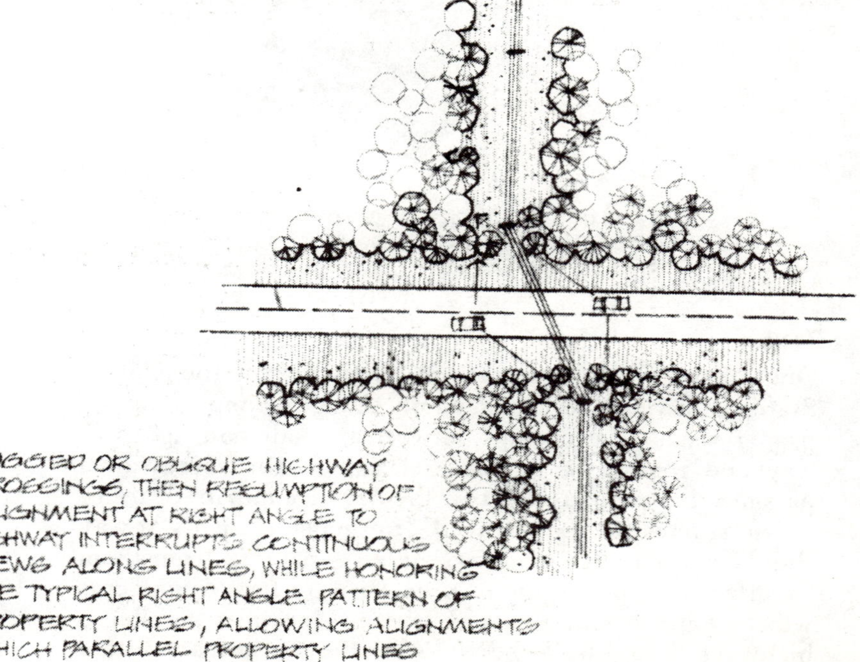

−

+

AVOID ALIGNMENTS WHICH RESULT IN LONG VIEWS OF TRANSMISSION LINES PARALLEL TO HIGHWAYS. LOCATE TRANSMISSION ALIGNMENTS AT SUFFICIENT DISTANCE FROM THE HIGHWAY THAT INTERVENING VERTICAL ELEMENTS WILL INTERRUPT THE VIEW DOWN THE TRANSMISSION LINES.

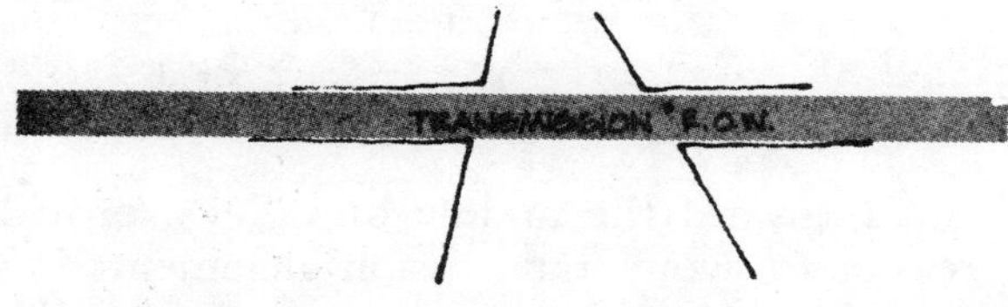

−

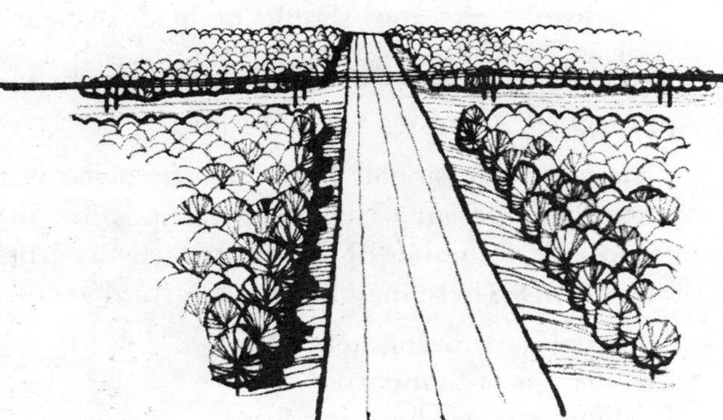

LOW TOWERS NECESSITATE REMOVAL OF VEGETATION IN RIGHT-OF-WAY AT ROAD CROSSING, ALLOWING VIEWS ALONG RIGHT OF WAY.

+

PRESERVATION OF FORESTED EDGE BLOCKS VIEW DOWN RIGHT-OF-WAY. HIGH, LONG SPAN TOWERS CARRY LINES OVER TREES

WHERE RIGHTS-OF-WAY CROSS ROADS AT AN OBLIQUE ANGLE, OR RUN TANGENT TO A SECTION OF HIGHWAY, PLANTINGS OF SUFFICIENT SIZE TO FOIL THE VIEW DOWN THE RIGHT-OF-WAY SHOULD BE RETAINED OR REPLACED. THESE NEED NOT BE SOLID PLANTINGS, BUT WOULD MORE APPROPRIATELY CONSIST OF LIMITED NUMBERS OR GROUPINGS OF TREES TO INTERRUPT THE OPEN VIEW.

Natural Areas

With the growing realization that the preservation of our relatively few remaining natural areas is critical to the future generations of our country, great emphasis should be placed on the proper placement of transmission structures and lines in relation to these assets. Wherever possible, these areas should be totally avoided or circumvented in order not to disrupt their natural qualities. Underground placement of lower voltage lines should be considered. Whether below ground or above, rights-of-way should be of minimum possible width in scenic areas with the width of the right-of-way sensitively articulated to reduce the corridor effect. Significant scenic, recreational and historical areas should not be intruded upon by high voltage transmission lines. Special emphasis should be placed upon preserving scenic forest areas from encroachment by transmission lines either above or underground. Transmission rights-of-way should not parallel rivers and streams nor should they be situated in close proximity to scenic lakes, ponds and shorelines. Rivers and streams tend to be located toward the middle of valleys, providing additional reason to avoid transmission alignments in this configuration. Lines should cross rivers, streams, and other bodies of water at low areas. Avoid crossings at high points in the landscape or at especially scenic points. As in the case of roadways, crossings should be made as near to perpendicular as feasible.

Approaches, Processes and Solutions

Design professionals have been involved in the location, development and design of electric transmission lines in a number of ways under a number of organizational structures and in a number of situations. These may be characterized as follows:

1. Guideline Development
2. Redesign of Standards
3. Alignment Analysis and Study
4. Integration of Activities on the Rights-of-Way and the Development of Secondary Uses of the Rights-of-Way
5. Beautification and Buffering

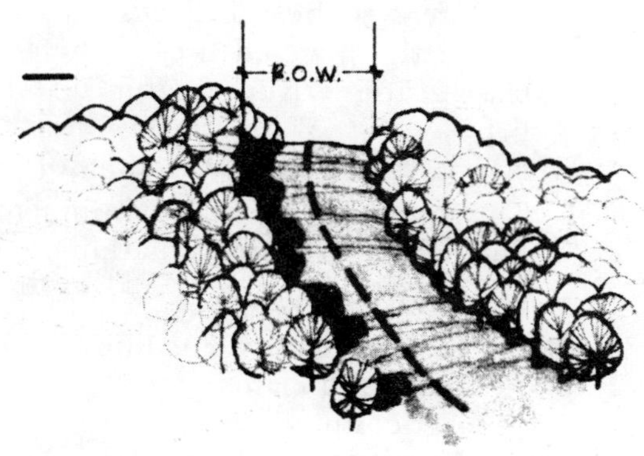

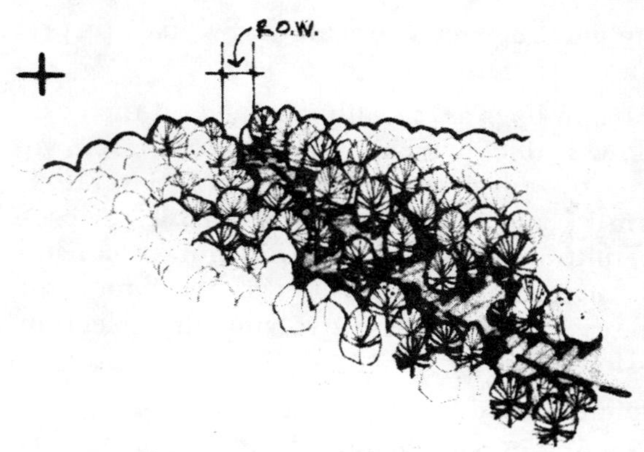

UNDERGROUND LINE RIGHTS-OF-WAY SHOULD BE OF MINIMUM POSSIBLE WIDTH WHEN CROSSING HILLS IN WOODED AREAS.
VARY THE WIDTH OF THE UNDERGROUND LINE RIGHT-OF-WAY TO REDUCE CORRIDOR EFFECT THROUGH WOODED ZONES.[38]

Essentially, design professionals of all types have been involved in all of these facets of positive solution exploration. Among the most important and complete guidelines which have been developed to assist in the location of transmission rights-of-way, are those which were prepared in the *Environmental Guidelines* publication prepared by the Western Systems Coordinating Council in December of 1971.

That publication outlines the following guidelines:

MAJOR TRANSMISSION LINES

1. Introduction

   Since the preservation of environmental quality has become a matter of national interest and priority, there is increasing national concern as evidenced by the actions by groups and private citizens to enjoin, postpone or alter certain projects on the basis that they may be damaging the natural environment. Clearly the public has an interest, individually and collectively, in protecting and enhancing the environment and minimizing adverse impact that transmission lines may have on the environment.

   Each transmission line project involves the following stages:

   (a) Planning and environmental impact analysis
   (b) Route selection
   (c) Rights-of-way acquisition
   (d) Design
   (e) Construction
   (f) Maintenance

2. Planning and Environmental Impact Analysis

   The objectives of transmission line planning are to develop system additions for at least a ten-year projection and to meet system reliability criteria. Current methods used for the timing of capacity additions are based on a 'most probable' load estimate.

   Alternative methods for planning new transmission line capacity must be considered to provide proper evaluation of the environmental impact in the areas affected. Obtaining new right-of-way corridors or utilizing existing corridors in a coordinated effort with other util-

38. Ibid., pp. 26-46.

ities, governmental agencies, private individuals or private groups are alternatives to be considered.

New transmission lines are usually planned on the basis of power flow studies which project the requirement for additional power in some specific area. Several alternative sources and routes may be available for providing this additional power and concomitant consideration of environmental factors is an important element of feasibility studies involving the selection of transmission routes.

Project planning should include:

(a) Environmental impact analysis.

(b) Cooperation and coordination with other utilities, where appropriate, in order to develop the transmission line route so that the visual impact on the landscape will be minimal.

(c) Consideration of joint use of lines by upgrading or replacing existing lines to avoid the need of acquiring additional rights-of-way.

(d) Consideration of transmission line corridors when long-range plans are coordinated with other utilities, governmental agencies and private organizations.

(e) In situations where minimum visibility conflicts with flight safety regulations, the safety regulations should govern.

3. Route Selection

Utility representatives must be fully apprised of the specific environmental concern of the people in the areas affected and have an awareness of the need to evaluate the visual impact and physical infringement on the landscape.

Locating lines in certain critical areas or in close proximity to parks and monuments, scenic, natural, historic, archeological, and recreational sites should be avoided, if possible.

The process of route selection requires that:

(a) A current land use and feasibility study be made of the areas to be traversed. This study should categorize the planned use of the areas into different classifications, such as rural, residential, business, light industrial and heavy industrial, and include estimates of real estate values of each classification.

(b) The appraisal of the environmental impact of the line necessitates an evaluation of the following:

(1) Industrial Values
Existing Utility Facilities
Industrial Plants
Agricultural Usage
Forests

(2) Social Values
Historic and Archeological
Schools and Institutions
Recreation
Cemeteries
Scenic Views
Residential Development
Commercial Development
City Streets, Highways and Freeways
Railroads
Rivers
Flood Control Channels

(3) Physiographic Obstructions
Foundation Soil Values
Swamps, Lakes and Inlets
Mountain Ranges

(c) Each possible route should be evaluated to determine its capabilities, such as number of circuits (lines), type of lines (tower lines on right-of-way and poles on streets and estimated costs, including undergrounding of other facilities.

(d) The route of transmission lines must meet the present and long-range objectives of local planning and be compatible with existing and future land uses in the area.

To minimize the environmental impact of a transmission line, the following criteria should be considered in the selection of the route:

(a) Locate the route in an area of minimum conflict with present and future planned land uses.

(b) Replace or upgrade existing lines.

(c) Add new lines parallel to existing lines provided that reliability criteria can be maintained.

(d) Develop joint use lines with other utilities.

(e) Avoid

(1) Line locations which create unusable spaces between lines.

(2) Heavily timbered areas, steep slopes, proximity to main highways and scenic areas.

(3) Crossing main highways in the vicinity of interchanges or bridges.

(4) Crossing highways at points where the structures can be seen from a long distance or silhouetting the structures against the sky, if possible.

(5) Wildlife concentration areas, such as nesting areas and flight corridors.

(6) Parks, monuments, scenic, natural, historic, archeological and recreational areas.

(f) Cross canyons up slope from roads which traverse the length of the canyon.

(g) Use the topography or natural cover to screen the line from highways and other areas of public view.

(h) Consider access and construction roads and locate the route for minimum damage from road building activities and for minimum use by unauthorized personnel. Permanent roads should by minimized.

(i) During field reconnaissance, guard against the natural tendency to favor the locations which are easily observed. Tangled underbrush and other areas which must be observed by rough foot travel may hide the best route. Find all practicable routes joining the termini.

4. Right-of-Way Acquisition

Each right-of-way agent who contacts property owners to acquire easements or fee title to property for a right-of-way must be:

(a) Knowledgeable concerning all phases of the planned transmission line;

(b) Able to discuss the fundamentals concerning the need for the line and its ultimate capacity; and

(c) Fully apprised of the proposed design of structures, location of structures, conductor size and configuration and construction schedules.

5. Design

Appearance is the prime environmental impact of transmission lines. Consequently, extreme care is required in the overall design and location of tower structures.

The fact that most esthetic tower structures stand out in bold fashion, emphasizes the importance of correct structure selection for a particular location and area. The shape of the structure should relate to the environment by blending, harmonizing or in some cases, contrasting.

Except in the areas where lattice transmission towers are environmentally acceptable, tower structures artistically designed should be used wherever possible.

Each of the following numbered designs is shown in the Edison Electric Institute publication dated 1967 entitled 'Electric Transmission Structures':

(a) A 'slingshot' pole (5/HS/2T2G) would create minimum impact in an open area with sparse tree growth.

(b) In areas of massive concrete structures, the use of bridge structures (19/HS/2T2G) might be more effective to lessen the environmental impact.

(c) When crossing a parking lot of modern design and construction, a contemporary tower design (14/HT/2T1G) might be the correct selection.

Where the towers are visible from close range, an esthetic type structure should be used. The design configuration should avoid irregularities. Collars and flanges for bolted connections are not artistic. Simple lines, straight or curved, are preferred. Color is also most important.

Additional costs will be incurred in construction of a transmission line which will have a minimum visual impact and physical infringement on the landscape. For example, the average total line cost of an esthetically designed line may be between 1.3 and 1.5 times, or greater, than for lattice construction. Most design concepts call for larger amounts of material, but experience has indicated that erection costs should be lower. The cost of footings for two-legged or single-pole structures is an important part of the total cost of the line and in some cases may amount to 20% of the total cost of the line.

The desired objective should be to blend the line into the particular surrounding by judicious use of shape, size, texture, color and location. In other words, the idea is to minimize contrast with the environment.

Contrast, however, cannot always be avoided and specific situations may occur in which contrast can be used somewhat harmoniously. For example, a 750 KV river crossing with towers several hundred feet high might be interesting enough to be environmentally self-sufficient, if the towers are of a modern and striking design.

Transmission line designers must visualize the finished product as they progress through the technical phases of design. Changes in the topography or life style along the route may divide the line into natural environmental style, and other relevant factors. Modern and striking design concepts may be appropriate for a new and growing architecturally designed industrial area, but could detract from the natural beauty of a wooded rural area.

Reducing the size of a transmission structure below optimum increases total line costs radically, but color and shape can be varied to segments. These segments should be considered separately for structure type, color, make them appear smaller to the eye.

The location of an individual structure can drastically alter the visual impact. As an extreme example, visualize the contrast formed if a steel lattice tower were to be located in the center of a small placid mountain lake. Analogous situations of varying visual impact are structures at highway crossings, river crossings, ridge crossings, small meadows, and other areas where structures located inside the surroundings may tend to degrade scenic values. When these areas must be crossed, structures should be placed outside and screened from the open area.

Visualization during the design phases is not an end in itself. Structure locations should be field staked and visually checked for environmental impact. Designers should also view the line after the construction is completed.

6. Construction

(a) Rural Areas

During the various construction phases it is imperative that the use of mobile heavy equipment be closely monitored so that damages to the environment can be minimized. As a general rule the communications between the field forces and those who are charged with responsibility for supervising construction of the project may need additional emphasis concerning environmental limitations involved and constant review of those limitations.

If the lines are to be built by contractors, the construction specifications should convey not only management's desire for environmental control, but separate specifications should be developed which will provide for preservation, or failing that, restoration of the environment. Some specifics are:

(1) Clear the right-of-way in a manner that will minimize marring and scarring of the landscape by using 'brush blades' instead of dirt blades on bulldozers.

(2) Do not

   i. 'Clear cut' the right-of-way. Instead, leave low growing shrubs and bushes.

   ii. Clear in canyons where trees are not a danger to the line.

   iii. Blast in or near streams without adequate protection of aquatic life.

(3) Locate construction road crossings of streams at sites where a minimum of silting of the water will occur.

(4) Leave a screen of vegetation between the right-of-way and highways or rivers.

(5) Time the clearing operation to minimize damage to critical areas.

(6) Install culverts, side drainage ditches and water bars to prevent erosion of roads.

(7) Avoid oil spills and littering.

(8) Develop fire prevention plans and a fire inspection program.

(9) Tension string conductors to reduce the amount of vegetation clearing.

(10) During the clean-up operation, obliterate ruts and scars, repair damage to ditches, terraces, roads and other features. Seed the right-of-way with appropriate grasses and other vegetation to restore the ground cover and prevent erosion. Remove all unused material and equipment from the construction area.

(b) Urban Areas

In an effort to lessen the impact on the landscape and open spaces and to minimize urban land requirements for transmission lines, joint use of air space and multiple purpose rights-of-way will be necessary in the future. The following are illustrative of opportunities for possible joint use of city streets, railroad rights-of-way, flood control channels and freeways:

(1) City Streets

   i. Single-Circuit Steel Pole Lines

   Numerous arm shapes have been used for single-circuit steel pole lines on streets. In most instances, the pole base diameter can be limited to 24" where the structure is between the curb and sidewalk. The color and dimensions of the poles should fit the environment. Tapered, welded, round or polysided, painted poles are most attractive.

   ii. Double-Circuit Steel Pole Lines

   For transmission voltages, double-circuit construction will, in most cases, require the line to be located in a median strip. This location might require special negotiations in case of state or federal funded highways. In general, median strips contain fewer substructures than shoulder or sidewalk areas. Pole diameters can, in most cases, be maintained at around 3′ at the ground line. Many examples of different arm shapes are available. Since structures cannot be placed in intersections, it is extremely difficult to turn corners from one street into another.

   Flying corners on mid-span taps are one possibility. For esthetic reasons, pole top design should be considered. New esthetic designs are needed.

   The span lengths for single- or double-circuit poles on streets should be short (400′-600′). The details of the structure become most important.

Recommended for consideration are:

Tapered vs. Straight Legs
Painted vs. Galvanized Structures
Welded vs. Bolted Connections

iii. Other Pole Lines on City Streets

Wooded, concrete or other materials may be used for some transmission lines on city streets.

(2) Railroad Rights-of-Way

Presently there are many miles of transmission lines on railroad rights-of-way.

Railroad rights-of-way often lend themselves to compatible uses by transmission lines and permit maximum usage of available corridors without devaluating surrounding property.

Transmission lines on railroad rights-of-way along streets may result in an upgrading of the street due to undergrounding of existing distribution or telephone facilities.

Using modern steel pole structures, two double-circuit 230 KV lines can be placed on a 90 foot railroad rights-of-way.

Bridge structures may be used to overbuild railroad tracks.

Short spans are recommended with simple esthetic structures, having straight galvanized legs with welded connections.

(3) Flood Control Channels

Many flood control channels are artificial, concrete lined structures that lend themselves to joint usage for utility corridors. Joint use of airspace, by overbuilding flood control channels, should be used if possible. Single poles or bridge structures are applicable. Lines should not be constructed along the banks parallel to natural rivers because of the adverse environmental impact on natural and scenic values.

(4) Freeways

Experience has indicated that new freeway routes often follow existing transmission line rights-of-way without serious impairment of environmental values. Governmental agencies and utilities should consider the mutual benefits of the joint use of such airspace when requirements for lines arise adjacent to existing freeways.

(5) Fundamentals in Urban Line Construction

i. Street intersections should be avoided and structures should be located at equal distances from crossings. Crossings with freeways and highways should be at 90° and accomplished with tangent towers if possible.

ii. When planning alignments parallel to highways, consideration should be given to placing the structures within the road right-of-way rather than several hundred feet away. This will lessen the visual impact and physical infringement on the landscape.

iii. The structures should be evenly spaced and of uniform height where practicable.

iv. Deadend structures are less attractive than tangent structures because they are heavier and have more insulators and hardware.

v. Careful consideration should be given to scale. The structure dimensions should not overpower the environmental dimensions.

vi. If possible and if compatible with line safety, the minimum number of trees should be removed as they serve to screen the lower and heaviest part of the structure. Replacing of excessively high trees with trees which are compatible with the line should be considered.

vii. At each location take a 360° view to determine all critical points from where the structure can be seen. Minor adjustments in location may solve an appearance problem.

The most esthetically desirable length of span will be determined by the location. The structure locations should be such that the line follows the terrain, without serious difference in height. It is recommended that pictures be taken every 200 to 300 feet along the route to determine if the contemplated structures would fit in against the background. Aerial photographs of 1 inch to 100 feet or other suitable scale prove to be very useful in locating structures.

Very often in urban areas only very narrow rights-of-way are available. After determining conductor sideswing in the usual manner, it may be necessary to consider special structures at specific locations where clearances are reduced.

A thorough search of city and county records is necessary to determine the existence of all underground facilities in franchise locations or on rights-of-way. Copies of street maps and engineering drawings showing locations of underground structures should be obtained. Gas and water pipes, oil lines, electric and communication cables must be accurately located to avoid damage during footing excavation and to assure no mechanical or electrical interference. There may be occasion when it will be advantageous to rearrange and relocate existing underground facilities to provide for the proposed new transmission structures.

7. Landscaping

Esthetic landscaping should be considered at road or highway crossings where line visibility is high and in conflict with scenic values of the general area. The location of all planting should be carefully considered so maximum advantage of screening qualities can be utilized without loss of the artistic values. Trees and vegetation indigenous to the area should be used whenever possible so as to minimize visual attraction to the area which might otherwise occur from the planting of unusual or contrasting vegetation or trees.

Access routes from public roads or highways should be developed so as to be unobtrusive or they should be screened, if practicable. Gates and fences should be neat, substantial and well maintained at all times.

8. Secondary Right-of-Way Uses

Utilities should make available to communities and others right-of-way land areas for compatible productive and recreational purposes such as parks, golf courses, equestrian trails, agricultural and nursery uses and similar beneficial uses. Such right-of-way uses must be carefully planned. Maximum use should be made of topography and the right-of-way maintained in such a manner as to be an asset to the neighborhood. Special treatment may be necessary at road or street crossings to minimize the impact of the line.

9. Radio and Television Interference

Prior to construction, ambient radio frequency noise level and received radio and television signal strength measurements should be made along the proposed transmission line rights-of-way.

After construction and energization of equipment and lines, additional ambient radio frequency noise and received radio and television signal strength measurements should be made along transmission line rights-of-way at the same locations used prior to construction. Further investigation of radio and television interference should be made if complaints are received. Approximately 70% to 85% of all complaints received are caused by customer owned devices and the remaining are usually traced to line hardware or contamination problems and are cleared by line crews.

Any problems located by the repeat measurements should be corrected at once.

10. Maintenance

Some of the transmission line maintenance problems confronting the utility are:

(a) Vehicular traffic is a major hazard to lines located on franchise locations in streets. Poles and structures are normally set behind curbs in accordance with local codes.

(b) Insulator contamination caused from industrial, automotive, salt spray and other natural pollution is frequently very high in urban areas. Frequent washing on insulators is a solution. However, washing becomes objectionable to the public because of the water spray and noise of the washing equipment.

(c) Transmission lines may emit audible and radio noises if not properly maintained. Proper design, construction and good maintenance will keep the noise at a low level.

(d) Buildings, structures or combustible materials, such as lumber, hay or flammable fluids located under transmission lines are a hazard to the line.

(e) Control of vegetation in areas of extreme high fire hazard.

(f) During the patrolling processes, whether by aircraft or land vehicles, special care must be exercised so as not to frighten wildlife, livestock or poultry.

(g) Access to maintain transmission lines in urban areas must be preserved. Line patrolling in urban areas is usually by patrol trucks because the use of helicopters is not permitted.

(h) Although transmission lines can be designed to minimize or prevent trouble caused by:

(1) Floods
(2) Earth slippage
(3) Unstable soils
(4) Avalances
(5) Earthquakes
(6) Winds
(7) Fire

their significant impact requires that they be given full consideration. Even with the best design, expert operation and maintenance are required to prevent hazardous conditions from occurring. The need for constant and regular maintenance procedures is of the utmost importance and their value to the reliability of lines cannot be over-emphasized.[39]

One other set of guidelines which was prepared by the Working Committee for Utilities of the President's Council on Recreation and Natural Beauty in December of 1968 dealt with guidelines for the protection of natural, historic, scenic, and recreational values in the design and location of rights-of-way and transmission facilities. This working committee stated at that time the following:

Guidelines for the Protection of Natural, Historic, Scenic, and Recreational Values in the Design and Location of Rights-of-Way and Transmission Facilities

The Working Committee on Utilities has found that the form and appearance of transmission facilities can be improved if these facilities are creatively designed and constructed with the imaginative use of colors and materials. Moreover, we have found that even with improved form,

39. Western Systems Coordinating Council, Environmental Committee, *Environmental Guidelines*, (Los Angeles, California, Western Systems Coordinating Council, 1971), pp. 25-36.

the appearance of rights-of-way and transmission facilities can be further improved if they are concealed in part by the natural features of the landscape, such as the existing vegetation and terrain, and newly planted vegetation. The guidelines set forth below have followed from these general findings and are applications of them to the kinds of circumstances frequently confronted by management decision-makers in the selection of rights-of-way routes. They are not simply theories, but are objects to be achieved by practice. It is with this in mind that we strongly urge management to make these guidelines in substance a field manual to govern the planning, locating, clearing and maintenance of rights-of-way and the construction of transmission facilities.

The Selection and Clearing of Rights-of-Way Routes

- Rights-of-way should be selected with the purpose of minimizing conflict between the rights-of-way and present and foreseeable uses of the land on which they are to be located. To this end, existing rights-of-way should be given priority as the locations for additions to existing transmission facilities, and the joint use of existing rights-of-way by different kinds of transmission facilities should be considered.

- Rights-of-way should avoid scenic, recreational and historic areas where possible. If rights-of-way must be routed through scenic, recreational or historic areas, they should be located in corridors least visible from areas of public view.

- Rights-of-way should avoid heavily timbered areas, steep slopes and proximity to main highways where possible.

- Long views of transmission lines perpendicular to highways, down canyons and valleys or up ridges and hills should be avoided. The lines should approach these areas diagonally and should cross them at a slight diagonal.

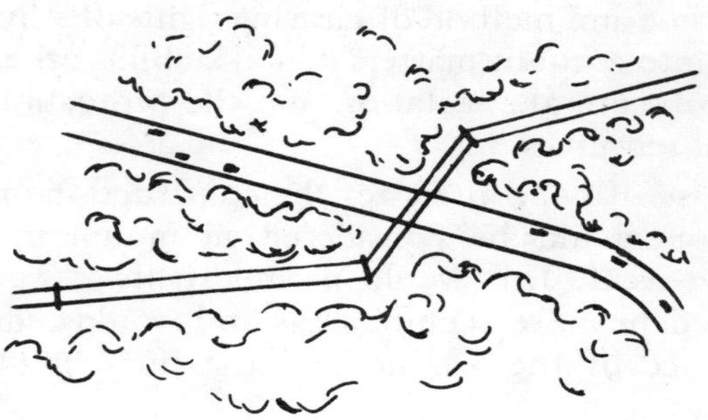

- Rights-of-way clearings should be kept to the minimum width necessary to prevent interference of trees and other vegetation with the proposed transmission facilities. Trees which would interfere with the proposed transmission facilities and those which could cause damage if fallen should be selectively cut and removed.

POOR EXAMPLE

PREFERRED

- The time and method of clearing rights-of-way should take into account matters of soil stability, the protection of natural vegetation and the protection of adjacent resources.
- The use of helicopters for the construction on rights-of-way should be considered in mountainous and scenic areas. This would permit rights-of-way to be located in more remote areas and would reduce disturbance of the ground and the number of access roads.
- Trees and other vegetation cleared from rights-of-way in areas of public view should be disposed of without undue delay. If trees and other vegetation are burned, local fire and air pollution regulations should be observed. Unsightly tree stumps which are adjacent to roads and other areas of public view should be cut close to the ground or removed.
- Trees, shrubs, grass and top soil which are not cleared should be protected from damage during construction.
- Rights-of-way should not be cleared to the mineral soil where possible. Where this does occur in scattered areas of the rights-of-way, the top soil should be replaced and stabilized without undue delay by the planting of appropriate species of grass, shrubs and other vegetation which are properly fertilized.
- Soil which has been excavated during construction and not used should be evenly filled back onto the cleared area or removed from the site. The soil should be graded to comport with the terrain and the adjacent land, and the top soil should then be replaced and appropriate vegetation should be planted and fertilized.
- Bulldozing generally should not be done on slopes which exceed 35 percent. Scars on the surface of the ground should be repaired with top soil and replanted with appropriate vegetation.
- Terraces and other erosion control devices should be constructed where necessary to prevent soil erosion on slopes on which rights-of-way are located.
- Where rights-of-way cross streams or other bodies of water, the banks should be stabilized to prevent erosion. Construction on rights-of-way should not damage shorelines, recreational areas or fish and wildlife habitats.
- Replacement of earth adjacent to water crossings should be at slopes less than the normal angle of repose for the soil type involved and sodding or seeding should be accomplished without undue delay.
- Rights-of-way should cross streams or other bodies of water at low lands rather than at high banks or wild areas where possible.
- Blasting should not be done within or near stream channels without adequate protection for fish and other aquatic life.
- The so-called "jacking" technique for laying pipe or cable beneath river beds should be used where approach elevations and soil conditions permit.
- Cofferdam techniques to lay pipe or cable across streams should be used in order to permit full flow in one part of the stream while construction work is being performed in another part.
- Care should be taken to avoid oil spills and other types of pollution while work is performed in streams.
- Water used for pipe testing purposes and taken from streams or other bodies of water should be limited to volumes which will not cause harm to the ecology or aesthetics of the area. When the testing water is released, it should be done at least one-fourth mile from the streams or bodies of water into vegetation on levels upland so as not to cause erosion and siltation.
- Rights-of-way strips through forest and timber areas should be deflected and should follow irregular patterns. This will prevent the rights-of-way from appearing as tunnels cut through the timber.

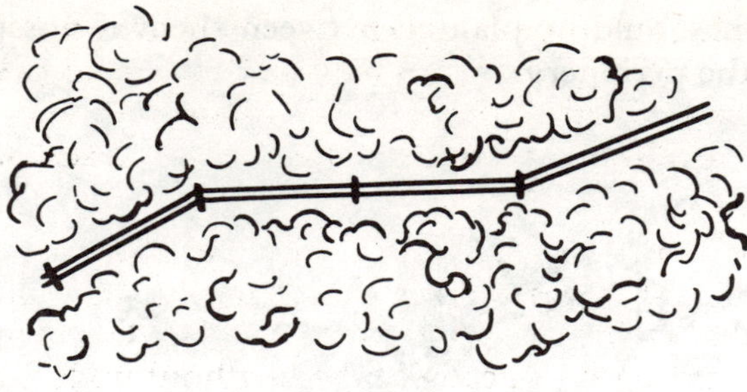

• Small trees and plants should be used to feather back the rights-of-way from grass and shrubbery to larger trees.

POOR EXAMPLE

PREFERRED

• Rights-of-way strips through forest and timber areas should be cleared with curved, undulating boundaries and trimmed to comport with the topography of the terrain.

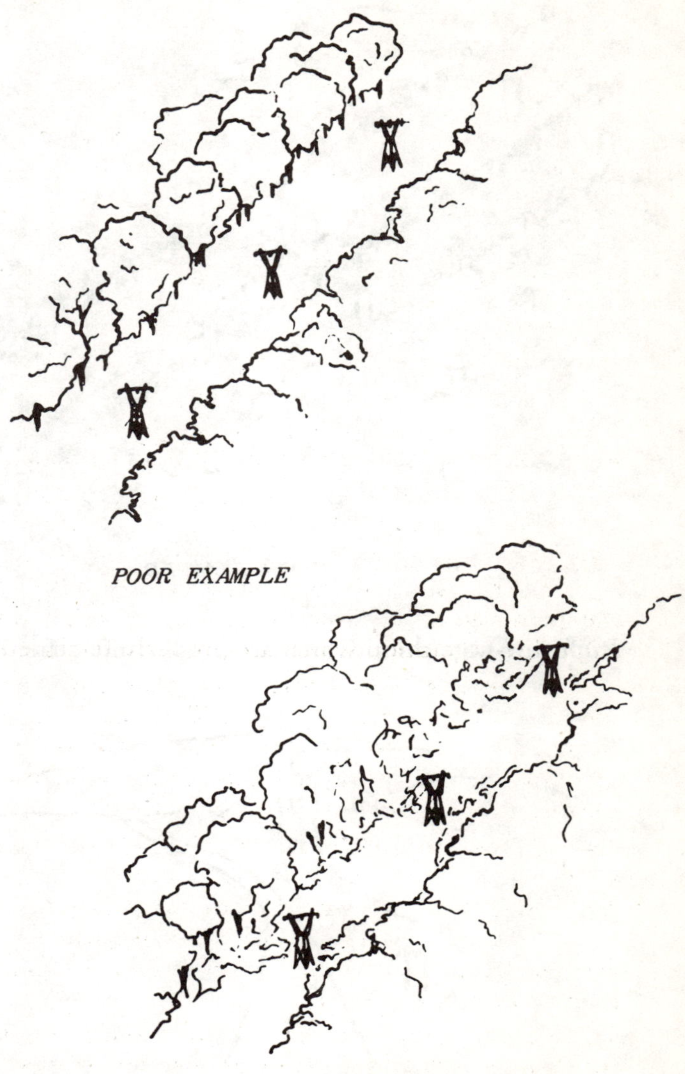

POOR EXAMPLE

PREFERRED

· Where there are several adjacent rights-of-way, low-growing vegetation should be planted between them if possible. Groves of trees should be maintained with lower growing shrubbery on the periphery.

POOR EXAMPLE

PREFERRED

· Rights-of-way should not cross hills and other high points at the crests.

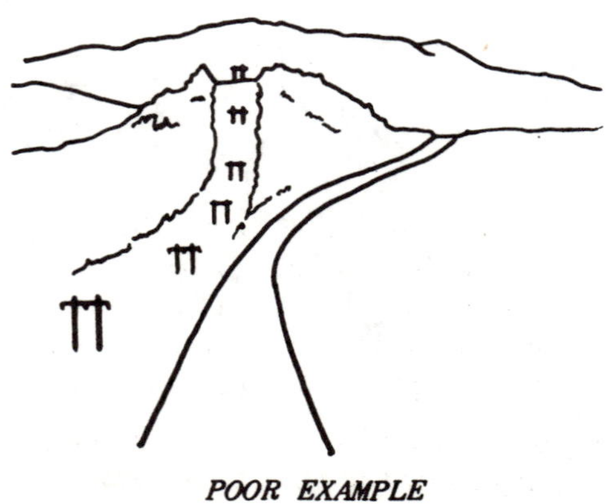

POOR EXAMPLE

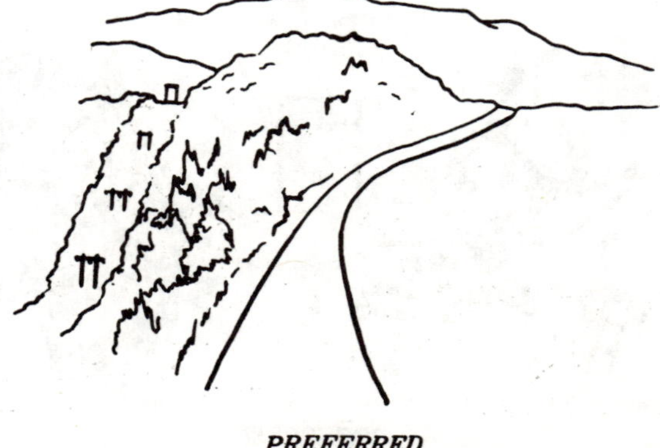

PREFERRED

- If underground transmission lines must be located near the crests of hills or other high points, trenching should be done with small equipment in order to minimize the width of the rights-of-way clearings.
- Roads used during construction should be stabilized without undue delay by erosion control measures and the planting of appropriate grass and other vegetation. These roads should be designed for proper drainage, and water bars to control soil erosion should be installed.
- Access roads should not be constructed on unstable slopes. Where feasible, service and access roads should be used jointly.

The Location of Transmission Towers and Overhead Lines

- If an overhead line must be routed across uniquely scenic, recreational or historic areas or rivers, the feasibility of placing the line underground should be considered. If the line must be placed overhead, it should be located on a right-of-way least visible from areas of public view.
- Transmission facilities should be located with a background of topography and natural cover where possible. Vegetation and terrain should be used to screen these facilities from highways and other areas of public view.

- Where transmission facilities must be placed on slopes which parallel highways or other areas of public view, they should be located approximately two-thirds the distance up the slopes where feasible. With the slopes as background, the presence of the facilities would be less noticeable.

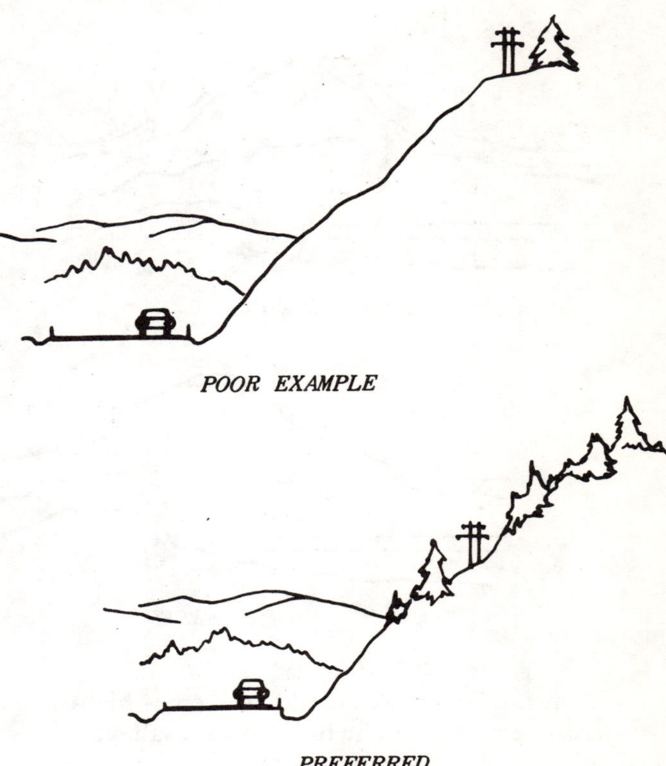

- Transmission line towers which are in areas of public view should not be placed in a straight-line pattern for long distances through forest or timber areas. They should follow an irregular pattern along rights-of-way alignments which have been deflected.
- Transmission line towers should generally follow the contours of the land.

- Transmission facilities should not cross the crests of hills and other high points. To avoid placing a transmission tower at the crest of a ridge or hill, towers should be spaced below the crest to carry the line over the ridge or hill, and the profile of the facilities should not be silhouetted against the sky.

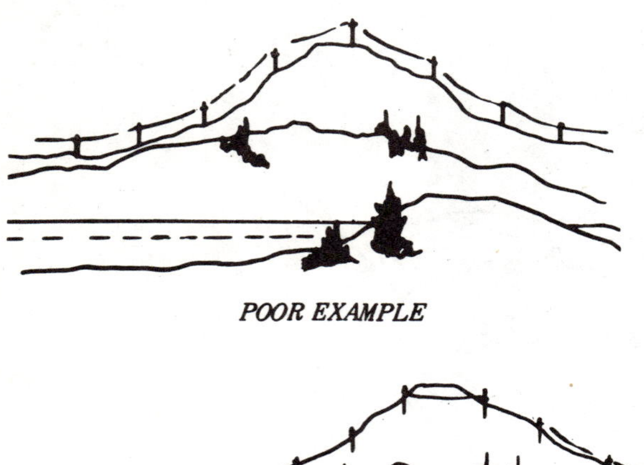

POOR EXAMPLE

PREFERRED

- Transmission lines should not cross highways at the crest of a road or the bottom of a valley.

POOR EXAMPLE

PREFERRED

- Long views of transmission lines parallel to highways should be avoided where possible.

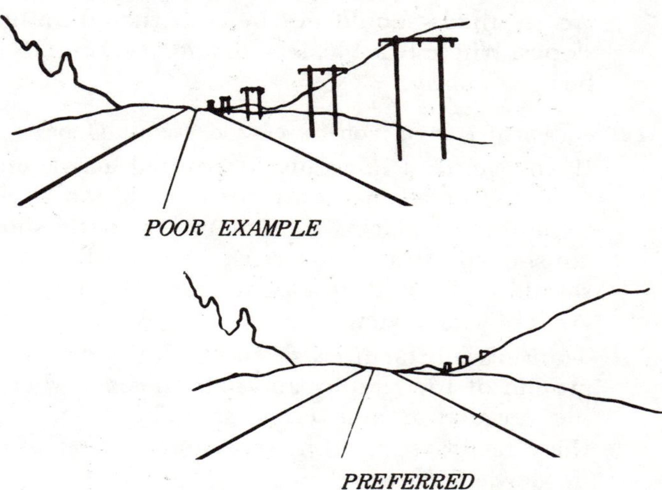

POOR EXAMPLE

PREFERRED

- Transmission lines should cross canyons up-slope from roads which traverse the canyon basins if the terrain permits.

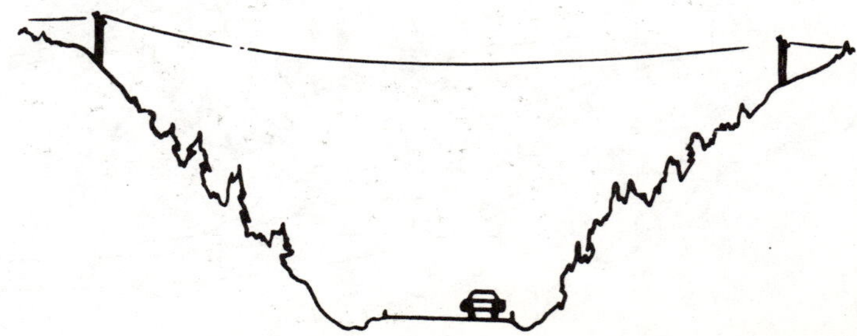

- When crossing valleys in a forest, high, long-span towers should be used to keep the power lines above the trees and to eliminate the need to clear all vegetation from below the lines. Only as much vegetation as is necessary to string the line should be cut.

POOR EXAMPLE

PREFERRED

- Where ridges or timber areas are adjacent to highways or other areas of public view, overhead lines should be placed beyond the ridges or timber areas.
- In forest or timber areas, high, long-span towers should be used to cross highways in order to retain much of the natural growth along the highways.

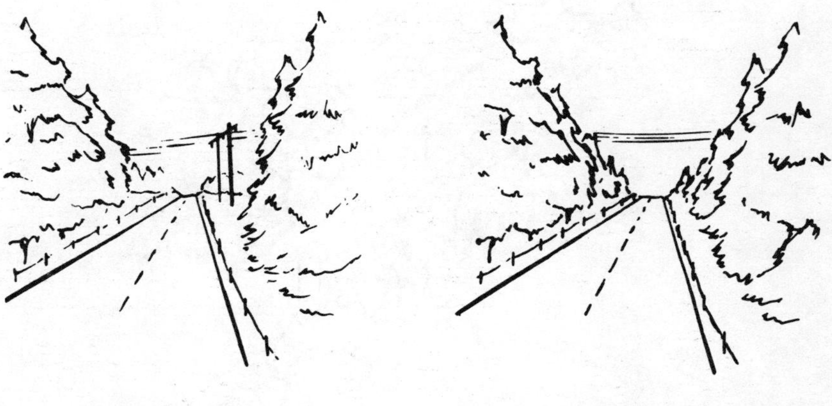

POOR EXAMPLE *PREFERRED*

- Rights-of-way should cross hills and mountains obliquely rather than straight up and down the sides.

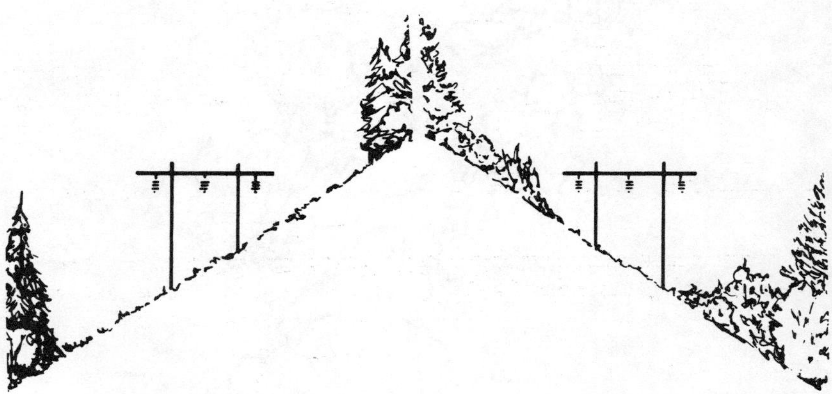

POOR EXAMPLE *PREFERRED*

- Where rights-of-way enter dense timber from a meadow or other clearing, trees should be feathered in at the entrance of the timber for a distance of 150-200 yards.

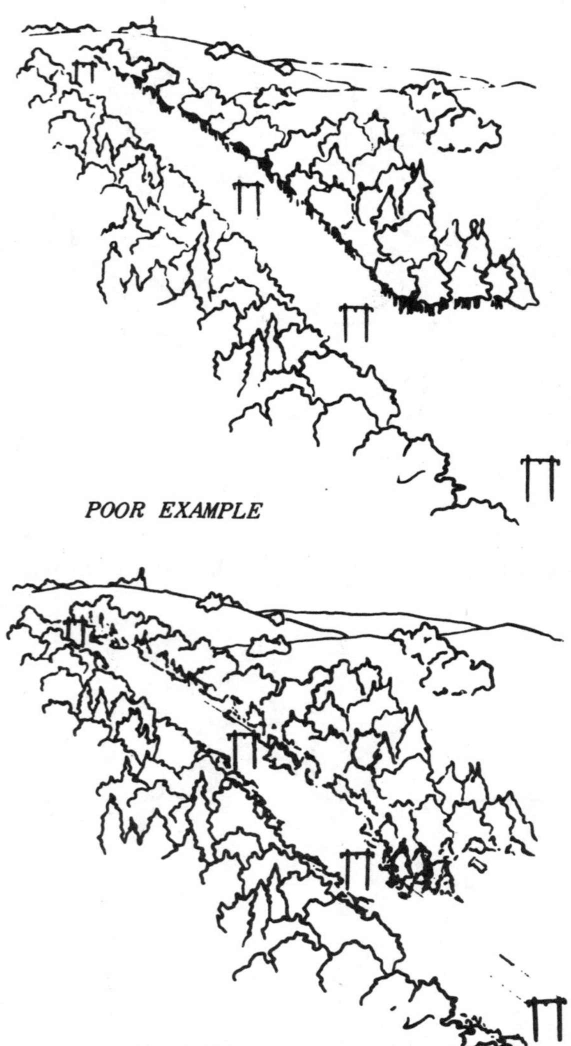

POOR EXAMPLE

PREFERRED

- Native shrubs and trees should be left in place or planted randomly, with the necessary allowance for safety, near the edges of rights-of-way adjacent to roads.

POOR EXAMPLE

PREFERRED

- Transmission lines should not be located or cross at road intersections or interchanges where possible.
- The Federal Highway Administration and the State Highway Department should be consulted with respect to any applicable guidelines or regulations they might have to govern transmission lines which cross highways.

The Design of Transmission Towers

- The size of transmission towers should be kept to the minimum feasible.

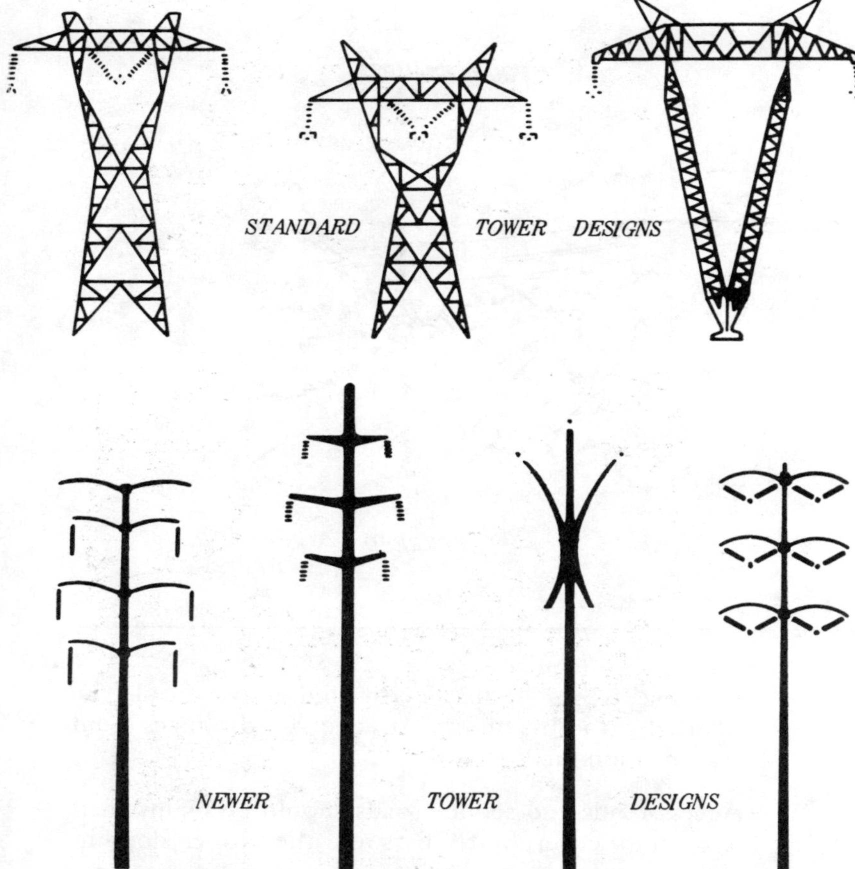

- Simple, but functional, designs of towers and poles should be used. Illustrations of these kinds of structures can be found in the book *Electric Transmission Structures,* sponsored by the Electric Research Council.

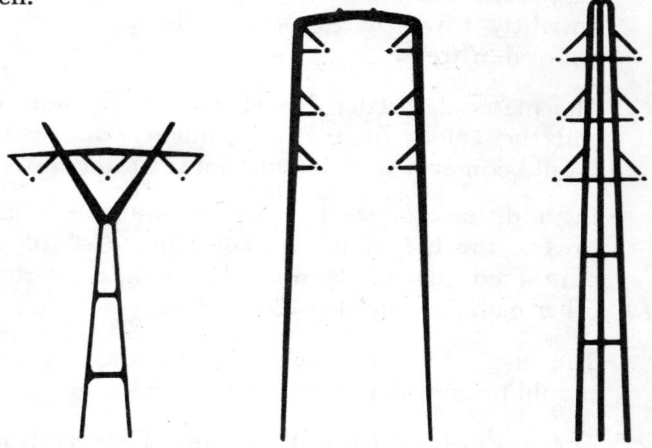

HENRY DREYFUSS & ASSOCIATES DESIGNS

- The use of poles designed without cross-arms for electric transmission lines of 138 kv and below and communications cables should be considered.

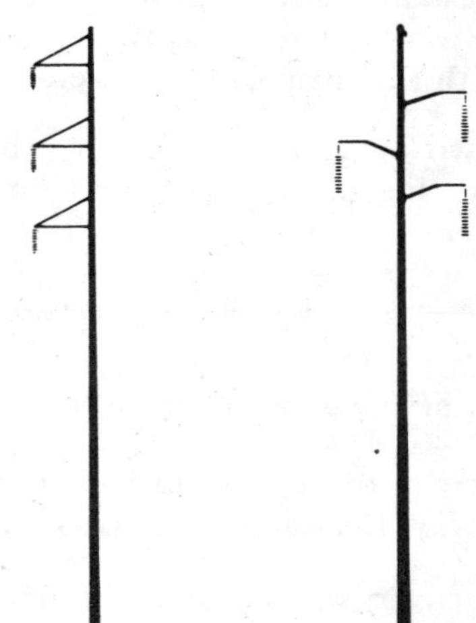

- The use of different tower designs at intervals along rights-of-way should be considered.
- Transmission towers should be spaced apart at the maximum feasible distance which would not cause unsightly tower height, conductor sag, or excessive removal of trees.
- The materials used to construct transmission towers and the colors of the components of the towers should comport with the natural surroundings.
- In addition to steel and aluminum transmission towers, the use of towers constructed of fiberglass, reinforced plastic, laminated wood, concrete, and other materials should be considered.
- The use of treated single or double wood poles should be considered in forest or timber areas.
- The use of weathered galvanized steel structures should be considered when transmission towers are to be silhouetted against the sky.
- The form of the insulator should be integrated with the form of the transmission tower. Where both a strain insulator and a post insulator are used on a single tower, the designs of each should be coordinated with the form of the tower.
- Where two or more circuits are required at highway crossings, the use of multiple circuit towers should be considered.

POOR EXAMPLE

PREFERRED

The Maintenance of Transmission Line Rights-of-Way

- Once a cover of vegetation has been established on a right-of-way, it should be properly maintained.
- A grass cover should be maintained in the areas immediately adjacent to transmission towers. Low growing trees, shrubs, herbs and grass should be planted on rights-of-way at adequate distances from transmission facilities.
- Access roads and service roads should be maintained with grass cover, water bars and the proper slope in order to prevent soil erosion.

- In scenic areas, vegetation should be treated with chemicals in late autumn or early spring in order to minimize the impact of the temporary discoloration of the foliage. Chemical growth retardants which would cause permanent discoloration of the foliage should not be used. Care should be taken to assure that chemicals used to control the growth of tree stumps do not damage other vegetation.
- When rights-of-way are inspected to check the operation of appurtenant facilities, attention should be given to locate gullies and fallen timber and to observe the condition of the vegetation. The use of aircraft to inspect and maintain rights-of-way should be encouraged.

Possible Secondary Uses of Rights-of-Way

One of the potential benefits of transmission line routes is that clearings at safe distances adjacent to transmission facilities may be used for secondary purposes. The following should be considered as possible secondary uses of rights-of-way:

Cultivation of Christmas trees, elderberry or huckleberry bushes, and other nursery stock.
Parks
Golf courses.
Equestrian or bicycle paths.
Orchards.
Picnic areas.
Bird sanctuaries.
Game refuges.
Hiking trail routes.
General agriculture.
Winter sports.

The Location of Appurtenant Aboveground Facilities

- The proposed designs and locations of electric substations, pipeline compressor stations, pumping stations, cooling stations, and other aboveground facilities, including meter and regulator stations and communication towers, should be made available to local agencies which have jurisdiction over these matters sufficiently in advance of construction deadlines to permit adequate review.
- Unobtrusive sites should be selected where possible for the location of substations and like facilities.
- Potential noise should be considered when the locations for gas turbines, substations and like facilities are being determined. Such facilities should be located in areas where sound will not be resonated.
- The size of substations and like facilities should be kept to the minimum feasible.
- The designs of the exteriors of substations and like facilities should comport with the surroundings and other buildings in the area. For example, if a substation is to be located in a residential area, its design should comport with the designs of nearby residences.
- If substations are located in residential or other scenic areas, the appurtenant transmission conductors and distribution conductors adjacent to the substations should be placed underground.[40]

Probably one of the more vital studies on transmission routing was done by the San Francisco landscape architectural firm of Eckbo, Dean, Austin & Williams, in conjunction with the Pacific Gas and Electric's study of the proposed Davenport power site. This total study was in three parts, as was mentioned previously.

The first part dealt with the land use of the surrounding area; the second part dealt with the location of the nuclear power generating plant and its appurtenances; and the third part dealt with the optimum alignment for the transmission lines from the Davenport Power Plant site on the Pacific Ocean south of San Francisco.

According to the consultants the study was outlined as follows:

40. Working Committee on Utilities, *Report to the Vice President and to the President's Council on Recreation and Natural Beauty,* (Washington, D.C., Working Committee on Utilities, 1968), pp. 5-21.

The scope of the transmission line study consists of analysis, feasibility, determination of routing, right-of-way treatment, and the design of transmission lines attendant to the proposed power plant at Davenport. The study works within the parameters of the proposed points of transmission origin and destination as established by P.G.&E.

Transmission requirements consist of two 230 kv lines from Davenport to Monta Vista, one 230 kv line from Davenport to Mount Hermon, and one 230 kv line from Mount Hermon to the Metcalf-Monta Vista line in the vicinity of Soda Spring Canyon. The Mount Hermon to Soda Spring Canyon line is needed to provide for normal electrical load growth in the Santa Cruz area. If a power plant is built at Davenport, the line can function as an integral part of the needed outlets for the generated power. It will thereby reduce the new line requirements for such a plant.

Power demand, location of origin and destination points, transmission of voltage, and structural engineering requirements for transmission towers are determined by P.G.&E. and are not within the scope of this study.[41]

The process utilized by the consultants in the study takes advantage of the capabilities of storage retrieval and graphic depiction capabilities of the computer land planning. The process is described by the consultants as follows:

As with any other development, transmission line development constitutes a modification of the landscape. The process followed, therefore, is one of identifying areas of relative sensitivity to landscape modification, the concept being to follow the paths of least sensitivity, and to develop special criteria for traversing unavoidable, highly sensitive areas.

The first step is to define a study area beyond which transmission line routes are impractical and within which the major effects of transmission line routes can be evaluated.

Within the study area, existing regional data relevant to landscape modification has been collected from various County, State and Federal agencies. The data is encoded for computer use at a scale of 1/2" = 1 mile and stored in the data bank.

The information contained on most of these maps is interpreted, sorted, and combined to produce interpretative maps relating specifically to the effects of landscape modification.[42] (See Six maps on pp. 181-186.)

A second level of interpretive mappings has been produced by grouping related elements in the first interpretive level to produce generalized maps of Physical, Economic, Visual and Socio-Cultural Constraints to Landscape Modification. The constraints implied by each grouping are then described in terms of transmission line development.

To optimize the determination of the best overall transmission corridors, a map of Suitability for Transmission Lines is established by combining and weighing all of the foregoing constraint information. Areas of greatest constraint are mapped in dark values. Test alignments which in general follow the lightest values have been drawn on clear plastic overlays and checked against all of the basic maps for correctness, detail, and special treatment criteria in sensitive areas. The highly sensitive areas requiring detail study along transmission routes are identified. Preliminary studies were made to determine probable feasibility and more detailed studies have been conducted to determine specific alignments and treatment within these areas.

Throughout the report visual impact of transmission development has been described in terms of *major*—readily apparent and conspicuous, *moderate*—readily apparent but not conspicuous, and *minor*—not readily apparent. Terms used to relate the effectiveness of concealment were: *substantially*—completely to 90% concealed and *partially*—50 to 90% concealed.[43] (See four maps on pp. 187-190.)

41. Garrett Eckbo, Francis H. Dean, Donald B. Austin, and Edward A. Williams, *Transmission Line Routing,* (San Francisco, California, Eckbo, Dean, Austin and Williams, for Pacific Gas and Electric Company, 1970), p. 1.
42. Ibid., p. 3.
43. Ibid., p. 4.

GENERAL VISIBILITY[44]

VISIBILITY

General:

A field survey was conducted to determine scenic view points, areas of visual accessibility from public thoroughfares, and visually degraded areas. The visually accessible areas and visually degraded areas were mapped and combined with computerized exposure beyond and within 4 miles of view from the scenic view points in order to produce the map of visibility.

Definitions:

Scenic view points: Significant view points within parks, view points along roads leading to parks, points along roads with a continuous display of scenic quality, view points along existing and proposed scenic highways, and accessible points from which the length and quality of view are noteworthy.

Visually accessible areas: Visible areas not categorized as scenic.

Visually degraded areas: Highly modified landscapes appearing unstable and inconsistent with their surroundings.

Based on field observations under good visibility conditions, four miles was chosen as the limit of high visual impact for transmission line corridors. Beyond four miles only long straight clear cut alignments running parallel to the line of sight would be major visual problems.

Use:

The visibility map depicts from dark to light, areas most sensitive to visual exposure. These areas should be avoided or corridors should be designed and constructed specifically to minimize visual impact.

Source:

1. Field mapping by Eckbo, Dean, Austin & Williams.

2. Exposures calculated from U.S. Geological Survey Maps, scale 1:24,000, 1955. Photo revised 1968.

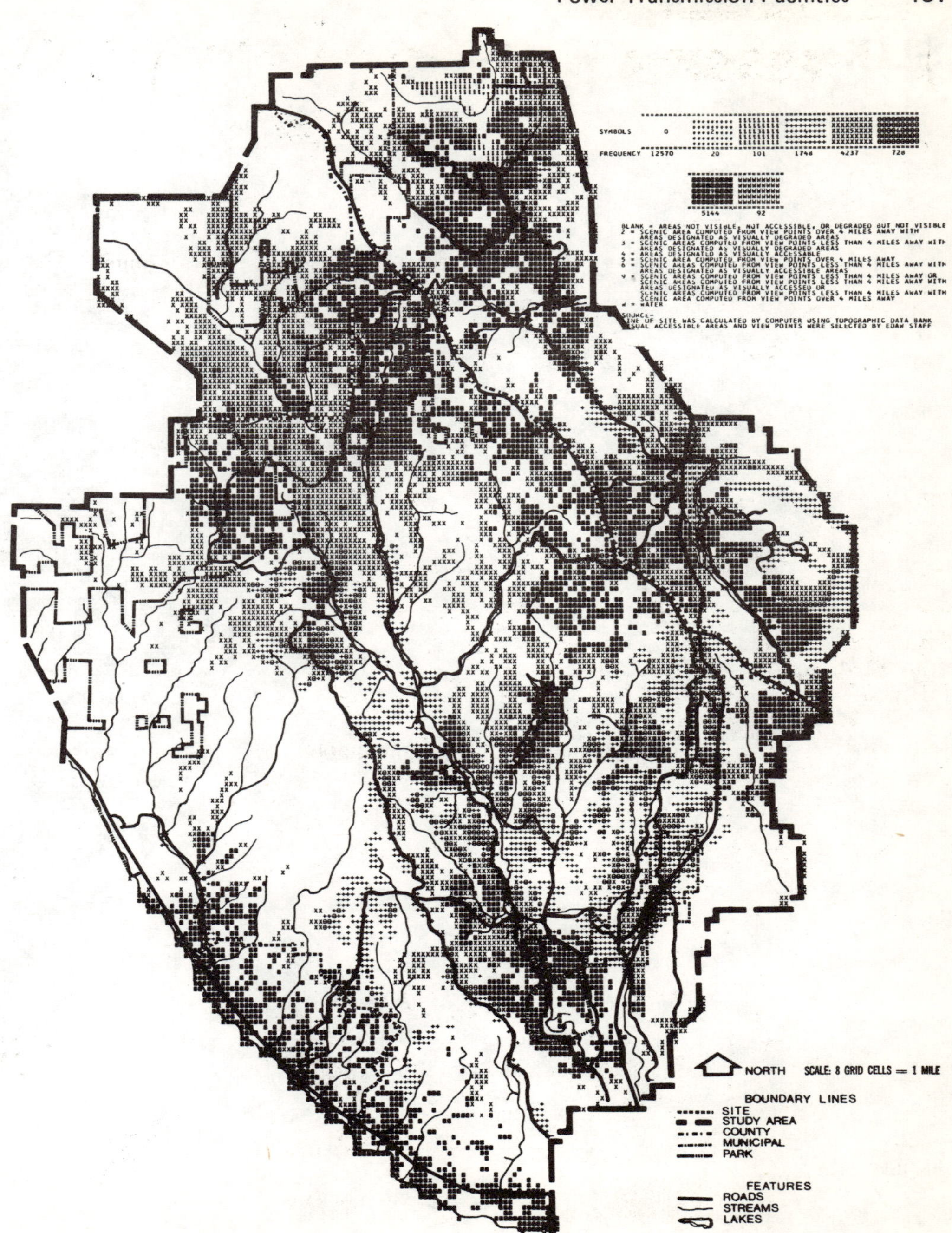

44. Ibid., p. 14.

SLOPE[45]

SLOPE

General:

The slope map depicts 4 categories of the average maximum gradient per grid cell. Slopes were calculated from U.S. Geological Survey maps.

Definitions:

0 - 10%
10'
0
100'
11 - 20%
20'
11'
0
100'
21-50%
50'
21'
0
100'
51-100%
100'
51'
0
100'

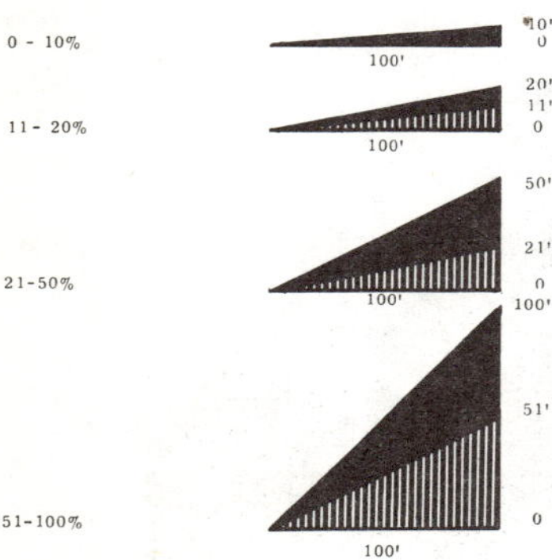

Use:

Generally the greater the slope the greater the difficulty in access road construction. Large cuts and fills result and become scars in the landscape. All slopes over 50% should be avoided as access road sites and slopes over 20% in discontinuous canopy or unstable areas should be avoided or designed and treated specifically to overcome the scarring problem. Slopes under 20% require less detailed attention unless they occur in highly unstable areas.
The slope map was combined with physiography and slope instability maps to produce a map of physical constraints to landscape modification.

Source:

U.S. Geological Survey Maps, scale 1:24,000, dated 1955, photo revised 1968.

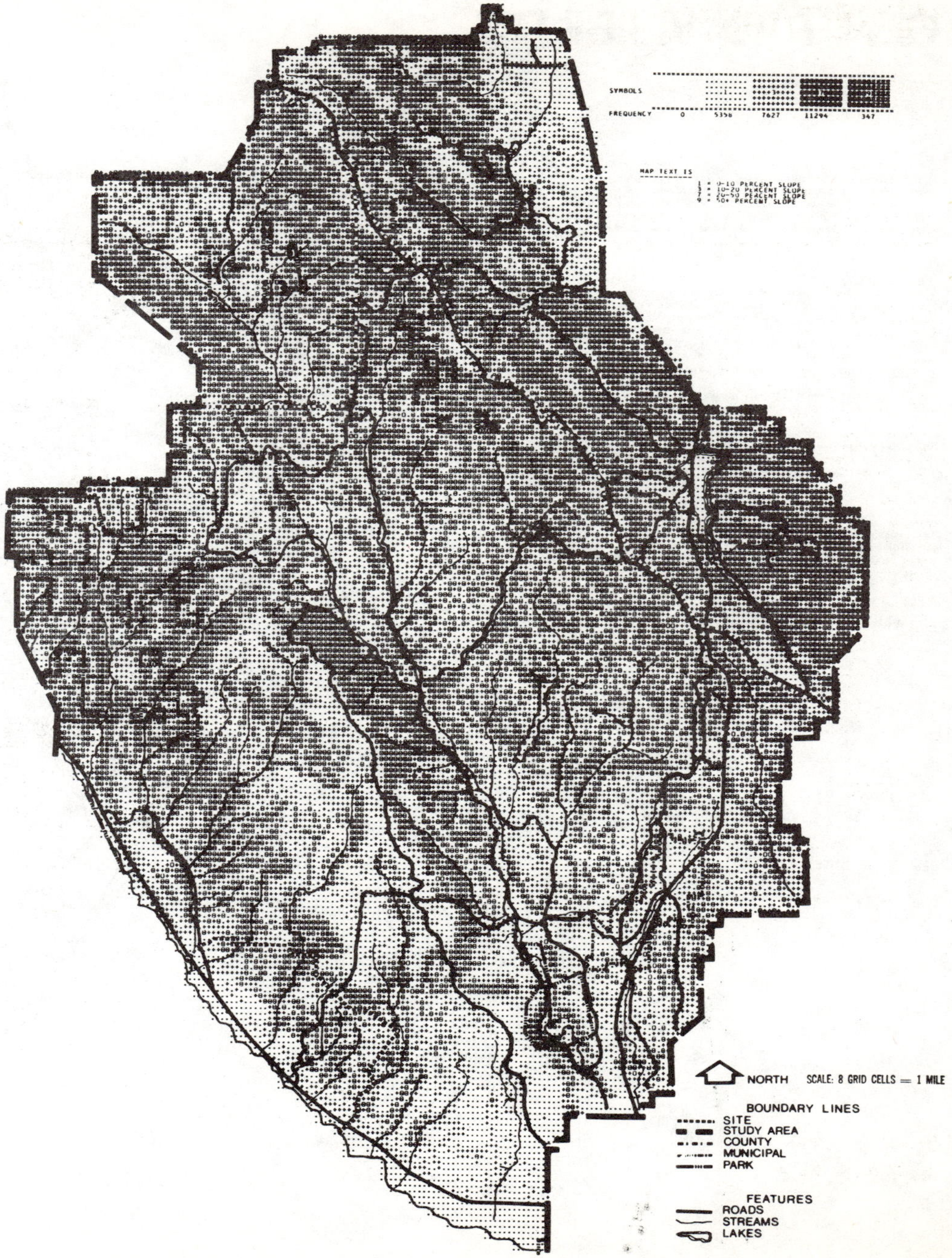

SYMBOLS
FREQUENCY 0 5356 7627 11294 347
MAP TEXT IS
1 = 0-10 PERCENT SLOPE
3 = 10-20 PERCENT SLOPE
7 = 20-50 PERCENT SLOPE
9 = 50+ PERCENT SLOPE
NORTH
SCALE: 8 GRID CELLS = 1 MILE
BOUNDARY LINES
SITE
STUDY AREA
COUNTY
MUNICIPAL
PARK
FEATURES
ROADS
STREAMS
LAKES

45. Ibid., p. 26.

VEGETATION [46]

VEGETATION

General:

The natural plant communities occurring within the study area are coastal scrub, pine forest, redwood forest, Douglas fir forest, mixed evergreen forest, chaparral, grassland, and riparian.

In order to clarify the canopy conditions these communities were grouped into the following categories: High canopy (timber trees); high-moderately high canopy (minor conifers); moderately high canopy (hardwoods); thin, low or no canopy (thin woodland, discontinuous canopy).

Definitions:

High canopy areas (redwood and Douglas fir forest), principal species are coast redwood and Douglas fir with many species of the mixed evergreen forest.

High to moderately-high canopy areas - (pine forests) - principal species are knobcone pine, Monterey pine, ponderosa pine.

Moderately-high canopy areas - (mixed evergreen forest and riparian). In the mixed evergreen forest principal species are tan oak, madrone, California laurel, bigleaf maple, black oak, blue oak, California live oak, highland oak, and valley oak. In the riparian areas typical species are white alder, red alder, California box elder, arroyo willow (Salix lasiandra), Coulter's willow (Salix coulteri), black cottonwood, sycamore, Oregon ash, brown dogwood (Cornus glabrato) and Douglas baccharis (Baccharis douglasii).

Thin, low and no canopy areas - (chaparral, coastal scrub, grassland). The principal species in the chaparral are chamise, eastwood manzanita, California scrub oak, scrub canyon live oak, birchleaf mountain mahogany, Christmas berry, California buckthorn, California huckleberry, poison oak, kidneywort (Baccharis pilularis var. consanguinea). The principal species in the coastal scrub are coyote brush (Baccharis pilularis var. consanguinea), sticky monkey flower, California blackberry, poison oak, California sagebrush.

In grassland the most dominant species are little quaking grass (Briza minor), California brome (Bromus carinatus), soft chess (Bromus mollis), ripgut grass (Bromus rigidus), San Francisco blue-grass (Poa unilateralis), pine blue-grass (Poa scabrella), Western melica (Melica californica), western six-weeks fescue (Festuca megalura), rattail fescue (Festuca myuros), six-weeks fescue (Festuca dentonensis) and red fescue (Festuca rubra). Due to heavy grazing the existing grassland is a highly unstable community. Most of the grasses are now annuals and will usually succeed to Baccharis if grazing or management are removed.

Use:

The basic vegetation information was interpreted and combined with other data to produce maps of timber tree area characteristics, unique vegetation forms, discontinuous canopy characteristics, timber productivity, and soil-vegetation contrast.

Source:

1. California Division of Forestry, Timber Stand-Vegetation Cover Maps, 15 Minute Ben Lomond Series, 1950.
2. The Santa Cruz Mountains Regional Pilot Study Early Warning System. T. Patri, David Streatfield and Thomas J. Ingmire, 1970.

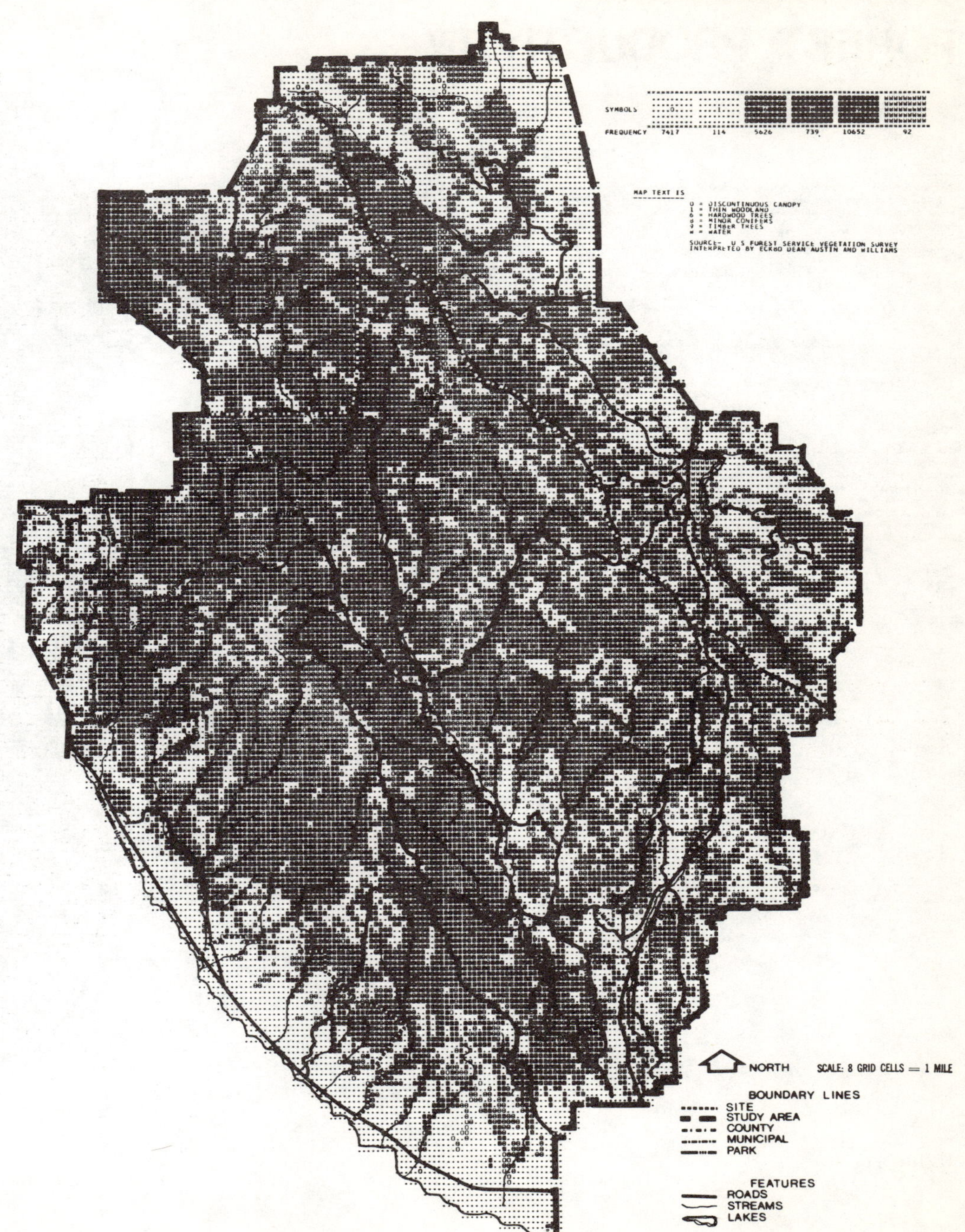

46. Ibid., p. 28.

FOREST PRODUCTIVITY[47]

FOREST PRODUCTIVITY

General:

The forest productivity map was interpreted from the Forest Service forest conditions map. The forest productivity map depicts those areas of greatest and least potential for commercial conifer production.

Definitions:

Site index 7 or greater - Estimated heights over 225 feet at 300 years of age.

Site index 6 - Estimated heights between 200 and 225 feet at 300 years of age.

Site index 5 - Estimated heights between 175 and 200 feet at 300 years of age.

Site index 4 or less - Estimated heights less than 175 feet at 300 years of age.

Use:

The higher the site index the greater is the potential for timber production. Areas where the site index is 7 or greater are the most valuable timber production areas and should be avoided where the terrain requires clearing of the right of way. Areas where the site index is 5 or 6 are valuable timber production areas, rights of way and access roads should be closely coordinated with timber production operations to insure minimum adverse impact on sustained yield of the forest and maximum benefits from access road construction.

Source:

U.S. Department of Agriculture, Forest Service, Forest Conditions Maps, 1944, Scale 1:62,500.

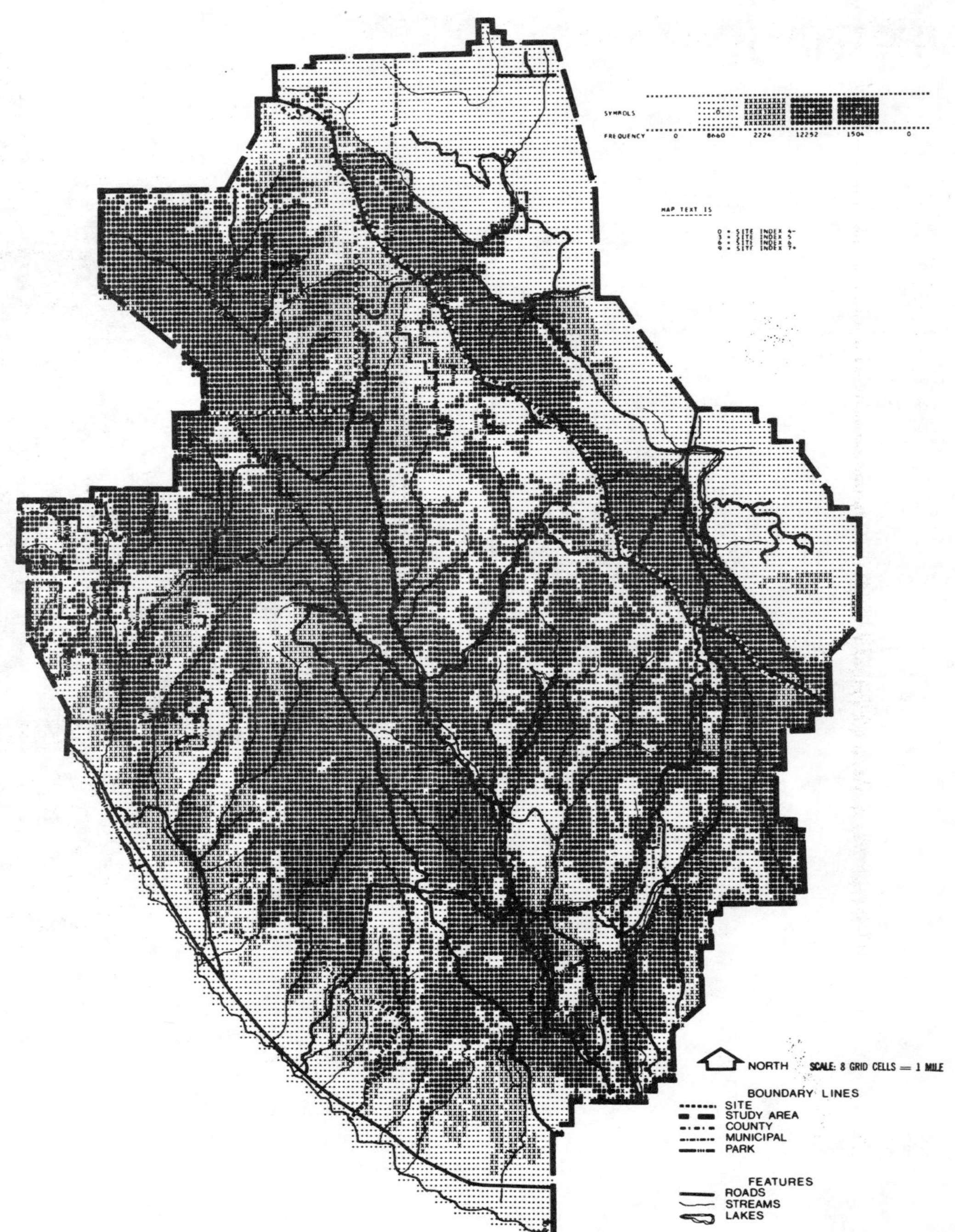

47. Ibid., p. 30.

DISCONTINUOUS CANOPY[48]

DISCONTINUOUS CANOPY

General:

The discontinuous canopy map is an interpretation from the basic vegetation data provided by the California Division of Forestry. Areas of low vegetation, or where the ground surface is visible from a distance were considered to be areas of discontinuous canopy. Depicted on the map are the specific elements comprising the discontinuous canopy.

Definitions:

Scrub - Tall,heavily branched shrubs, such as the manzanitas, scrub oaks, and chamise, and other low slenderly branched shrubs, such as the sagebrushes.

Grassland - Grasses, bushy herbaceous plants and marshes.

Bare Ground and Rock - Bare or litter covered soil devoid of vegetation and rugged areas devoid of soil such as talus slopes or cliffs.

Cultivated - Land under cultivation, irrigated pastures, and fallow land.

Urbanized - Residential, urban, and industrial areas.

Thin Woodland - Minor conifer areas with canopies covering less than 50% of the understory.

Use:

The elements of the discontinuous canopy were combined with soil color to produce a map of soil vegetation contrast.

Source:

California Division of Forestry, Timber Stand - Vegetation Cover Map. 15 minute, Ben Lomond Series, 1950.

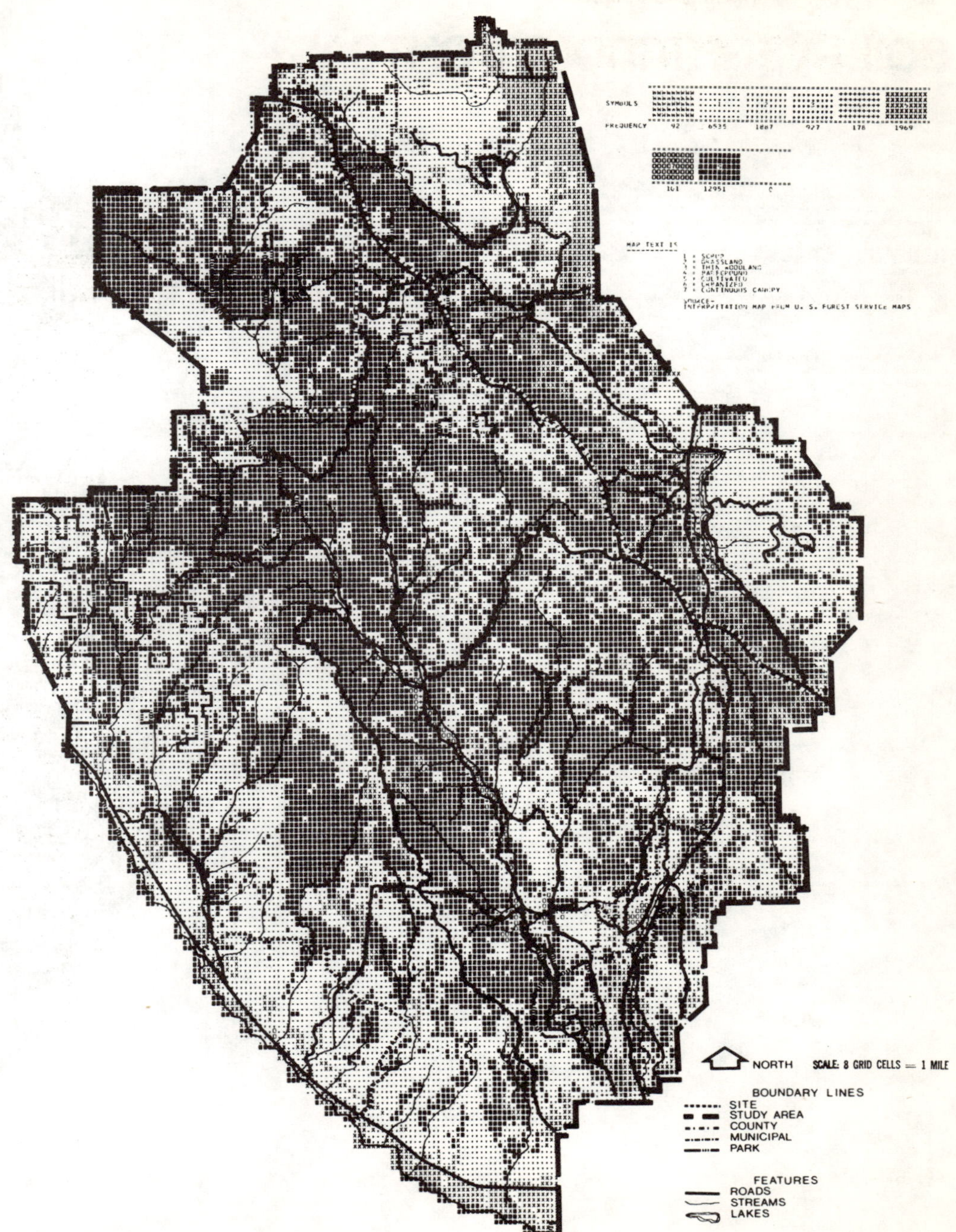

48. Ibid., p. 32.

SOIL-VEGETATION CONTRAST[49]

SOIL-VEGETATION CONTRAST

General:

The soil color map was combined with the discontinuous canopy map to produce the soil-vegetation contrast map. A light soil in scrub was considered to be a light-dark contrast. Colorful soils in scrub, grassland or thin woodland were considered to be a color contrast. Soils with both light and colorful properties in scrub, grassland, or thin woodland were considered to be a light-dark and color contrast. Neutral soils in grassland were considered to be a dark-light contrast during the summer months. Other areas of discontinuous canopy were considered to have little contrast. In areas of continuous canopy no contrasts were recognized.

Use:

The soil-vegetation contrast map indicates those areas where access road construction should be avoided or designed and treated specifically to minimize visual impact.

Source:

1. Soil color map.

2. Discontinuous canopy map.

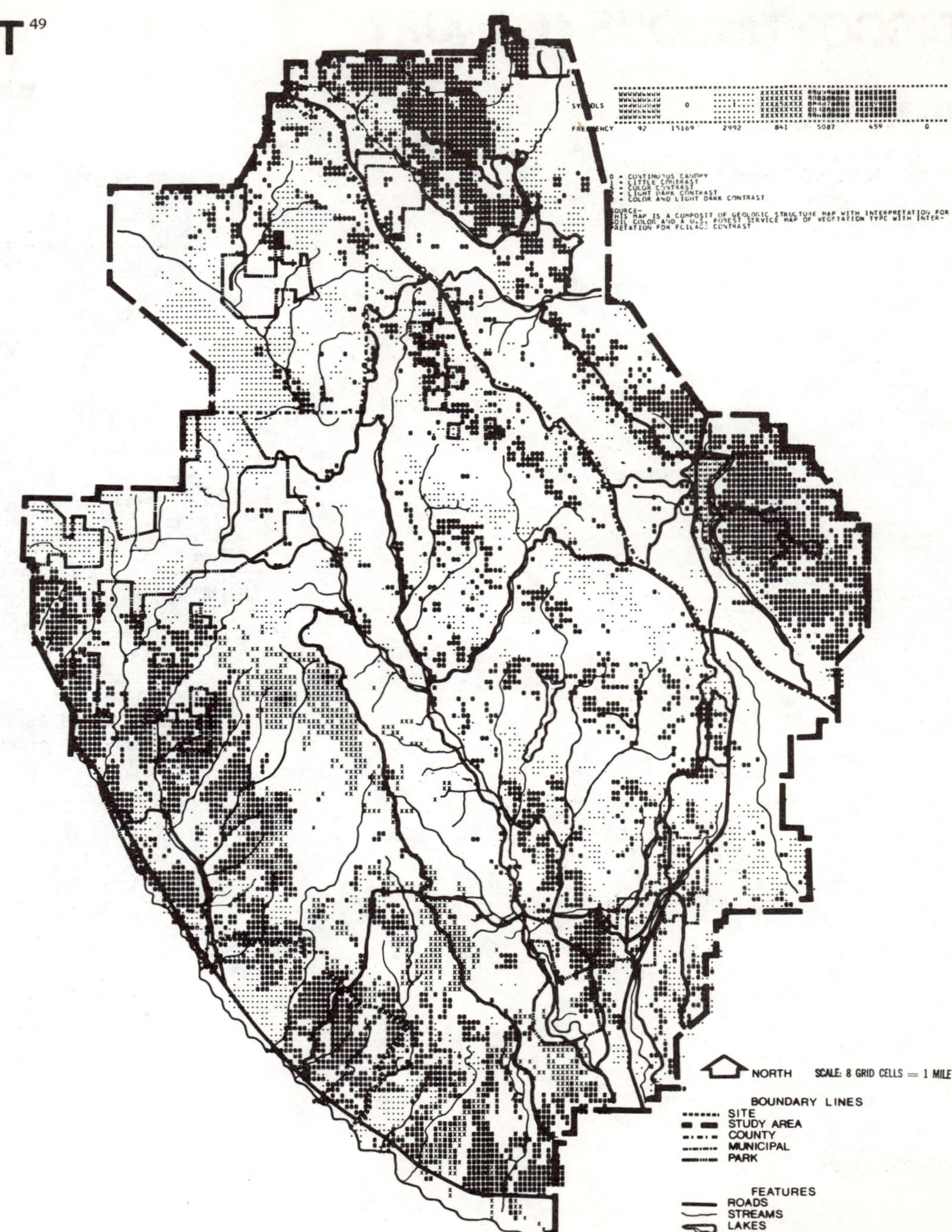

49. Ibid., p. 33.

The following maps show some of the constraints to landscape modification:

PHYSICAL CONSTRAINTS TO LANDSCAPE MODIFICATION[50]

PHYSICAL CONSTRAINTS TO LANDSCAPE MODIFICATION

General:

The map of physical restrictions to landscape modification was derived from the slope, slope instability and physiography maps. The map indicates zones of the degree of physical restrictions to landscape modification.

Definitions:

Extensive - Slopes over 50% or landslide areas, or course lines.

Major - Slopes of 21%-50% or areas of high risk of slope instability.

Moderate - Slopes of 11%-20% or areas of moderate risk of slope instability.

Minor - Slopes under 10% and areas of low risk of slope instability.

Use:

The physical restrictions to landscape modification map when related to transmission line development implies the following actions for each zone of constraints.

Extensive - The necessity for extensive grading operations, the very high risk of further landsliding, or the high risk of stream siltation within this zone makes it a consideration for transmission line development only when no other opportunities exist and when detailed designs assure that the problems related to the zone can be overcome.

Major - Cuts and fills should be engineered and planted to insure slope stability, fills should be compacted, and access roads should be designed to minimize the concentration of runoff.

Moderate - Access roads should be designed to minimize the concentration of runoff.

Minor - No special actions.

Source:

1. Slope Map.
2. Slope Instability Map.
3. Physiography Map.

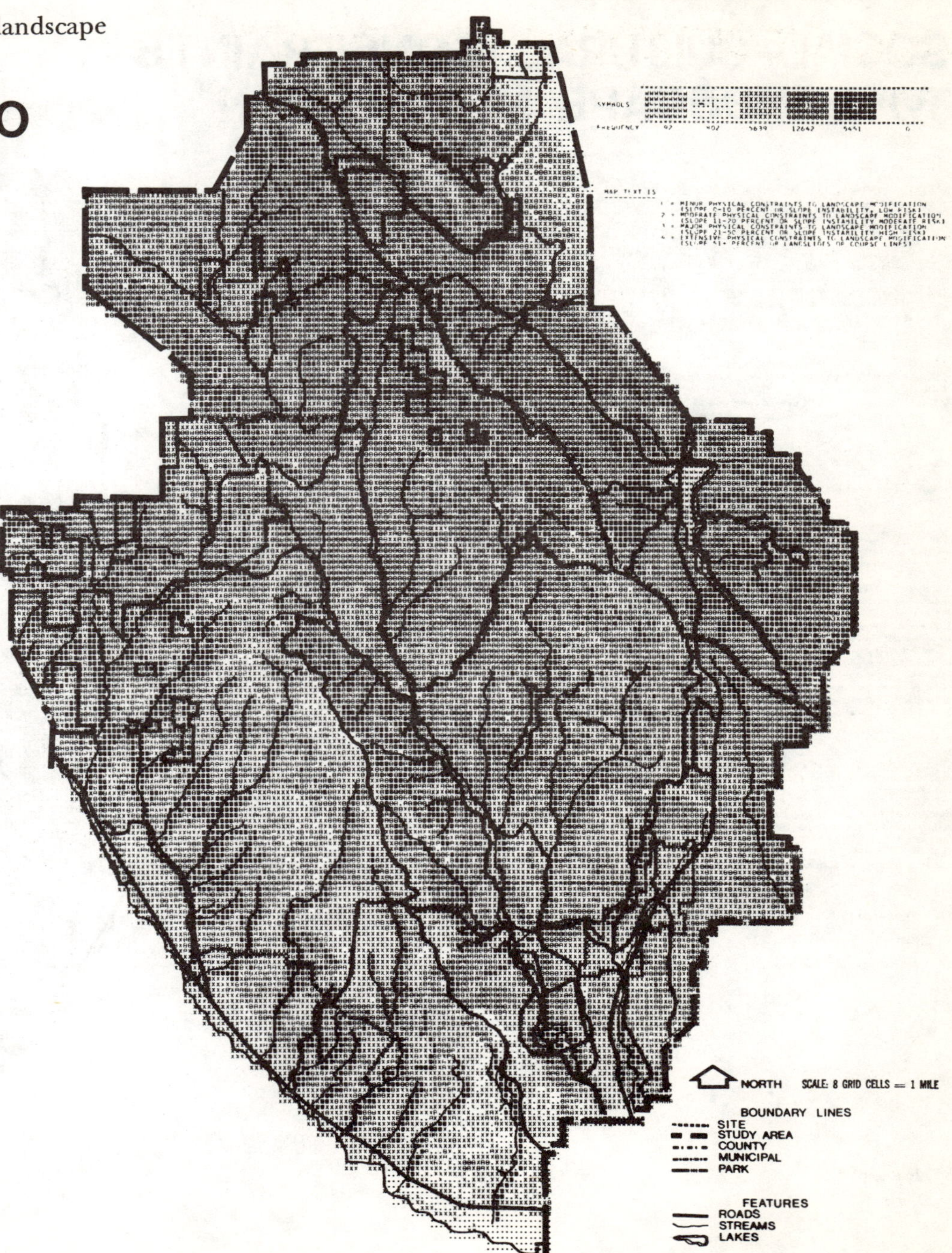

50. Ibid., p. 10.

SOCIAL-CULTURAL CONSTRAINTS TO LANDSCAPE MODIFICATION[51]

SOCIAL-CULTURAL CONSTRAINTS TO LANDSCAPE MODIFICATION

General:

The map of social-cultural constraints to landscape modification was derived from the existing and anticipated development and recreation map and the maps of historical and archaeological sites. The constraints map depicts the areas where the greatest and least potential social and cultural disruptions would occur if the existing landscape were to be changed.

Definitions:

Extensive - Built-up areas of densities greater than one dwelling unit per acre, historical or archaeological sites.

Major - Built-up areas of densities of one dwelling unit per 2-5 acres or anticipated built-up areas of greater than one dwelling unit per acre.

Moderate - Anticipated built-up areas of densities of one dwelling unit per 2-5 acres.

Minor - Areas not included in the above.

Use:

The social-cultural constraints to landscape modification map when related to transmission line development implies the following actions for each zone of constraints.

Extensive - The necessity to relocate individuals, businesses, industries or institutions within this zone makes it a consideration for transmission line development only when no other opportunities exist. When there are no other opportunities, development in this zone should follow the path of least social disruption and plans for making the right of way useful to the community should be implemented. Historical or archaeological sites should be avoided. When the sites are difficult to avoid, special detailed designs should be developed to insure that excavation, the potential for improved public access, or the appearance of towers, rights of way or access roads do not degrade the value of the sites.

Major - This zone should be avoided. When better opportunities are unavailable rights of way and access roads should be designed to minimize interference with the existing or anticipated use of the landscape.

Moderate - Rights of way and access roads should be developed to minimize conflicts and provide opportunities for the anticipated land use.

Minor - No special actions.

Source:

1. Existing and Anticipated Development and Recreation Map.
2. Historical Sites Map.
3. Archaeological Sites Map.

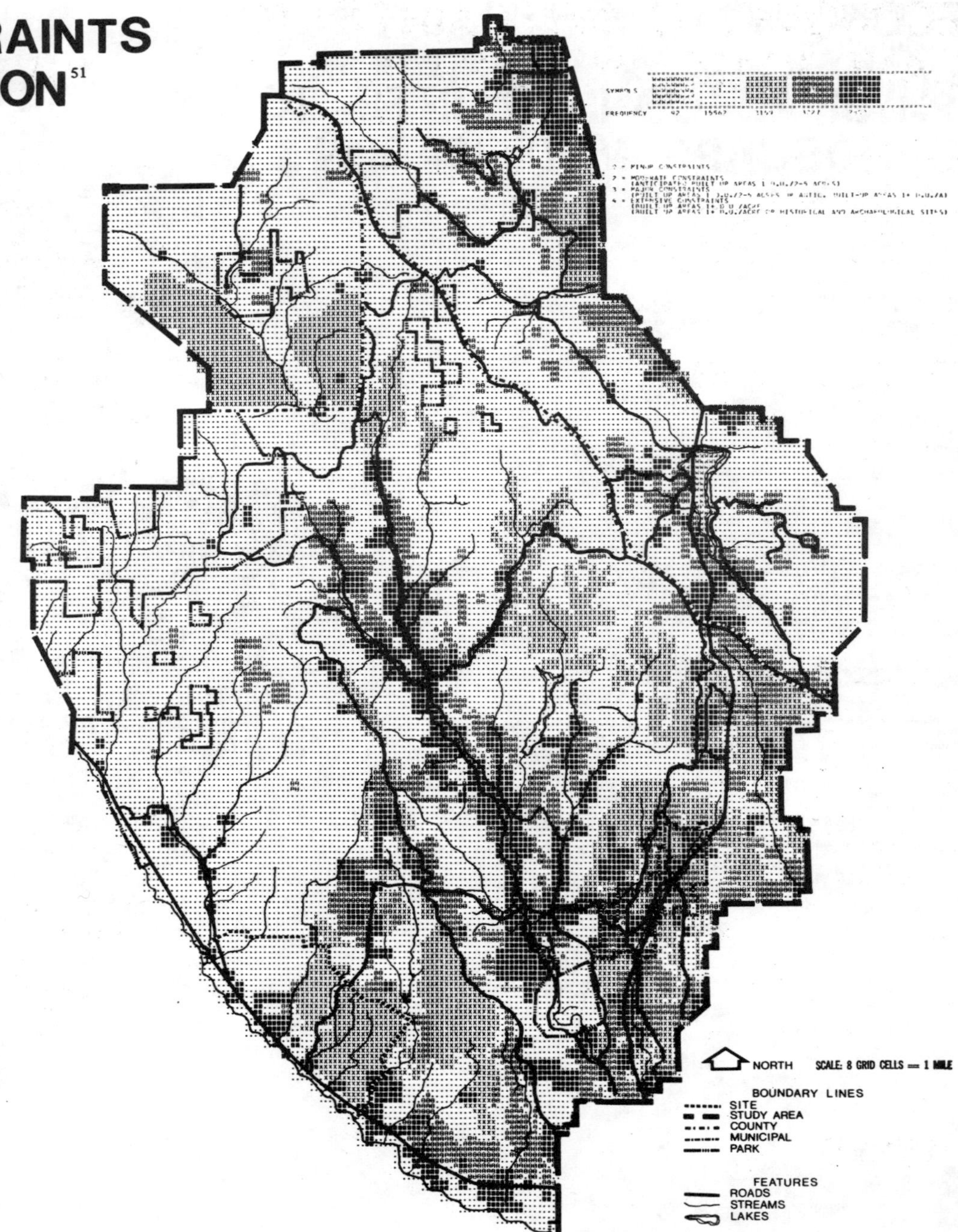

51. Ibid., p. 12.

ECONOMIC CONSTRAINTS TO LANDSCAPE MODIFICATION[52]

ECONOMIC CONSTRAINTS TO LANDSCAPE MODIFICATION

General:

The map of economic constraints to landscape modification was derived from the existing and anticipated development and recreation map, the land capability for cultivation map, the county general plans and the timber productivity map. The map indicates land value zones in which a change of use could cause a general economic loss to the community.

Definitions:

Extensive - Existing built-up areas of densities greater than one dwelling unit per acre.

Major - Existing built-up areas of densities of one dwelling unit per 2-5 acres, or anticipated built-up areas of densities greater than one dwelling unit per acre, or existing and potential reservoir sites, or quarry sites, or areas where the land capability for cultivation is high, areas where the timber site index is 7 or greater.

Moderate - Anticipated built-up areas of densities of one dwelling unit per 2-5 acres, or areas where the land capability for cultivation is moderate, or areas where the timber site index is 6.

Minor - Areas where the land capability for cultivation is low, or where the timber site index is 5.

Very Minor - Areas where the land capability for cultivation is very low, and areas where the timber site index is 4 or less.

Use:

The economic constraints to landscape modification map when related to transmission line development implies the following actions for each zone of constraints.

Extensive - The necessity to remove buildings and structures and the inhibition of building construction in this zone by transmission line development constitutes the highest relative economic loss to the community. This zone should be considered for transmission line development only when no other opportunities exist. Should it become necessary to develop within this zone, the path of least economic loss to the community should be followed, and plans for offsetting the loss by providing public amenities should be implemented.

Major - This zone should be avoided. When better alternatives are unavailable, rights of way and access roads should be closely coordinated with existing and anticipated land uses in order that maximum economic benefits and minimum economic penalties for adjacent land uses can be derived. When the agricultural value of the land is the only determinant of the zone the actions of the moderate constraints zone should be applied.

Moderate - Rights of way and access roads should be coordinated with existing and potential land use plans.

Minor and Very Minor - No special actions.

Source:

1. Existing and Anticipated Development and Recreation Map.
2. Land Capability for Cultivation Map.
3. County General Plans.
4. Timber Productivity Map.

SYMBOLS

FREQUENCY 92 4418 2321 11102 4335 2372 0

MAP TEXT IS

1 = VERY MINOR
(LAND CAPABILITY - VII & VIII OR TIMBER SITE INDEX 4-)
2 = MINOR CONSTRAINTS
(LAND CAPABILITY - VI OR TIMBER SITE INDEX 5)
3 = MODERATE CONSTRAINTS
(LAND CAPABILITY - III & IV OR TIMBER SITE INDEX 6 OR ANTICIPATED BUILT UP AREAS 1 D.U./2-5 ACRES)
4 = MAJOR CONSTRAINTS
(LAND CAPABILITY - I & II OR TIMBER SITE INDEX 7+ OR BUILT UP AREAS 1 D.U./2-5 ACRES OR ANTICIPATED BUILT UP AREAS 1+ D.U./ACRE OR QUARRIES OR RESERVOIR SITES)
5 = EXTENSIVE CONSTRAINTS
(BUILT UP AREAS 1+ D.U./ACRE)

NORTH SCALE: 8 GRID CELLS = 1 MILE

BOUNDARY LINES
SITE
STUDY AREA
COUNTY
MUNICIPAL
PARK

FEATURES
ROADS
STREAMS
LAKES

52. Ibid., p. 13.

VISUAL CONSTRAINTS TO LANDSCAPE MODIFICATION[53]

VISUAL CONSTRAINTS TO LANDSCAPE MODIFICATION

General:

The map of visual constraints to landscape modification was derived from the maps of existing and anticipated development, recreation, visibility, slope instability, timber trees, discontinuous canopy and physiography. The constraints map depicts the areas where visual quality is of greatest and least importance to the general public.

Definitions:

Extensive - Scenic areas within 4 miles from the viewpoint,or existing and proposed parks, orexisting and proposed scenic highways, or existing and potential reservoir sites, or built-up areas of densities greater than one dwelling unit per acre.

Major - Private recreation areas or potential recreation areas,or built-up areas of densities of one dwelling unit per 2-5 acres,or anticipated built-up areas of densities greater than one dwelling unit per acre.

Moderate - Scenic areas beyond 4 miles from the viewpoint,or visually accessible areas,or anticipated built-up areas of densities of one dwelling unit per 2-5 acres,or soil vegetation contrast areas,or areas of a high risk of slope instability in discontinuous canopy,or ridges,or course lines, or slopes over 50% in areas of discontinuous canopy, or timber tree areas.

Minor - Visually degraded and other areas not included in the above.

Use:

The visual constraints to landscape modification map when related to transmission line development implies the following restrictions for each zone of constraints.

Extensive - The visual quality of this zone is an irreplaceable resource of great importance to the public and must be preserved. From significant points of viewing,transmission line development within this zone should not appear to have cleared rights of way, continuous alignments of over 3 towers, or ground modification incompatible with the surrounding landscape. Towers and conductors should be designed and treated to blend with or enhance the existing landscape. There should be no highly visible towers within 1000 feet of significant view or activity points.

Major - The visual quality of this zone is of special importance to many people. From significant viewpoints, transmission line development should not appear to have cleared rights of way.

Moderate - The visual quality of this zone is of general importance and/or highly susceptible to visual degradation. Within scenic areas beyond 4 miles from the viewpoint, transmission line development should not appear to have continuous cleared rights of way for more than 2 tower spans. In the areas of soil-vegetation contrast, steep slopes and high risk of slope instability in discontinuous canopy, the impact of access road construction should be minimized by special treatments of cuts, fills, and road beds. In timber tree areas,cleared right of way should be designed not to appear continuous. Within visually accessible areas and anticipated built-up areas of densities of one dwelling unit per 2-5 acres,towers should be screened from view. Background should be used to prevent silhouetting of towers on ridges. Course lines should be spanned to prevent clearing of vegetation.

Minor - No special actions.

Source:

1. Existing and Anticipated Development and Recreation Map.
2. Visibility Map.
3. Slope Instability Map.
4. Timber Tree Map.
5. Discontinuous Canopy Map.
6. Physiography Map.

53. Ibid., p. 11.

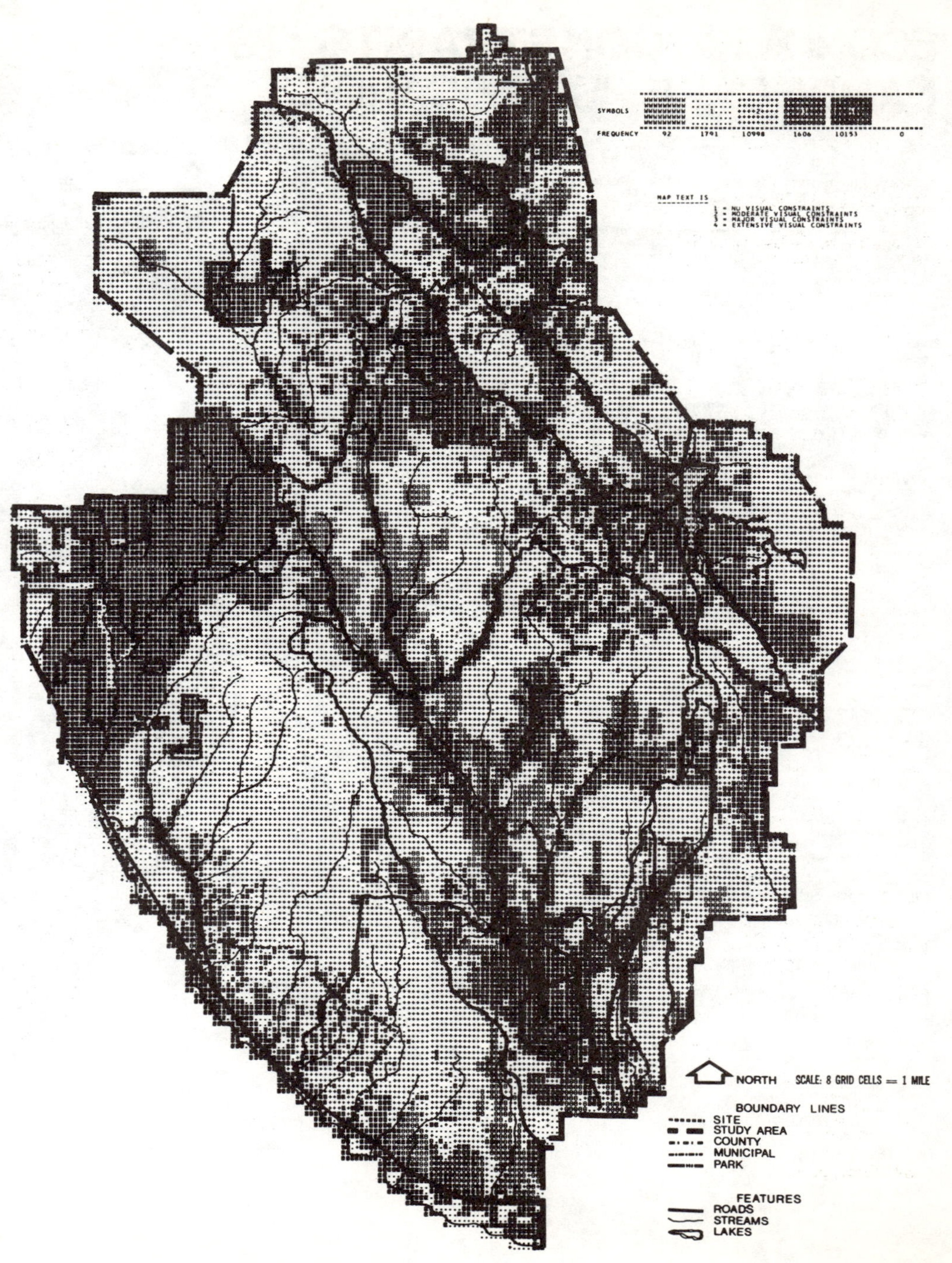

As a result of the analysis conducted through the review of the information which had been mapped by the computer, one composite map indicating the suitability for transmission lines was developed and is shown on the following illustration:

SUITABILITY FOR TRANSMISSION LINES[54]

SUITABILITY FOR TRANSMISSION LINES MAP

General:

The constraints from the foregoing maps were weighted, combined and accumulated to produce the map of suitability for transmission lines. The suitability map depicts the relative environmental resistance of areas for transmission line development.

Definitions:

Weighting	Constraints
9	Existing and proposed parks.
9	Existing and proposed scenic highways.
9	Existing, future, and potential reservoirs.
9	Built-up areas 1 + D. U. /acre.
9	Public institutions.
5	Potential recreation areas.
5	Private recreation areas.
5	Built-up areas 1 D. U. /2-5 acres.
5*	Scenic areas within 4 miles of viewpoint.
4	Anticipated built-up areas 1 D. U. /acre.
3*	Scenic areas beyond 4 miles from viewpoint.
2	Anticipated built-up areas 1 D. U. /2-5 acres.
2*	Visually accessible areas.
2	Timber tree areas.
1	Course lines and low lands.
1	Ridge lines and uplands.
1	Soil vegetation contrast areas.
1	Areas of high risk of slope instability.
1	Slopes of greater than 50% in discontinuous canopy.
*/2	Visually degraded areas.

Accumulated Weighting	Suitability	Constraints
0	Very high	None recognized.
1-2	High	Physical.
3-4	Moderate	Major physical.
5-8	Low	Visual or social, or physical and visual, or physical and social.
9+	Very low	Extensive social or visual, or high occurrence of others.

Use:

The suitability map together with field inspections was used to evaluate and determine the best transmission line routes. As far as practical, the transmission corridors should follow the path of least environmental resistance. Areas of high environmental resistance or low suitability are environmentally very sensitive and should be avoided or given special treatment as described on the constraints to landscape modification maps.

Source:

1. Existing and Anticipated Development and Recreation Map.
2. Counties' General Plans.
3. Visibility Map.
4. Physiography Map.
5. Vegetation Map.
6. Soil-Vegetation Contrast Map.
7. Slope Instability Map.
8. Slope Map.

54. Ibid., p. 9.

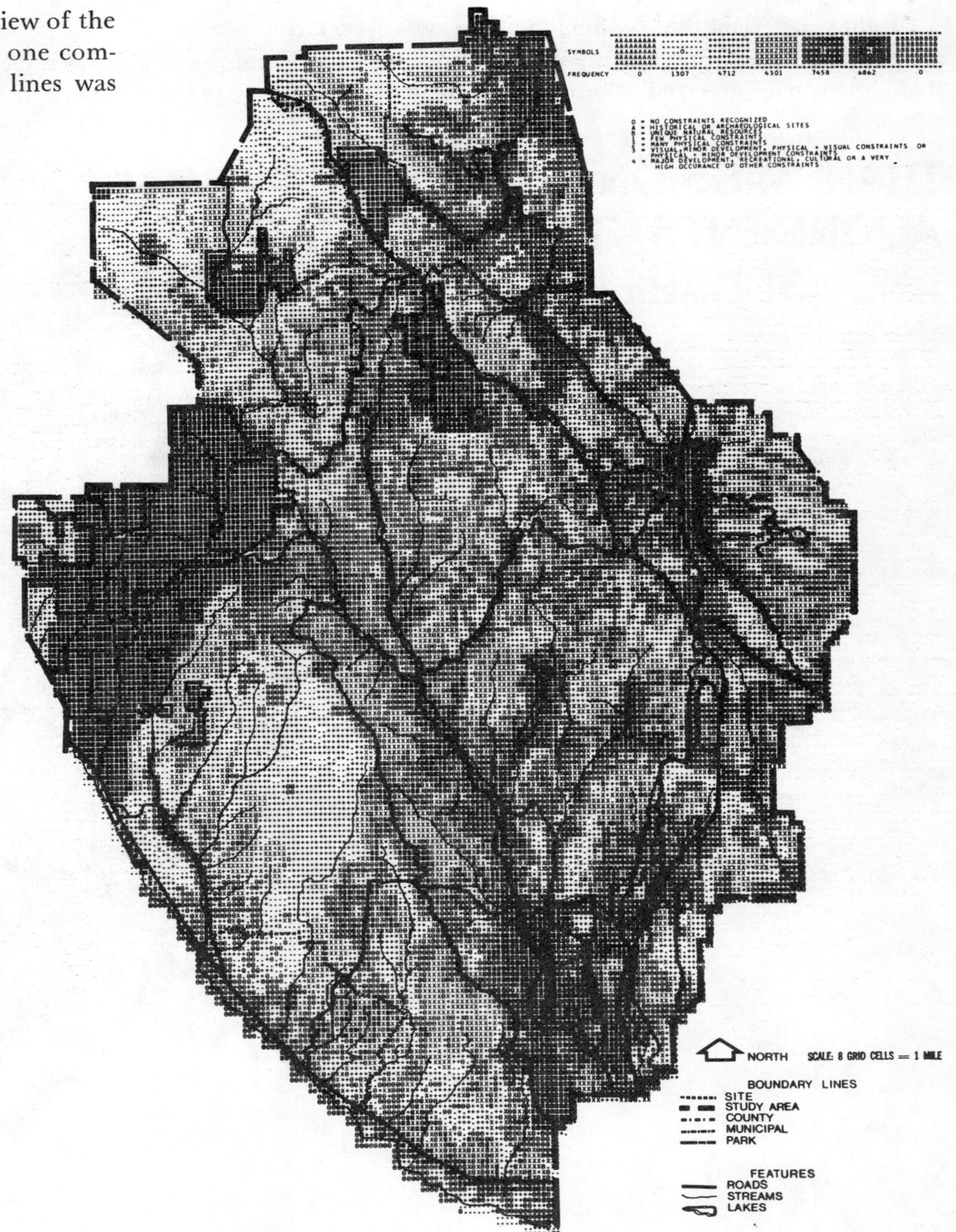

From that, another computer map was produced which showed transmission alignments studied and suitability. From this, then, the final preferred transmission alignments map was prepared.

TRANSMISSION ALIGNMENTS STUDIED AND SUITABILITY[55]

PG&E/D/T

DAVENPORT TO MONTA VISTA

C	=	Central Alignment
NW	=	Northwest Variation
NE	=	Northeast Variation
SW	=	Southwest Variation
SE	=	Southeast Variation
E	=	Eastern Alignment

DAVENPORT TO MT. HERMON

N	=	Northern Alignment
S	=	Southern Alignment

MT. HERMON TO SODA SPRING CANYON

W	=	Western Alignment
E	=	Eastern Alignment

FACILITIES

PP	=	Power Plant
MVS	=	Monta Vista Substation
MHS	=	Mt. Hermon Substation
LS	=	Lexington Substation
M-MV	=	Metcalf-Monta Vista Tap

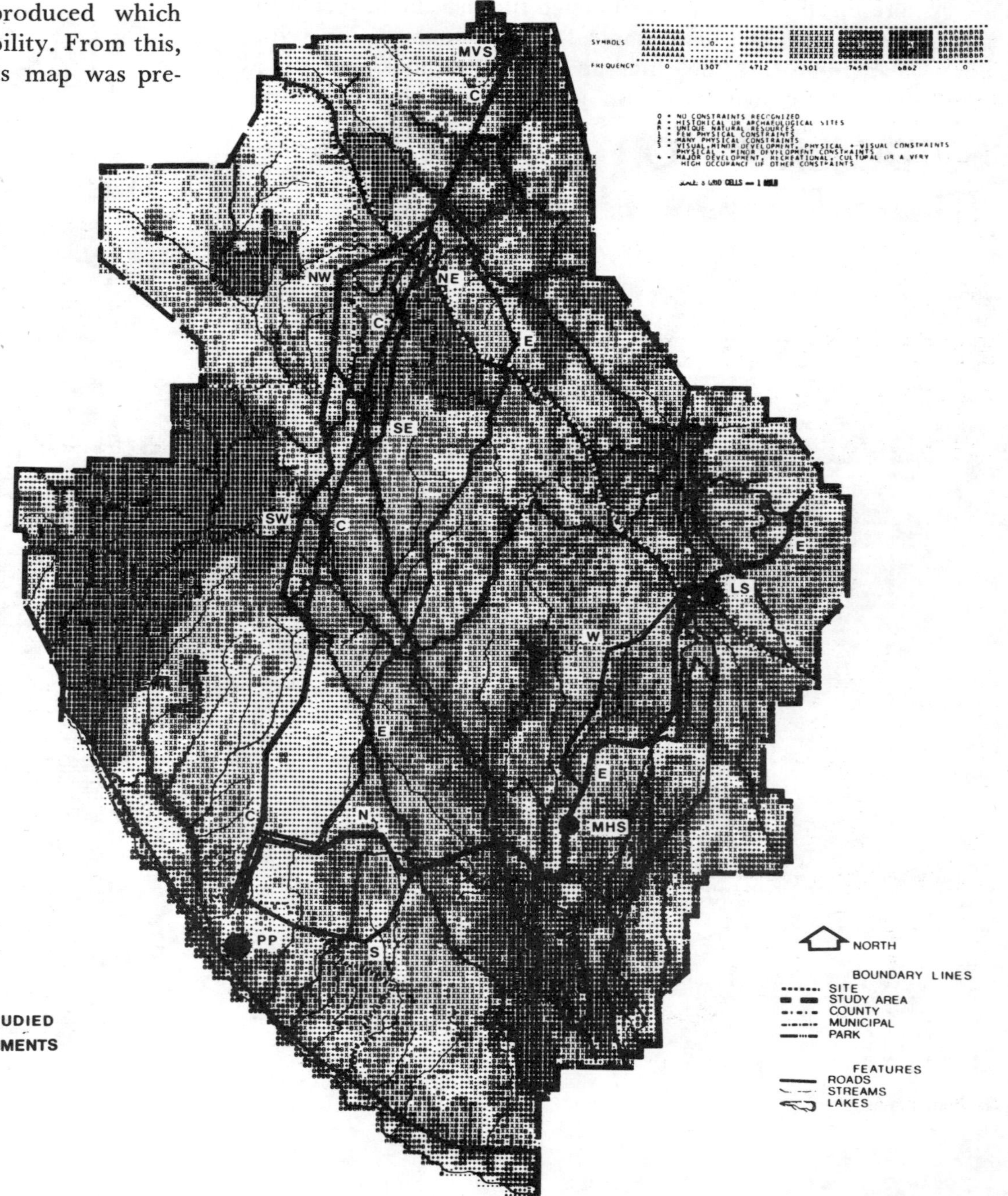

55. Ibid., p. 5.

Detailed alignment studies were developed and placed on finite contour maps. Each of the critical areas were then studied in greater detail, and a series of diagramatic sketches and eye-level perspectives were developed to indicate the treatment, effect and view at these various places on the overall site. Some of these illustrations are shown as follows:

EFFECTS of POSSIBLE FUTURE SKYLINE BLVD. ALIGNMENT at SUMMIT RD.[56]

A. DRIVING EAST

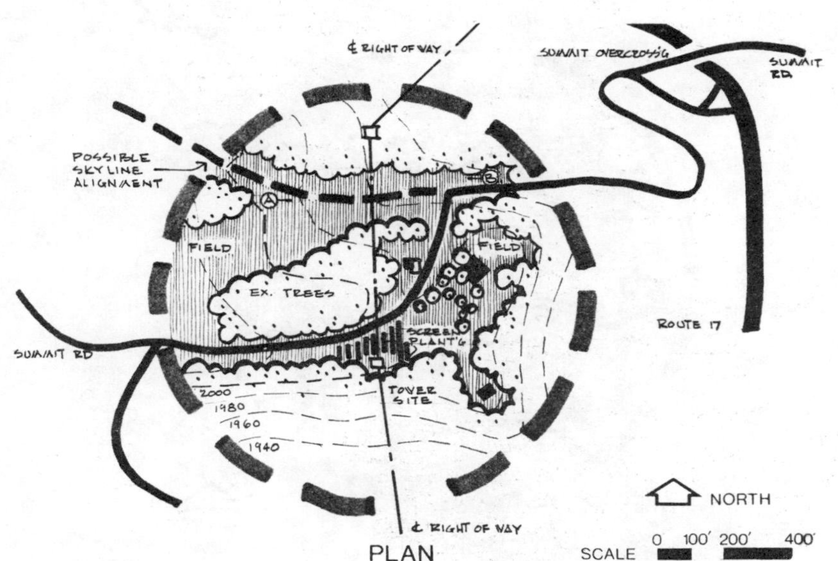

PLAN

B. DRIVING WEST

56. Ibid., p. 66.

TRANSMISSION SCHEME at SUMMIT RD. OVERCROSSING[57]

B. DRIVING SOUTH

C. DRIVING NORTH

A. DRIVING SOUTH

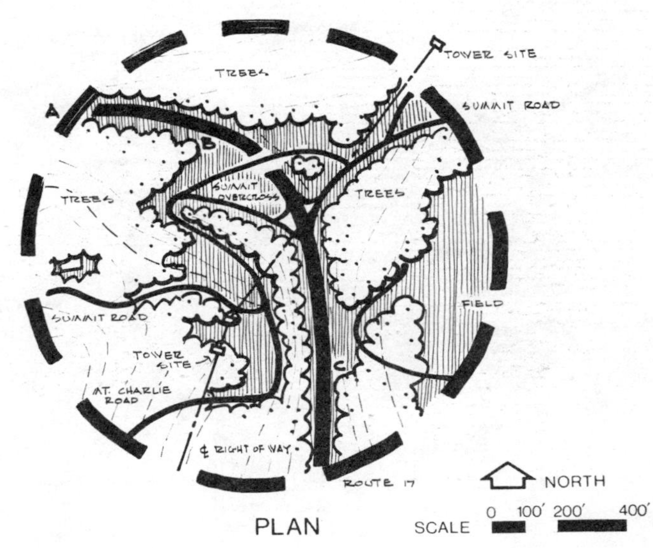

PLAN

57. Ibid., p. 63.

TRANSMISSION SCHEME at RT. 17[58]

E. DRIVING NORTH

TRANSMISSION LINES ABOVE VISUAL FIELD

G. DRIVING SOUTH

D. DRIVING NORTH

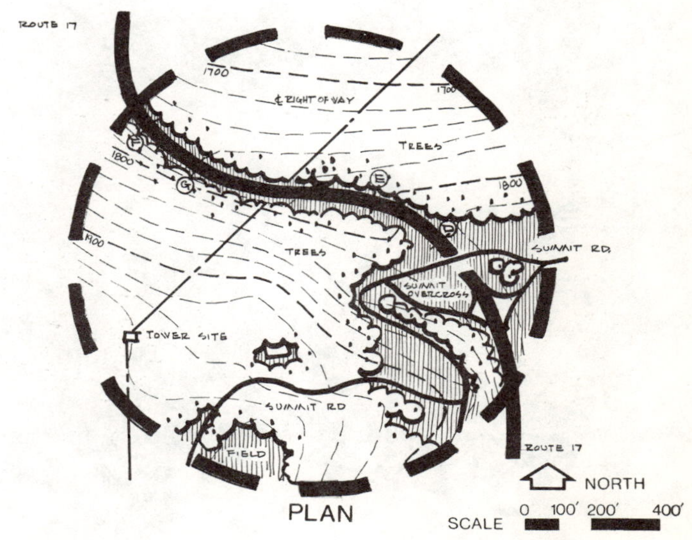

58. Ibid., p. 64.

TRANSMISSION SCHEME at SUMMIT RD.[59]

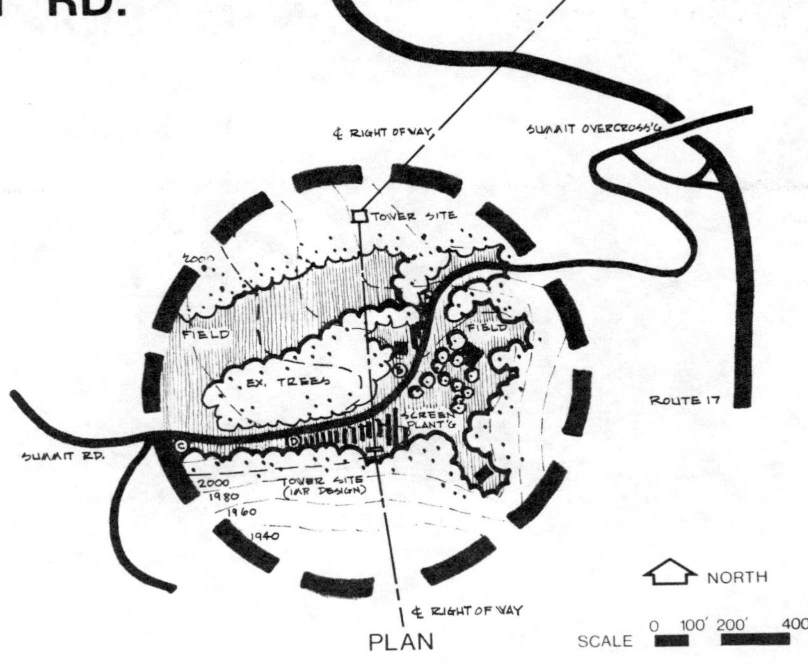

59. Ibid., p. 65.

One other extremely significant study conducted by an environmental design firm for an electric utility was a visual impact study conducted for Northern States Power Company by InterDesign, Inc., of Minneapolis, St. Paul. This study was completed in September of 1971.

The introduction to the study gives some of the background and the purpose of this particular study and its somewhat unusual approach.

> During the spring of 1971, the Department of Environmental Affairs of Northern States Power Company determined the need to analyze the visual impact of its facilities upon the seven county metropolitan region of the state. InterDesign was subsequently commissioned to undertake the study, focusing upon defining key problems and developing priorities for action to improve the aesthetic character of the various facilities.[60]

The approach to the overall study is unique among such projects for members of the electric utility industry. Therefore, it bears some discussion at this point as an introduction to InterDesign's findings concerning power transmission facilities. Other references will be made later in this book to the work and the processes utilized in this study for Northern States Power Company.

The following is a verbal description to the approach and the process utilized by the consultants:

> The study was organized into three phases as indicated on the accompanying charts. Phase I focused upon defining criteria and procedures for approach to the analysis and evaluation of the facilities. The second phase focused upon carrying out the analysis of as many facilities as possible within a two-month time period, utilizing the criteria and procedures defined in Phase I. Phase III then focused upon summarizing the analysis and documentation, defining critical problems and related recommendations, and presenting the study's findings.[61]

The following chart shows the process and the phases used by the consultants:

60. InterDesign, Inc., *Visual Impact Study*, (St. Paul, Minnesota, for Northern States Power Company, 1971), p. 3.
61. Ibid., p. 13.

Process[62]

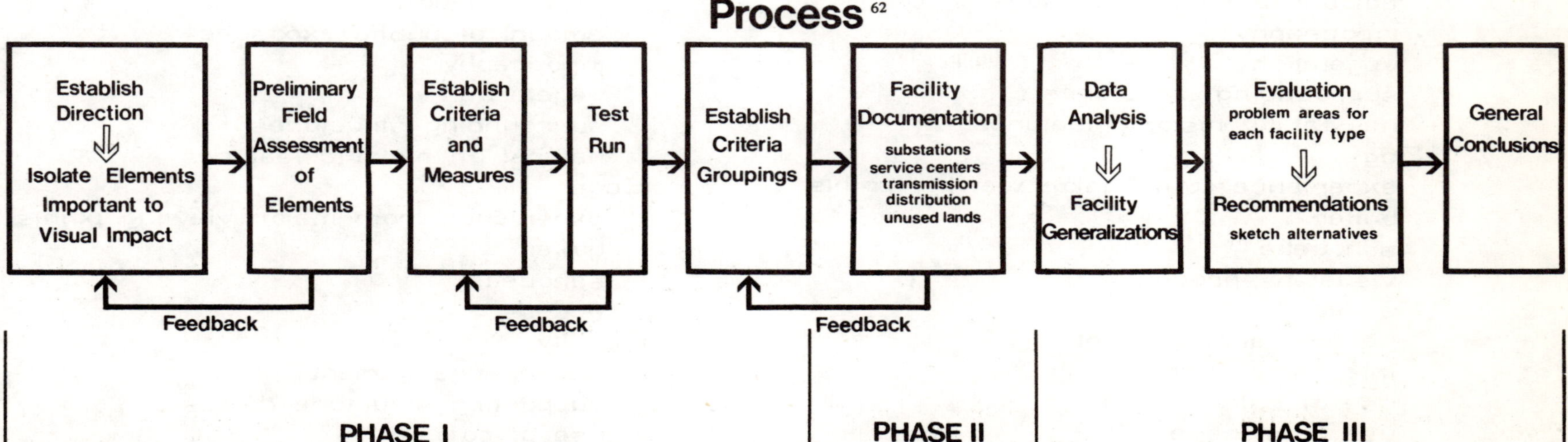

62. Ibid., p. 11.

The following illustrations show the approach to developing criteria used by InterDesign.

APPROACH TO DEVELOPING CRITERIA

I. FOCUSING ON FACILITY FROM SURROUNDINGS

A. SPOT FACILITIES

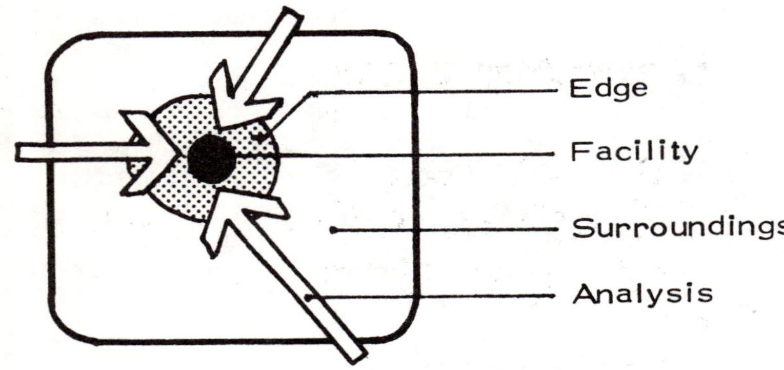

Important Considerations

1. Surroundings
   - surrounding land use and public circulation
   - amount of public exposure
   - topography
   - vegetation
   - surrounding structures
   - natural or historic features
2. Edge
   - experience from major viewing points
   - buffer
   - silhouette
   - visual by-products
3. Facility
   - general arrangement
   - internal elements
   - use of color
   - use of signage

B. LINEAR FACILITIES

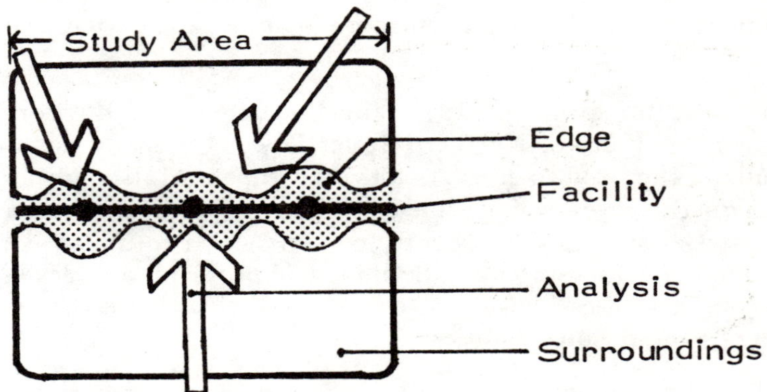

Important Considerations

1. Surroundings
   - adjacent land use and public circulation
   - amount of public exposure
   - topography
   - vegetation
   - surrounding structures
   - natural or historic features
2. Edge
   - experience from major viewing points
   - buffer
   - silhouette
   - access
3. Facility
   - linear arrangement
   - supporting structure details
   - use of color[63]

63. Ibid., p. 17.

II. OVERALL IMPACT OF FACILITY ON SURROUNDINGS

A. SPOT FACILITIES

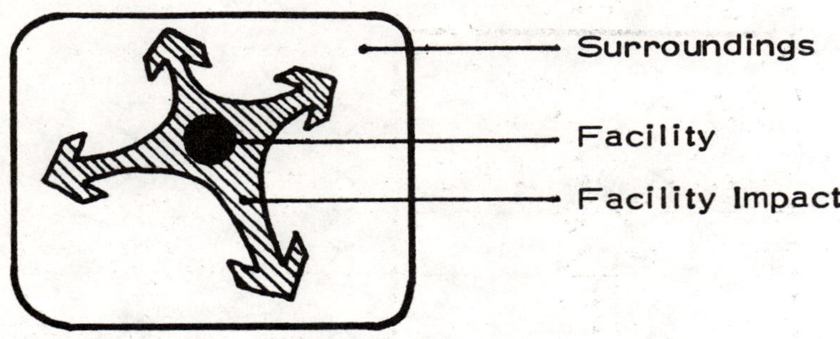

Important Considerations

Night continuity
Impact on adjacent land uses
Maintenance of existing character of an area

B. LINEAR FACILITIES

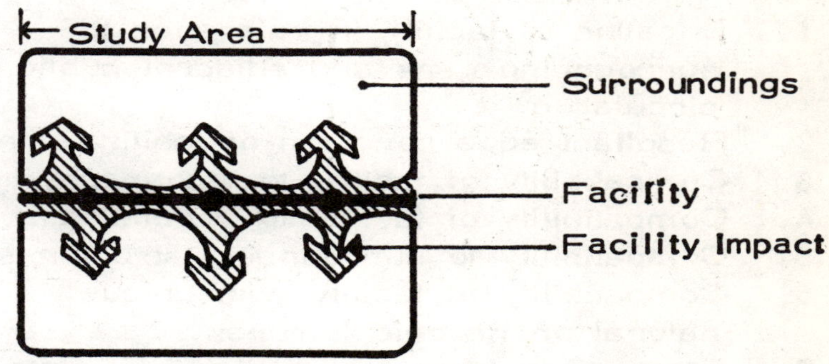

Important Considerations

Impact on adjacent land uses
Impact on surrounding character of an area[64]

64. Ibid., p. 18.

The following chart shows the facility analysis criteria established at the outset of this study:

FACILITY ANALYSIS CRITERIA[65]

Criteria	All kv Sub Stations	All kv Transmission and Distribution	Office/ Service facilities	Unused land & Facilities
1. Location of facility in reference to surrounding uses and adjacent public circulation	X	X	--	--
2. Resultant edge condition of facility siting	X	X	X	X
3. Compatibility of facility to topography	X	X	X	X
4. Compatibility of facility to vegetation	X	X	X	X
5. Compatibility to surrounding structures	X	--	X	X
6. Compatibility of facility with unique natural or historic features	X	X	X	X
7. Experience from major viewing points to the facility	X	X	X	--
8. Visual arrangement of the general facility	X	X	X	--
9. Visual arrangements of internal elements	X	X	--	--
10. Amount of public exposure	X	X	X	X
11. Results of visual by-products, e.g. parking and storage	X	X	X	--
12. Night Experience of facility	X	--	X	--
13. Use of Color	X	X	X	--
14. Use of Signage	X	--	X	--
15. Resultant impact of facility on adjacent land use patterns	X	X	X	X
16. Resultant effects of facility siting within existing character of the area	X	X	X	X

X: Criteria Applicable to Facility
--: Criteria Not Applicable to Facility

65. Ibid., p. 19.

The following indicates the facility and documentation file elements. This is an excellent means of data gathering, selection, organization and graphic depiction, and is certainly unusual among similar studies by environmental design consultants:

FACILITY DOCUMENTATION FILE ELEMENTS[66]

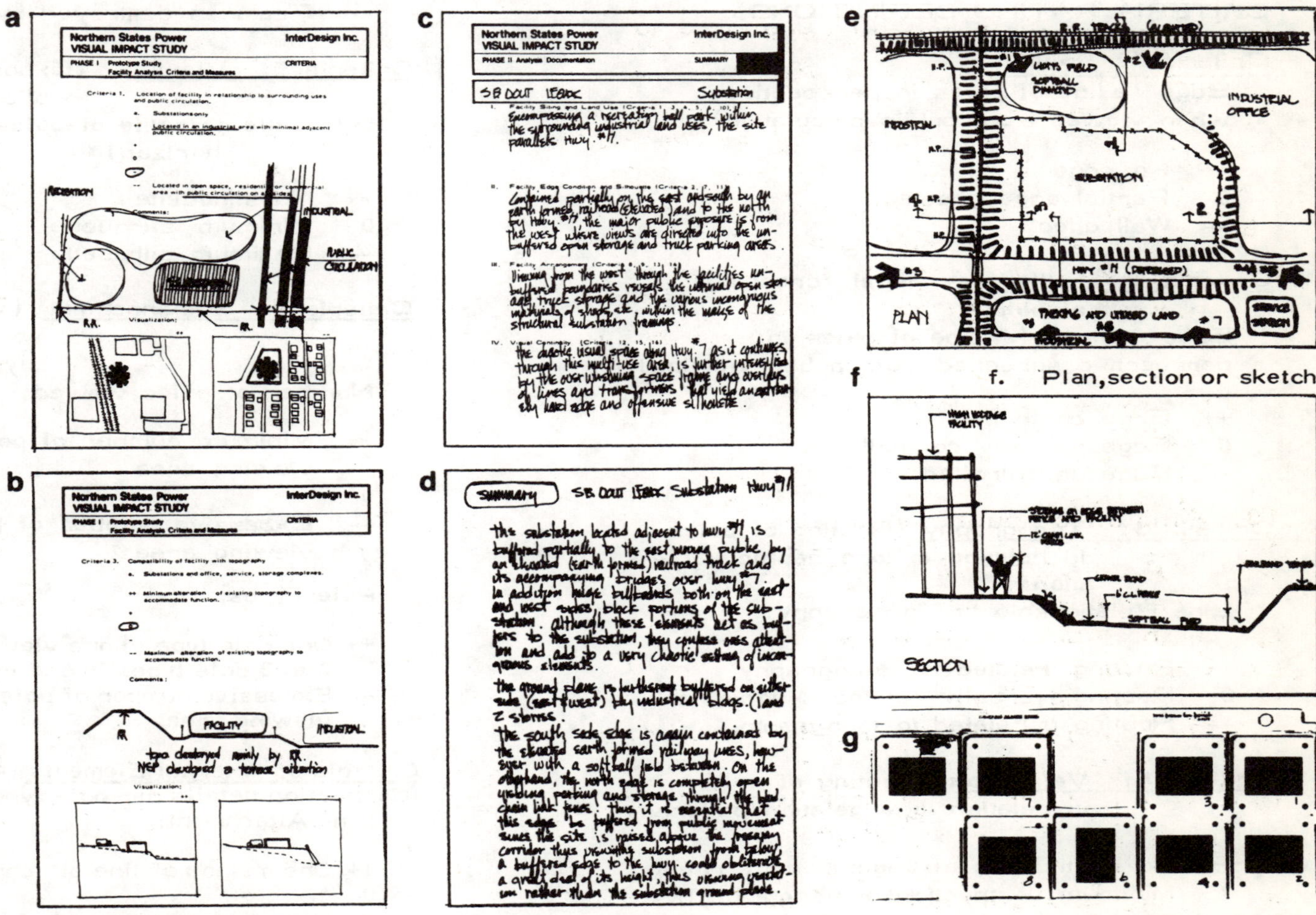

a. Completed criteria sheet
b. Completed criteria sheet
c. Summary of criteria evaluations.
d. Intuitive reactions to facility.
e. Plan,section or sketch
g. Slides: major views

66. Ibid., p. 25.

More will be said later concerning some of the other aspects of substation development and of the distribution facilities. However, at this point in the discussion of transmission distribution it seems appropriate to outline the criteria rating definitions utilized by the consultants in transmission and distribution facilities.

CRITERIA RATING DEFINITIONS

Criteria 1: Siting
Edge (a line of poles in perspective, when viewed from public circulation)

++ No edge
0 Partial edge created
-- Wall effect

Criteria 2: Buffer (A visual screening of poles)
Edge Control (A line of poles in perspective screened from public viewing)

++ Edge controlled
0 Edge partially controlled
-- Edge uncontrolled

Criteria 3: Topography (Routing of poles in relation to topographic variations)
Line Relationship to Topography:

++ Routing related to topography
0 Routing partially related to topography
-- Routing unrelated to topography

Criteria 4: Vegetation (Routing of poles in relation to vegetation patterns)

++ No destruction of vegetation by routing
0 Minimum destruction of vegetation by routing
-- Maximum destruction of vegetation by routing.

TRANSMISSION AND DISTRIBUTION

Criteria 6: Natural or Historic Features (Proximity of poles to natural or historic features)

++ No proximity of poles
0 Potential of area destroyed by proximity
-- Features destroyed by proximity

Criteria 7: Viewing (Exposure of poles to public)
Silhouette (Profile of poles exposed above horizon)

++ No silhouette
0 Minimum silhouette
-- Maximum silhouette

Criteria 8: Arrangement (Visual order of numbers and types of poles)
Number of poles viewed:

++ Minimum number of poles in one viewing area
0
-- Excessive number of poles in one viewing area[67]

Pole Types:

++ One pole type in one viewing area
0 2 to 3 pole types in one viewing area
-- Excessive number of pole types in one viewing area.

Criteria 9: Internal Elements
(Connection details per pole type)
Line Attachment:

++ One method of line attachment per pole
0
-- More than one method of line attachment per pole.

67. Ibid., p. 49.

Criteria 10: Public Exposure
(Viewing from major circulation routes)
Exposure Routes:

++ Located along minor roads
0 Located along major streets
-- Located along freeways or highways

Criteria 11: Visual By-Products
(Visibility of storage, maintenance and accompanying debris)
Service Roads:

++ Existing road used for service access
0
-- Additional roads required for service access

Criteria 13: Creative Use of Color
(Color used to blend, explain or accentuate design)

++ Color used positively
0 Color not used
-- Color used negatively

Criteria 15: Land Use Continuity

++ Pole routing does not disrupt existing land use patterns.
0
-- Pole routing disrupts existing land use patterns.

Criteria 16: Day Visual Continuity
(Siting and design are compatible with surrounding character)

++ Day visual continuity
0
-- Day visual discontinuity[68]

68. Ibid., p. 50.

The InterDesign staff then prepared a chart giving criteria ratings for 18 locations of transmission facilities throughout the entire system. The following chart shows the ratings and the application of the development criteria to the transmission facilities:

Criteria Ratings[69]

++ Desired Condition
o Neutral Condition
-- Undesired Condition

Transmission	1 Land Use / Substations	1 Edge / Trans. & Distr.	2 Vegetation / Substations	2 Topography / Substations	2 Structure / Substations	2 Major Edge Control / Service Centers	2 Minor Edge Control / Service Centers	2 Edge Control / Transmission	3 Topography Destruction / Subst. & Serv.	3 Line Relation To Topography / Trans. & Dist.	4 Vegetation Destruction / All Facilities	5 Scale Compatibility / Subst. & Serv.	6 Natural or Historic / All Facilities
1. Bloomington N - S		--						--		++	--		++
2. Broadway Plymouth		--						--		++	++		++
3. Brookdale Shopping		o						--		++	++		++
4. County Road #10		o						o		++	--		++
5. Gleason Lake		--						--		--	o		++
6. Hiawatha Avenue		--						--		++	++		++
7. Highway 100		o						--		o	++		++
8. Highway 280		--						--		++	++		++
9. I-494 E-W		o						--		++	++		++
10. I-494 N-S		--						--		o	o		++
11. Inver Grove		o						--		++	o		++
12. Kellogg Blvd.		o						o		--	++		++
13. King Plant		o						--		o	o		++
14. Minnehaha Park		o						o		++	o		++
15. Scott & Dakota Co		o						++		--	o		++
16. St. Louis Park		++						++		o	o		++
17. Theodore Wirth Park		o						o		o	o		++
18. University Avenue		--						--		++	++		++

Transmission	7 Viewing Level / Subst. & Serv.	7 Silhouette / Trans. & Distr.	8 Organization / Subst. & Serv.	8 Number of Poles Viewing / Trans. & Distr.	8 Pole Type / Trans. & Distr.	9 Clarification / Substation	9 Line Attachment / Trans. & Distr.	10 Exposure Routes / All Facilities	11 Ground Plane Exposure / Substations	11 Service Roads / Trans. & Distr.	11 Parking & Storage / Service	12 Night Visual Continuity / Subst. & Serv.	13 Creative Use of Color / All Facilities	14 Use of Signage / Subst. & Serv.	15 Land Use Continuity / All Facilities	16 Day Visual Continuity
1. Bloomington N - S		--		--	++		++	o		++			o		--	--
2. Broadway Plymouth		--		--	++		++	o		++			++		++	--
3. Brookdale Shopping		--		--	--		--	--		++			o		++	--
4. County Road #10		o		--	++		++	o		++			o		--	--
5. Gleason Lake		o		--	o		--	--		++			o		++	--
6. Hiawatha Avenue		--		--	++		--	o		++			o		++	--
7. Highway 100		--		--	++		++	o		++			o		++	--
8. Highway 280		--		--	--		++	--		++			o		++	--
9. I-494 E-W		--		--	o		++	--		++			o		++	--
10. I-494 N-S		--		--	o		++	--		++			o		++	--
11. Inver Grove		--		--	--		++	o		++			o		++	--
12. Kellogg Blvd.		--		--	o		++	o		++			o		++	--
13. King Plant		--		--	--		++	--		++			o		++	--
14. Minnehaha Park		--		--	++		++	o		++			o		++	--
15. Scott & Dakota Co		--		--	o		++	--		++			o		++	--
16. St. Louis Park		--		++	++		++	o		++			o		++	--
17. Theodore Wirth Park		o		++	o		++	o		++			o		++	--
18. University Avenue		--		--	++		++	--		++			o		++	--

69. Ibid., p. 51.

The following chart shows the satisfaction of criteria with regard to the transmission facilities within the system:

Satisfaction of Criteria[70]

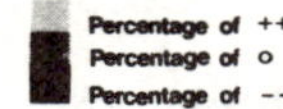

Transmission

1: Land Use / Substations; Edge / Trans. & Distr.

2: Vegetation / Substations; Topography / Substations; Structure / Substations; Major Edge Control / Service Centers; Minor Edge Control / Service Centers; Edge Control / Transmission

3: Topography Destruction / Subst. & Serv.; Line Relation To Topography / Trans. & Dist.

4: Vegetation Destruction / All Facilities

5: Scale Compatibility / Subst. & Serv.

6: Natural or Historic / All Facilities

7: Viewing Level / Subst. & Serv.; Silhouette / Trans. & Distr.

8: Organization / Subst. & Serv.; Number of Poles Viewing / Trans. & Distr.; Pole Type / Trans. & Distr.

9: Clarification / Substation; Line Attachment / Trans. & Distr.

10: Exposure Routes / All Facilities

11: Ground Plane Exposure / Substations; Service Roads / Trans. & Distr.; Parking & Storage / Service

12: Night Visual Continuity / Subst. & Serv.

13: Creative Use of Color / All Facilities

14: Use of Signage / Subst. & Serv.

15: Land Use Continuity / All Facilities

16: Day Visual Continuity

70. Ibid., p. 53.

The next chart, then, shows the high priority criteria for the transmission facilities:

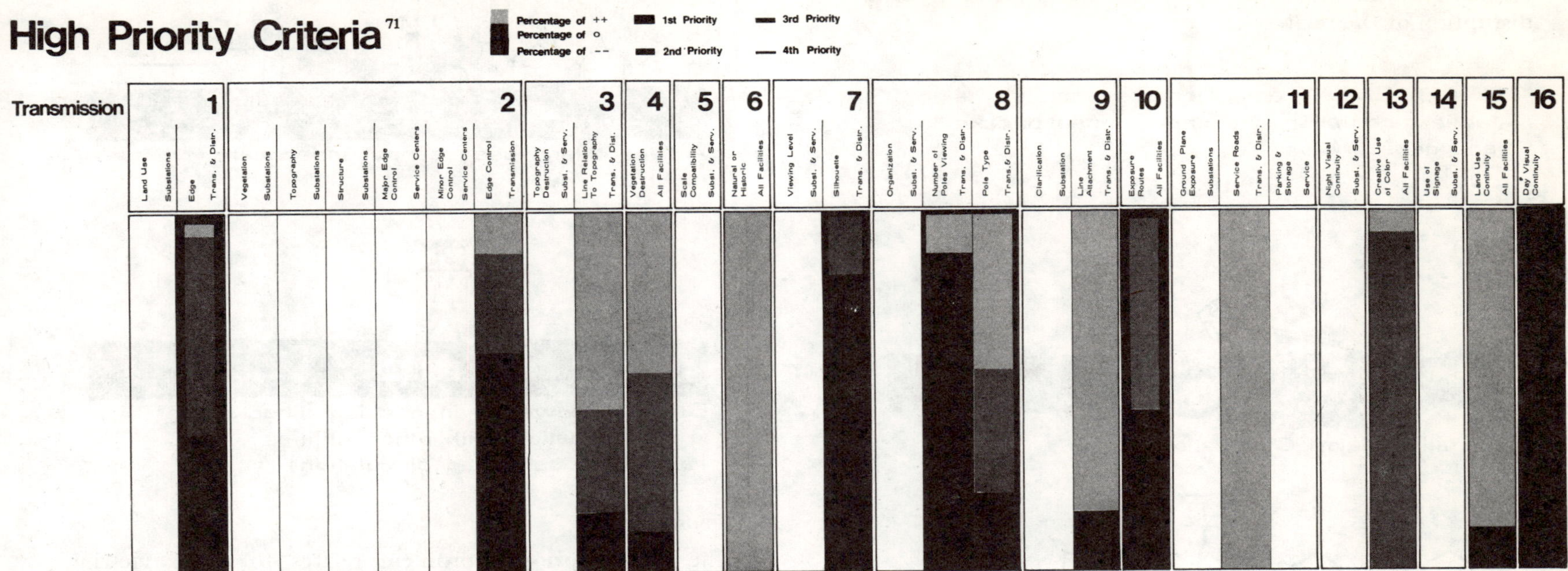

71. Ibid., p. 55.

From these analysis and data gathering steps, the consultants then prepared evaluation and recommendations for substations, service centers, and transmission lines. The consultants, then, had this to say concerning the transmission line aspect of this study:

> **Transmission**
>
> The first priority problem related again to the issue of location of the facility in relation to major public circulation routes. It was found that most visual problems occurred when these facilities paralleled or crisscrossed major roadways. If these facilities were located away from circulation corridors, possibly coupled with other utility right-of-ways, their visual impact could be more easily handled. In this regard, the right-of-way for various utilities such as sewer, pipelines, etc., could be amalgamated with power, resulting in a significant open space corridor. These utility corridors in natural landscape situations could, through careful siting, utilize vegetation and topography to effectively screen the elements from pedestrian view.
>
> The issue of disrupting adjacent land uses was also found to be most crucial related to Transmission facilities. The tower right-of-way cutting through land use areas tended to create visual walls between areas by the way in which the right-of-way edge reinforced the linear pattern of the lines. The purchase of a more varied right-of-way allowing a more varied development of the edge in structure and plantings would tend to reduce this negative effect. Plantings of controlled size and some fireproof structures might also be allowed under the wires themselves to help destroy the swath cut through areas.
>
> The rural Transmission facilities reviewed in Scott and Dakota counties presented a positive example of how these

elements could move through agricultural areas without disruption of the related land uses.[72]

PROBLEM: Facility routes parallel major public circulation and create open breaks in land uses.

Parallel Major Circulation

Cut swaths through use areas

RECOMMENDATION I: Locate facility routes away from major public circulation without disrupting adjacent land uses and correlated with open space systems and other utility right of ways.

Distance

Adjacent land uses carry underneath right of way.

Electric power transmission directed to consolidation with other utilities (excluding vehicular circulation).[73]

Transmission

The second priority problem relates to major viewing points and the detail location of the tower elements. It was determined that the viewing of the tower elements in total silhouette with exposed ground connections provided a most negative impact. Careful siting to hide ground connections behind hill forms rather than location on top of knolls would go far toward reducing negative impact on major viewing areas. In the same regard, location behind or inside major vegetation would help reduce the exposed silhouette. The facilities located near Brookdale effectively illustrate the effect of tower elements whose ground connections are hidden by these natural elements.

Exposing long vistas of the towers over one-half mile in length provided a negative visually dominating effect. An effort should be made to reduce long vistas of the towers, especially when related to major roadways.[74]

72. Ibid., pp. 92-93.
73. Ibid., p. 93.
74. Ibid., p. 94.

PROBLEM: From major viewing points the public sees maximum silhouettes of structures, and long views of the system.

RECOMMENDATION 2: Use buffers between major viewing points and facility to minimize silhouette, eliminate long views of the system, and conceal ground connections.

Vegetation Buffer

Creative siting of system relative to topography.

Topography Buffer

Structure Buffer

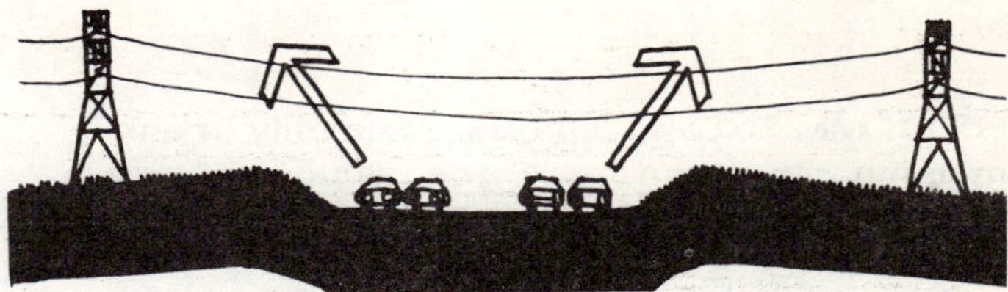

Cross circulation at right angle.[75]

The Transmission structure itself was the focus of the third priority problem defined. Most crucial in this regard was the use of many pole types within any one viewing area. The visual chaos of various line attachments and different silhouettes created by this practice could be reduced if a single pole type were used within viewing areas.

The steel truss structural system towers were found to produce the greatest negative impact. Their complex silhouette, especially when overlayed one with another in a viewing area, produced a high negative impact. The use of the simple steel tubular poles with insulator connections would provide the most transparent system and reduce the impact appreciably. Light colors, such as light blue and off-white, were found most effective on these pole structures, as the majority of the forms were often seen in silhouette with the sky as a background. These light colors, as a result, tend to minimize effectively the overpowering silhouette of the elements.[76]

PROBLEM: Facility presents a complicated changing pattern of structure.

75. Ibid., p. 95.
76. Ibid., p. 96.

RECOMMENDATION 3: Simplify transmission structure, use one structure type in any one area.

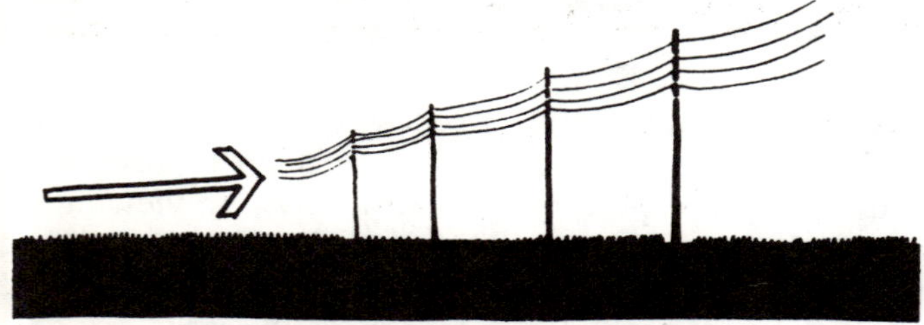

Create a more transparent system.[77]

The issue of visual continuity of the towers with the surrounding environment was judged a most crucial, but fourth priority, problem. The immense scale of the elements, even when simplified in silhouette, are most difficult to deal with as they overwhelm and dominate the view. It is almost impossible to eliminate the negative impact in these terms, but it can be reduced by accomplishing many of the recommendations defined for the first three problems.

Although burying these large voltage lines at this time is economically unfeasible, it may be worthwhile to consider, for the future, a system of smaller kilovolt lines in a grid within the metropolitan area which could be buried more economically. The technical and economic problems of replacing this larger kilovolt ring with a smaller net may, however, be insurmountable.[78]

The following pages indicate the prototypical studies for facility analysis criteria and measures as prepared by the consultants for Northern States Power Company.

77. Ibid., p. 97.
78. Ibid., pp. 92-98.

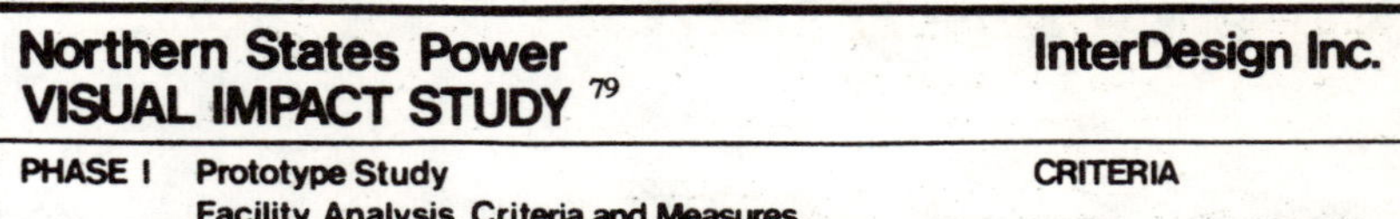

Northern States Power VISUAL IMPACT STUDY[79] — InterDesign Inc.

PHASE I Prototype Study — Facility Analysis Criteria and Measures — CRITERIA

Criteria 1. Location of facility in relationship to surrounding uses and public circulation

b. Transmission and distribution

++ Use of land area continuous with adjacent uses with minimal visibility from major public circulation ways.

\+

0

\-

-- Disruption of continuous adjacent land uses and/or right of way creates a ridgid edge in relation to adjacent uses and/or visibility from major public circulation ways.

Comments:

Visualization:

79. Ibid., p. 126.

Northern States Power
VISUAL IMPACT STUDY [80] **InterDesign Inc.**

PHASE I **Prototype Study** **CRITERIA**
Facility Analysis Criteria and Measures

Criteria 2. Resultant edge condition of facility siting.

a. All facilities

++ Visual edge control resolves facility siting

+

0

-

-- Visual edge uncontrolled (no resolution of siting)

Comments:

Visualization:

++

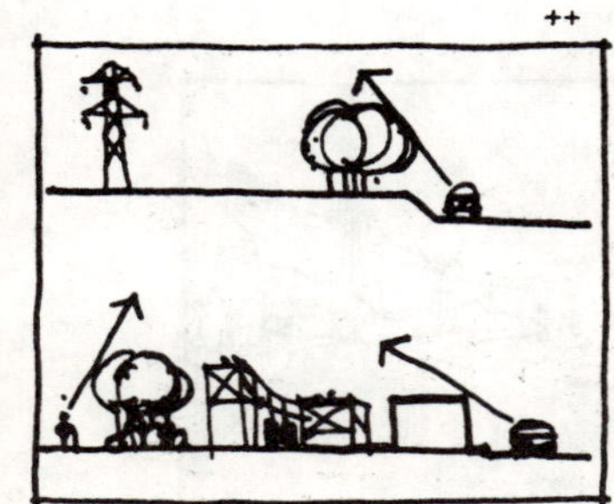

--

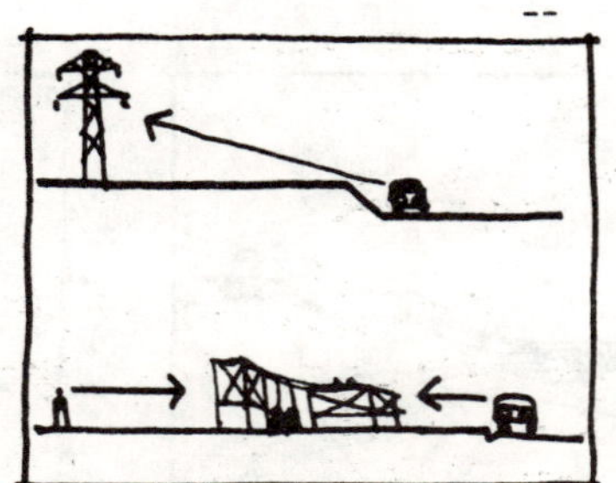

80. Ibid., p. 128.

Northern States Power
VISUAL IMPACT STUDY [81] **InterDesign Inc.**

PHASE I **Prototype Study** **CRITERIA**
Facility Analysis Criteria and Measures

Criteria 3. Compatibility of facility with topography

b. Transmission and distribution

++ Routing poles and lines in relationship to topography forms to conceal or minimize visual impact.

+

0

-

-- No relationship of routing or pole location to topography.

Comments:

Visualization:

++

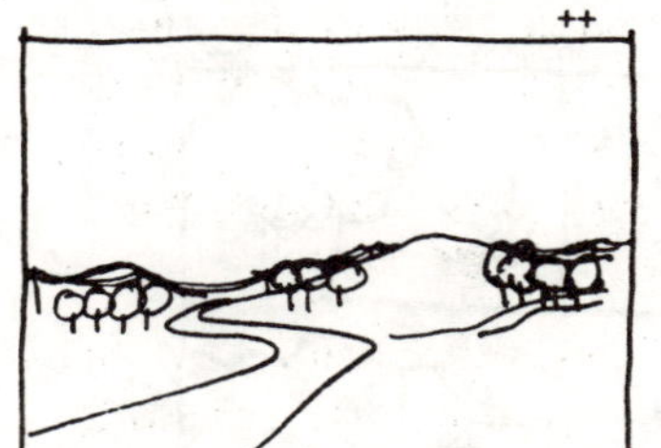

--

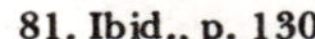

81. Ibid., p. 130.

Northern States Power **InterDesign Inc.**
VISUAL IMPACT STUDY[82]

PHASE I Prototype Study **CRITERIA**
Facility Analysis Criteria and Measures

Criteria 4. Compatibility of facility with vegetation patterns

a. All facilities

++ Minimum alteration of existing vegetation pattern and maximum useage of vegetation to minimize negative visual impact.

+

0

-

-- Maximum alteration of existing vegetation pattern and minimum use for screening.

Comments:

Visualization:

++

--

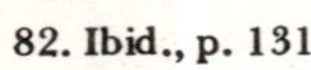

82. Ibid., p. 131.

Northern States Power **InterDesign Inc.**
VISUAL IMPACT STUDY[83]

PHASE I Prototype Study **CRITERIA**
Facility Analysis Criteria and Measures

Criteria 7. Experience from major viewing points to the facility.

b. Transmission and distribution.

++ Minimum silhouette, no long views of poles, simple or concealed ground connections.

+

0

-

-- Maximum silhouette, long views of poles, complex ground connections.

Comments:

Visualization:

++

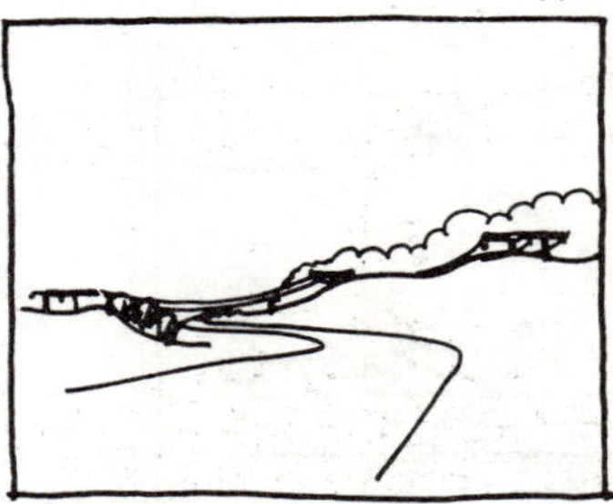

--

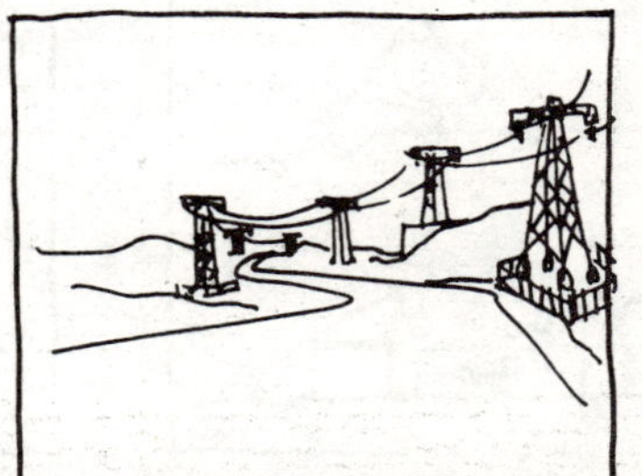

83. Ibid., p. 135.

Northern States Power **InterDesign Inc.**
VISUAL IMPACT STUDY [84]

PHASE I Prototype Study **CRITERIA**
Facility Analysis Criteria and Measures

Criteria 9. Visual clarification of internal elements.

b. Transmission and distribution.

++ Unified method of line attachment on a pole type, simplification of communications lines and attachment in vocabulary of power line attachment, one type of line sag for communications lines and power lines.

+

0

-

-- Various methods of line attachment on a pole, chaos of communication line and attachment, many types of line sag for communications lines and power lines.

Comments:

Visualization:

++

--

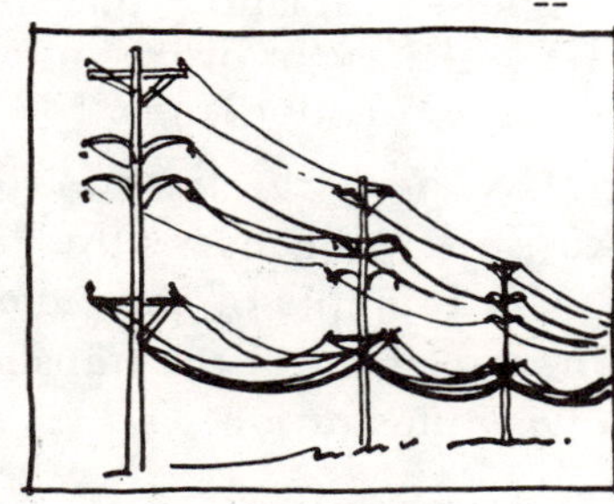

84. Ibid., p. 140.

Northern States Power **InterDesign Inc.**
VISUAL IMPACT STUDY [85]

PHASE I Prototype Study **CRITERIA**
Facility Analysis Criteria and Measures

Criteria 11. Results of visual by-products (parking and storage)

b. Transmission and distribution

++ Utilize existing roads and access points with minimum impact on topography and vegetation.

+

0

-

-- Provision of additional service roads with maximum impact upon topography and vegetation.

Comments:

Visualization:

++

--

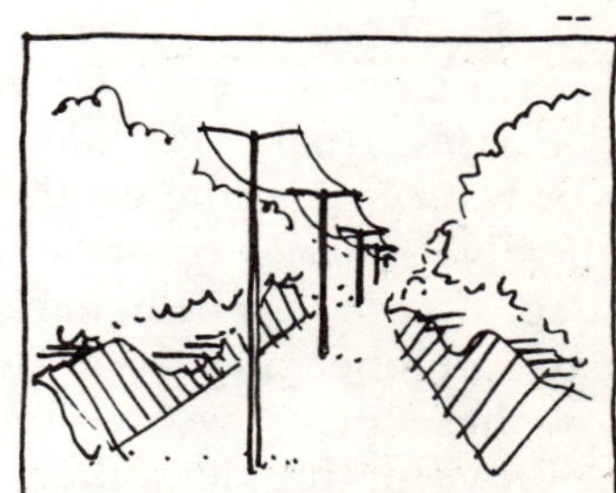

85. Ibid., p. 143.

Integration of Activities on Rights-of-Way

One other approach to dealing with the extensive rights-of-way required for electric transmission lines is to integrate a number of activities on the rights-of-way or to develop a series of secondary uses of the rights-of-way.

The statement in the publication entitled, "Electric Power and the Environment" in regard to such regional corridors is as follows:

> Effective regional planning requires the early acquisition and withdrawal of land for public service rights-of-way well in advance of the need—perhaps as much as 20 years. Changes in the planning techniques, rate making policies, and the attitudes of the utilities and others concerned may be required to motivate such investment.
>
> There are very real problems in resolving the conflicting requirements of the various utilities for optimum routing and the respective rights of the utilities to make use of the corridor itself. This process will require the detailed participation by the utilities in a public forum where all interested persons can be heard. This suggests that a land-use planning agency at the state or regional level may be required. However, one of the more important tasks facing the utilities in corridor planning is that of working and cooperating with the planning commissions or zoning boards in the various areas which they serve. The key to success in working with the zoning commissions and boards is planning corridors far in advance so that all interested parties can be included in the early stages of planning. Advance planning on the part of the utility will actually involve a series of projections with those 10 years in advance of completion being most definitive and those for a greater number of years in advance being more tentative. All of these plans will require careful thought and adequate provision for flexibility.
>
> If regional planning for EHV transmission lines is to be projected to the year 2000, broad and imaginative consideration must be given to how the rights-of-way can best be fitted to various land uses in ways that will keep environmental damage to a minimum. Joint use may require very wide rights-of-way to include regionally forecast needs for transportation movements of all kinds—power, rail, highway, pipeline, sewer, water and others. Other uses for these rights-of-way such as strip recreation and forest plantations will have to be considered in the early stages of planning. The question of whether a mixture of uses can be compatible so that a few major corridors can be made to serve a variety of needs must be explored. To accomplish this a team, including planners, landscape architects, engineers, industrial economists, and sociologists will be required. The acquisition or control of corridors of sufficient width to accommodate all present or foreseeable needs by a consortium of those public and private agencies using the corridor may require new laws and a 'corridor corporation' or other device with appropriate powers.
>
> The steps toward a good regional corridor include:
>
> (a) a forecast of needs for all transportation and communication sectors;
> (b) a layout of approximate alignment locations for all future needs;
> (c) delineation of potential corridors by grouping alignments;
> (d) an interdisciplinary team to study and plan the corridors;
> (e) close coordination with appropriate federal, state, and local land management agencies;
> (f) purchase or provision of other means of controlling the corridor lands.[86]

The Davenport transmission routing study prepared by Eckbo, Dean, Austin & Williams for the Pacific Gas and Electric Company contained an example of the recreational potential of the right-of-way on one segment of the transmission corridor. This is shown in the following illustration.

86. The Energy Policy Staff, Office of Science and Technology, *Electric Power and the Environment,* (Washington, D.C., U.S. Government Printing Office, Superintendent of Documents, 1970), p. 24.

RECREATIONAL POTENTIAL of the RIGHT of WAY MT. HERMON – SODA SPRINGS[87]

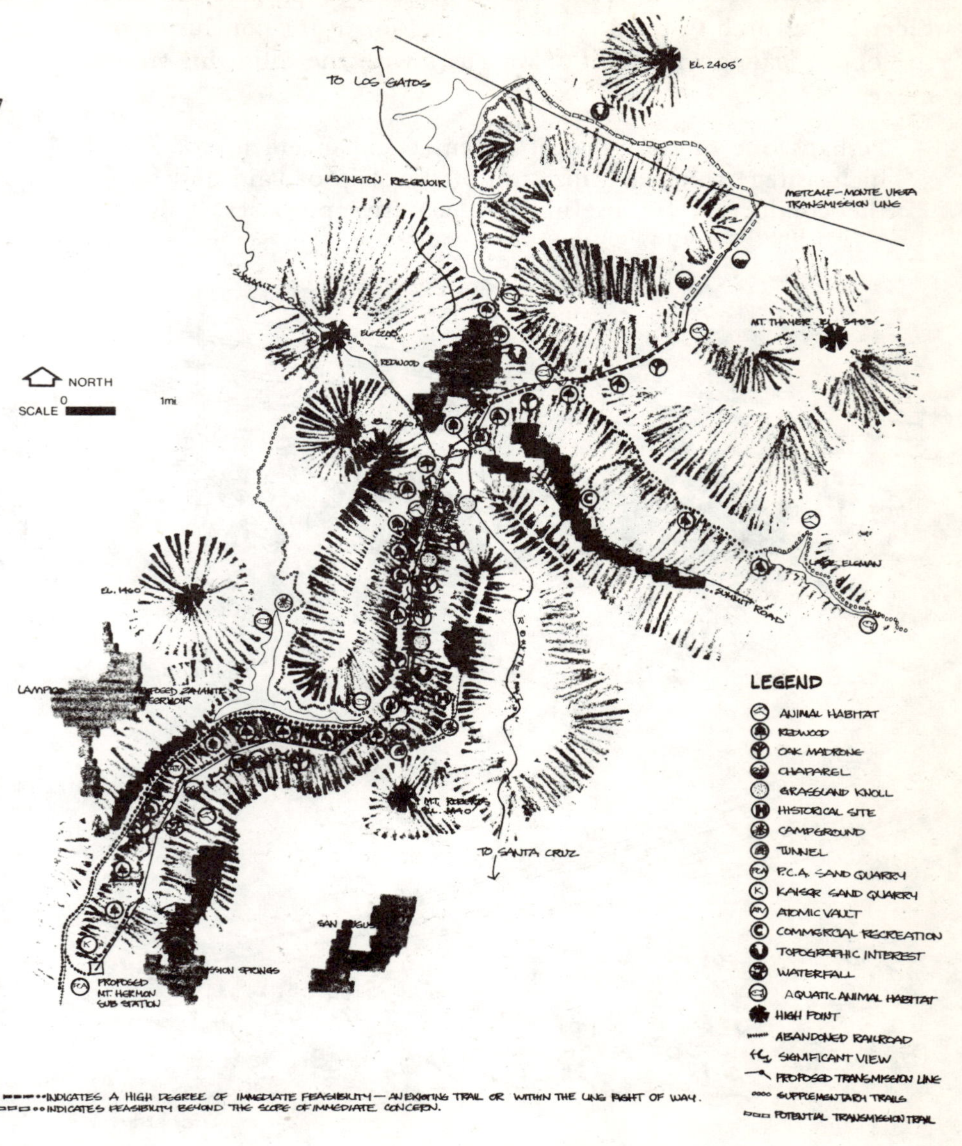

87. Garrett Eckbo, Francis H. Dean, Donald B. Austin, and Edward A. Williams, *Transmission Line Routing,* (San Francisco, California, Eckbo, Dean, Austin and Williams, for Pacific Gas and Electric Company, 1970), p. 70.

The book "Electric Transmission Structures" prepared for the Electric Research Council contained a section at the conclusion of the book dealing with right-of-way uses with the following statement:

> Perhaps one of the greatest potential enhancements of a high-voltage route is utilization of the strip of land under the conductors for useful and decorative purposes. This has been tried in some communities and, in many cases, it gives a handsome setting for the towers, a major benefit being the cultivation of trees and ground vegetation that does not endanger the electric conductors. It would, of course, be extremely beneficial for the type of designs in this collection. While imagination and experience could undoubtedly invent many uses for the right-of-way land, a few suggestions are illustrated here as examples.

Golf course

Park

Orchard

Equestrian or bicycle path [88]

88. Electric Research Council, *Electric Transmission Structures, A Design Research Program,* (New York, Edison Electric Institute, EEI Publication, No. 67-61, 1967), p. 78.

A number of counties and regional authorities throughout the United States have studied extensively the possible secondary uses of rights-of-way. Probably, Santa Clara County in California has done the most significant work in exploring the possibilities for hiking and bridle trail systems throughout the county utilizing the land under the utility transmission lines.

Illustrations from an excellent brochure prepared by Santa Clara County are shown in the following figures:

89

89. Santa Clara County Planning Department, *Greenways,* (San Jose, California, Santa Clara County Planning Department, 1969), p. 1.

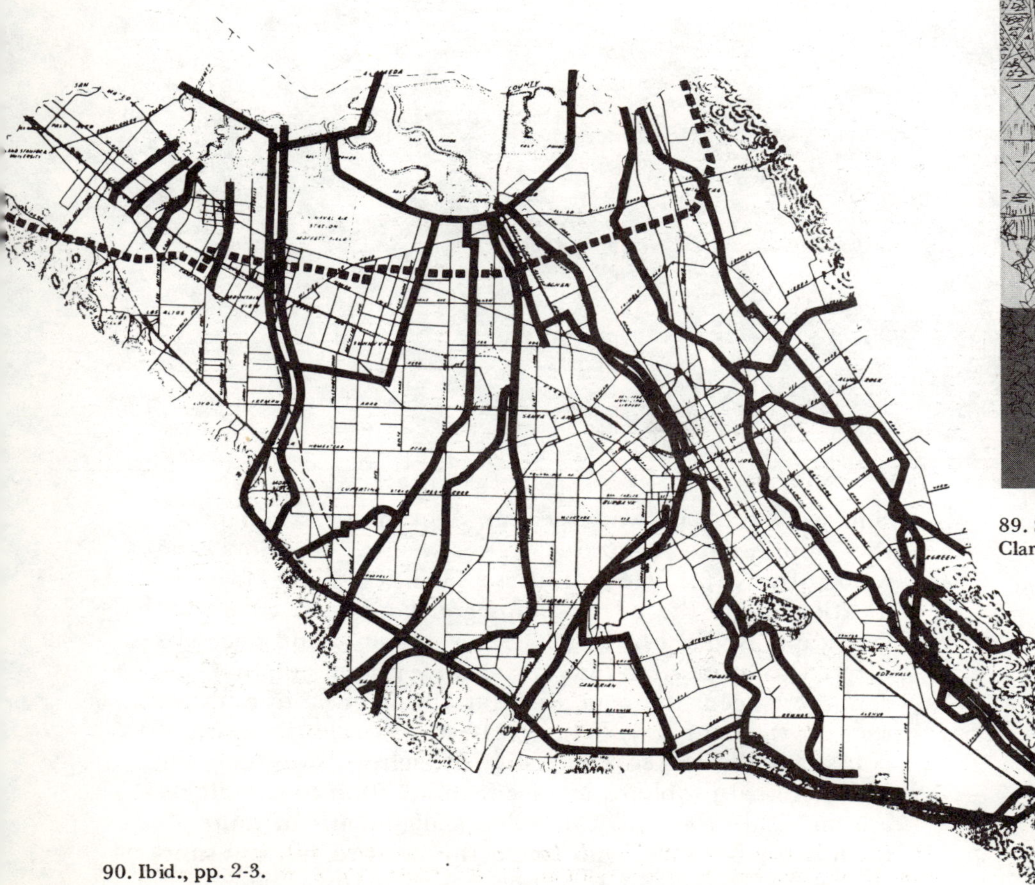

POWER LINES
WATERWAYS —
— streams flood control channels, water conservation canals and percolation ponds.
HETCH-HETCHY WATER LINE [90]

90. Ibid., pp. 2-3.

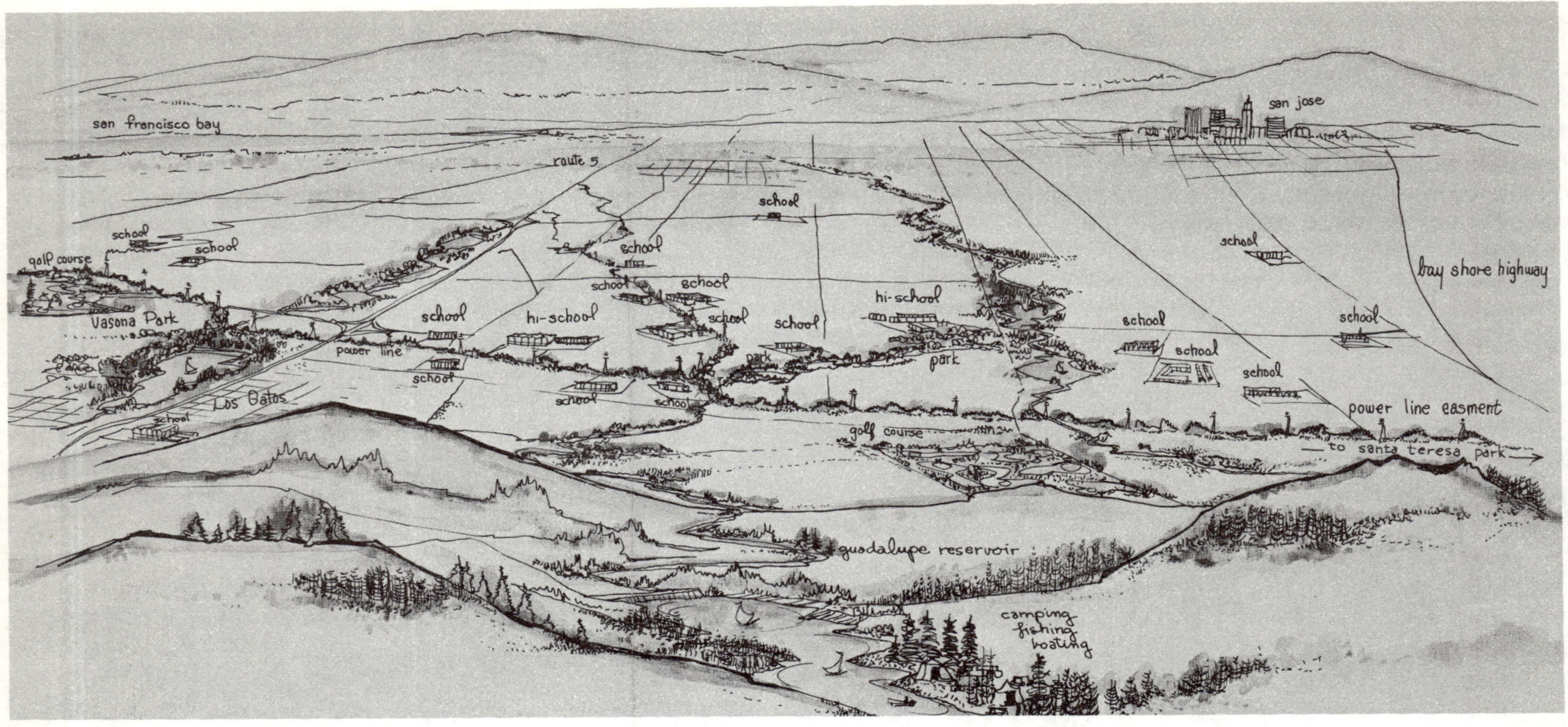

91. Ibid., p. 4.

91

Johnson, Johnson and Roy in their study for Consumers Power Company in Michigan on "Transmission and Distribution Rights-of-Way Selection and Development" made the following statement and showed an illustration of multiple use of rights-of-way.

Multiple Use of Rights-of-Way

Electrical transmission is one of many types of facilities that require lineal rights-of-way for their construction and maintenance. Other utilities—water, gas, communication, sewage—as well as transportation systems—highways, waterways, railroads, rapid transit systems, etc.—demand a similar corridor network system. A prime determinant for the appropriate location of easements, corridors and rights-of-way for each of these facilities is the logic of location along the edges of integral land units where the landscape is disturbed as little as possible and access is offered to more than one unit at any point. For that reason, there needs to be increased attention given to multiple use of rights-of-way. The greatest obstacle to this, of course, is the mixed ownership of rights-of-way and the jurisdictional problems involved. In addition to that, there are physical problems. Curvi-linear alignments of railroads do not for instance conform to the need to run transmission wires in straight lines between structures. There are, however, instances where relatively straight corridors are needed for each and could be combined into one line through the landscape. Also, in situations of scenic or historic interest, a curving electric utility alignment adjacent to a railroad requiring frequent turns is often an acceptable compromise between the conventional straight line transmission route and the more expensive alternative of putting the line underground. The concept of the development of multiple uses of utility corridors containing different types of utilities and transportation facilities needs to be studied further in areas of technology, compatibility between uses, jurisdiction and safety.

In addition to using transmission line corridors for other utilities or transportation systems, other uses within parts of the rights-of-way are possible and often feasible. Generally these are considered to be recreational or agricultural and include horseback riding trails, bike trails, hiking trails, parks, golf courses, picnic areas, general agriculture, orchards, Christmas tree and other nursery stock cultivation, bird sanctuaries and game refuges.[92]

93

Screening, Beautification And Buffering

There are numerous examples of the most common approach in dealing with environmental destruction caused by electric transmission lines in the environment. This is the approach to screening; to camouflage, or to "beautify." This, of course, is a less satisfactory approach to dealing with the basic problem of environmental destruction by transmission lines or facilities. However, it is exemplary, and since there are so many instances throughout the United States, it seems unnecessary to chronicle these any greater extent. It might be worth mentioning, however, a reference from the book *The Electric Utility Industry in the Environment,* a report to the Citizen's Advisory Committee on Recreation and Natural Beauty by the Electric Utility Industry Task Force on the Environment in 1968.

This document, in speaking of beautification facilities, makes the following statement:

> Fortunately, there has been a general trend among utility companies to beautify their facilities. Structure design, architecture, landscaping, and general appearance have been improved in most places during recent years.
>
> - Plants have been carefully sited, thoughtfully designed, and often surrounded by parks and picnic areas.
> - Substations have been located where they are unobtrusive.
> - Distribution substations have been vastly improved, in some instances being contained within structures indistinguishable from others in the environment.
> - New types of poles have been developed and transformers have been removed from sight.
> - It is not unusual for large utilities to have or retain architects, decorators, and landscape artists, a practice that is gaining, and is doing much to give utilities a leadership role in the field of industrial beautification. Sixty percent of those replying to the question-

92. William J. Johnson, Carl Johnson and Clarence Roy, *Transmission and Distribution Rights-of-Way Selection and Development,* (Ann Arbor, Michigan for Consumer's Power Company, Jackson, Michigan, 1970), pp. 51-52.
93. Ibid., p. 52.

naire (distributed by the Task Force) state that they use outside consultants.[94]

Unquestionably, there is much more to be done, but the progress already made would indicate that the industry generally assumes a large measure of responsibility and further programs can be anticipated.

Undergrounding

One of the areas receiving the greatest discussion and containing the most misunderstandings is the undergrounding of transmission facilities.

The 1970 National Power Survey, published by the Federal Power Commission had this to say about underground transmission lines:

> In 1970 there were about 2,000 miles of underground transmission lines in the United States, mostly in densely populated areas where overhead rights-of-way are not available or are prohibitively expensive, these high-voltage underground lines represent less than 1% of the nation's total transmission system. From an aesthetic point of view, the underground construction of transmission facilities might seem to be desirable. In wooded areas, however, undergrounding would require a completely cleared swath. Moreover, the earth disruption of burial could heighten erosion, and under certain soil and moisture conditions the heat dissipated into the earth by underground conductors could adversely affect vegetative growth. In such areas, the visual impact of undergrounding could be as pronounced as for overhead installations.
>
> Another aesthetic consideration in undergrounding alternating current transmission lines is the bulky equipment required to compensate for charging current. Providing such equipment along the underground transmission lines creates siting problems comparable to those involved in locating and designing substations.
>
> Fewer technical and environmental problems are encountered in undergrounding direct current transmission lines than for alternating current systems. However, esthetics must be an important consideration in designing and locating the expensive converted equipment required to change from alternating to direct current at the sending end and changing back to alternating current at the receiving end.[95]

The publication "Electric Power and the Environment" made the following statement in regard to underground utility tunnels.

> The concept of underground utility tunnels has been successfully and extensively employed at universities and government installations. It is not commonly used in cities in the United States. Substantial developments of such type are being constructed abroad, notably in Japan. Such construction might offer a unique solution to utility problems in built-up urban areas. Long-range master plans for future utility requirements in cities should consider the feasibility of having all utilities in one tunnel, where maintenance could be performed without street obstruction. Planning of this type should be coordinated with long-range master plans for mass transportation systems.[96]

It seems probable that it will be much easier to underground distribution facilities in the near future than it will to place underground major transmission lines to alleviate environmental disruption. A further dimension will be explored in a discussion of underground distribution lines in a later section of the book on that subject.

A great deal of publicity has resulted from some spectacular visual attempts at undergrounding distribution lines. Only a few truly significant instances exist.

94. Electric Utility Industry Task Force on Environment, *The Electric Utility Industry and the Environment,* A Report to the Citizens Advisory Committee on Recreation and Natural Beauty, (New York, New York, Electric Utility Industry Task Force on Environment, 1968), p. 27.

95. Federal Power Commission, *The 1970 National Power Survey: Guidelines for Growth of the Electric Power Industry,* (Washington, D.C., U.S. Government Printing Office, Superintendent of Documents, 1971), p. I-12-9.

96. The Energy Policy Staff, Office of Science and Technology, *Electric Power and the Environment,* (Washington, D.C., U.S. Government Printing Office, Superintendent of Documents, 1970), pp. 24-25.

Power Transformation Facilities

Next to the transmission lines associated with the electric power industry the transforming facilities are the items or physical accoutrements in the electrical energy generation and distribution intrastructure that have caused the most environmental concern and disruption. They are usually located in a variety of neighborhoods and districts where they are immediately visible to passersby.

Their purposes have been described in a number of references. The publication entitled, "Environmental Criteria for Electric Transmission Systems" gives the following statement of purpose for switchyards and substations.

> The purpose of these facilities is to receive power, to transfer power from circuit to circuit, to protect circuits and equipment, to transform power to other voltage levels and to provide meters for billing system power scheduling and control and to monitor system performance. The installation often serves as a location for maintenance crew headquarters, warehouses and storage facilities.[1]

The publication entitled, "Substation Site Selection and Development" prepared for Consumers Power Company of Michigan by the landscape architectural firm of Johnson, Johnson and Roy makes the following statement in regard to the purpose of substations for transforming facilities:

1. U.S. Department of the Interior, U.S. Department of Agriculture, *Environmental Criteria for Electrical Transmission Systems,* (Washington, D.C., U.S. Government Printing Office, Superintendent of Documents, 1970), p. 31.

The primary purpose of an electric distribution substation is to reduce the voltage of the bulk power source to a lower voltage for distribution to specific zones of the community. Installed in the basic substation are electrical equipment items such as transformers, voltage regulators, protective devices and switches. It also contains steel structures to terminate the sub-transmission and distribution lines and also to support protective devices, switches and other smaller components which make up the substation.

The function of the transformer is to reduce the voltage of the supply source to a lower voltage for distribution. The size and number of transformers depends on the electrical needs of the community.

The voltage regulator automatically maintains the voltage on the distribution system throughout the community within prescribed limits. It compensates for variations in voltage on the supply source to the substations and for variations in loading on the distribution system which would otherwise affect the voltage available on the system. This ensures each user of electric power of a supply voltage that conforms to the design limitations of his electrical appliances and equipment.

Two types of protective devices are generally installed, one of which is a fuse on the high voltage side of the transformer, for protection of the transformer from possible damage due to overload. When required to operate, it ensures continuity of service to other distribution substations connected to the same bulk power supply since operation of a protective device for the supply source should result in disruption of service to all connected substations. The other protective device installed is an automatic circuit breaker. This device senses troubles or overload conditions on a distribution circuit and operates automatically to interrupt the flow of power. Thus it prevents damage to the circuit and ensures continuity of service to users of electric power connected to other distribution circuits emanating from the same substation by removing, from service, the troubled circuit only.[2]

The basic purpose, then, of transformation facilities and substations is to terminate input, to exchange levels of power, to originate output, and to transform various levels of the essential incoming power to the optimum outgoing levels of power. It is basically a place for the lowering of power levels.

The transformer facilities are among the first stages in the distribution network. In essence, the electric power distribution network can be compared to the steps going down from a superhighway, to a highway, to a road, to a lane, and to a driveway.

The transformer is the point of interchange. As such, it must be placed in all types of neighborhoods which, at times, causes environmental destruction in many ways.

Basically, the types of transforming facilities may be divided into:

Major Transforming Facilities,
Bulk Power Stations, and
Sub-Transmission and Distribution System Transforming Facilities.

The location of transforming facilities is:

- Usually at points of change in power level
- As close as possible to the center of the electric power distribution load
- Near a high-voltage transmission line
- Accessible from a street or highway

In the simplest possible breakdown the components of electrical transforming facilities consist of:

- Input lines
- Transformer
- Enclosure
- Service drive
- Regulator and recloser structure
- Output lines
- Guying structures for the transmission lines

2. William J. Johnson, Carl Johnson and Clarence Roy, *Substation Site Selection and Development,* (Ann Arbor, Michigan, Johnson, Johnson and Roy, for Consumer's Power Company, Jackson, Michigan, 1969), p. 4.

ELECTRIC TRANSMISSION AND DISTRIBUTION SYSTEM[3]

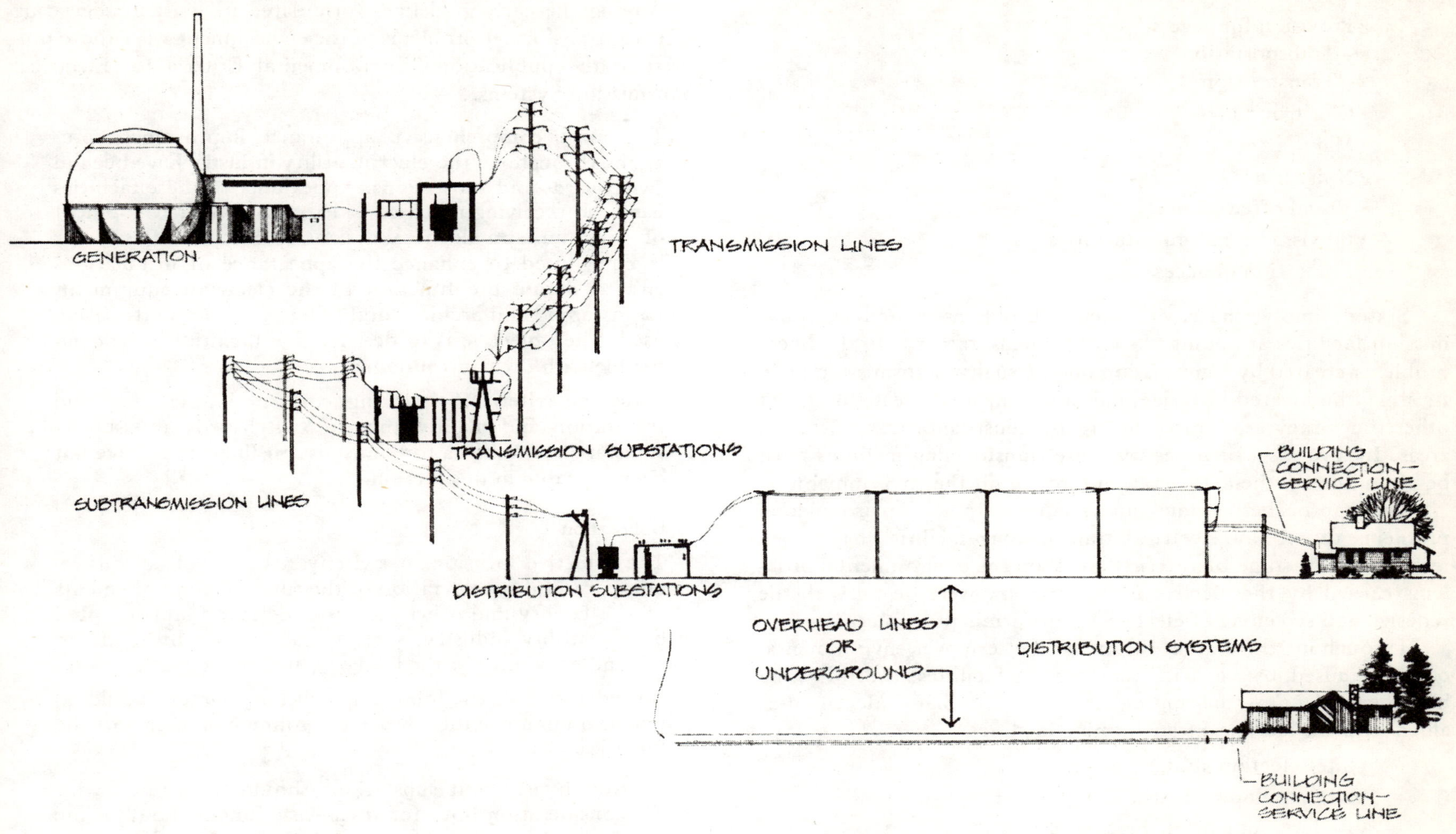

3. William J. Johnson, Carl Johnson and Clarence Roy, *Transmission and Distribution Rights-of-Way Selection and Development,* (Ann Arbor, Michigan for Consumer's Power Company, Jackson, Michigan, 1970), p. 1.

Storage facilities necessary on the site:

In essence, the environmental problems created by electrical transforming facilities may be categorized as follows:

- Provision for access
  - —Transportation
  - —Transmission
- Construction
- Maintenance
- Noise levels
- Visual offensiveness
- Provision for storage facilities
- Limitation of access

Stated simply, the environmental problems caused by transmission facilities are point or spot problems, rather than the linear problems created by transmission lines. Usually, transmission facilities may be located in residential or commercial areas, though at other times they are located in largely industrial or transportation areas. In any case, of necessity these transforming facilities must be ubiquitous in their scatteration throughout the environment.

The approaches, studies and solutions to the environmental problems caused by electrical transforming facilities have been many. In fact, some of the first work on the environmental problems caused by the electric utility industry were begun with the redesign and screening of electrical transforming facilities.

Throughout the current history of concern over environmental damage caused by electric transforming facilities, the following have been some of the approaches to the problems identified as allied with the electrical transforming facilities:

- Site selection studies
- Site development studies
- Redevelopment of the components
- Screening
- Camouflaging
- Hiding of transforming facilities
- Guideline development
- Evaluation techniques

Among the early guidelines formulated to assist in alleviating electrical transformer problems in the environment were those prepared in the publication "Environmental Criteria for Electrical Transmission Systems."

> The most comprehensive appearance improvement programs throughout the electric utility industry have been in switchyards and substations. Appearance of these facilities has been receiving much consideration in the development of structural and electrical designs by many utilities. Color is being used to enhance the appearance of installations and to define the function of the electrical equipment. Landscaping and architectural fences have been effectively used. The objective is to design these facilities to be compatible with their surroundings.
>
> Since the criteria for clearing, construction, cleanup, and restoration and maintenance of switchyards and substations are the same as for transmission lines, they have not been repeated in this section.
>
> **B. Location**
>
> The effective location of switchyards and substations requires careful consideration of the functional requirements of the facility and other factors understood and accepted by the utility industry, such as good access by road and rail and proximity to the load or generation station.
>
> Consideration of the following additional factors would do much to minimize the adverse environmental impact of the facilities:
>
> 1. Switchyards and substations should be located with consideration both for their basic function and for the preservation of public views of scenic, historic, natural and recreation areas, parks, monuments, etc.
> 2. The proposed location, layout, and design parameters should be coordinated with appropriate local planning

agencies to assure maximum compatibility between the facilities and present and future land use.

3. The location should be coordinated with the needs of utilities delivering power into or receiving power from the station. This is particularly important in the development of the site's electrical layout to minimize costly, unsightly, transmission line crossovers and unnecessary duplication of facilities.
4. If possible, locations should avoid populated areas, parks, scenic areas, wildlife refuges, hilltops and natural or man-made structures.
5. Locations near existing or proposed interstate or state primary highways should be avoided.
6. Where possible, substations should be located where they may be naturally or artificially screened.
7. Potential noise should be considered when the locations for high-voltage substations are being determined. The facilities should be located in areas where sound will not be resonated.
8. Multiple level, terraced substations may be used to minimize excavation and provide a facility that will blend effectively with sloping terrain.

C. Design

The following factors incorporated into the design of substations and switchyards will significantly contribute to the environmental compatibility of the facilities.

1. Facilities should be designed to be compatible with the area in which they are located and constructed from materials which are native to the area or which harmonize with their surroundings.
2. The size of switchyards and substations should be kept to the minimum which will accommodate the ultimate functional and aesthetic requirements.
3. Cut and fill slopes should be designed to achieve maximum compatibility with the surrounding natural topography.
4. Simplified, functional structures should be considered in preference to conventional lattice types.

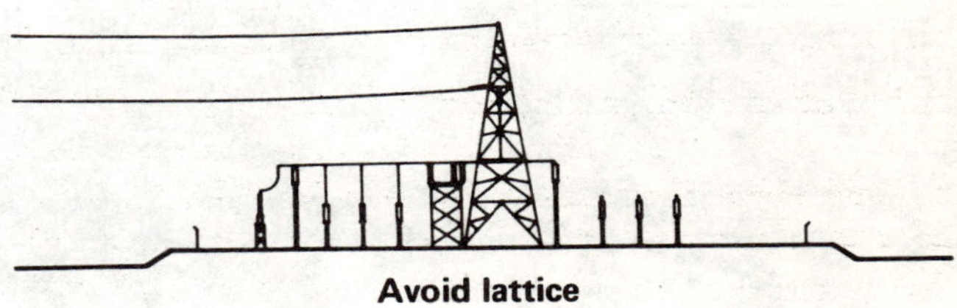

Avoid lattice

5. In general, the use of low profile concepts and simplified structures within switchyards and substations will enhance the overall appearance of the facility.
6. Buildings should be designed to be architecturally compatible with their surroundings.

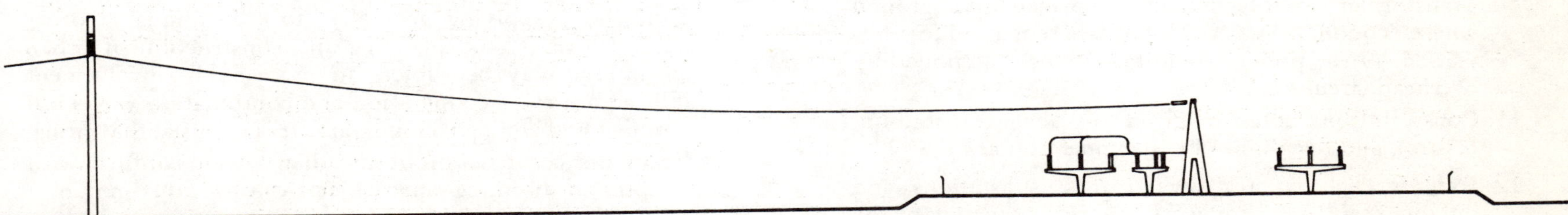

An uncluttered substation

7. Distinctive colors should be used wherever they appropriately enhance the appearance of substation structures and identify or express the function of electrical equipment; provided, however, that such colors are in harmony with the surrounding area.
8. Architectural fences or architectural features should be used along the fence line to complement the surrounding landscape or other features. Architectural fences may in some instances complement the architecture of service or control buildings within the substation area.

9. Creative landscaping practices should be developed and used to enhance the appearance of substation installation. However, the design may in some cases be compelled to abide with existing motifs, traditional forms and materials.
10. Transformer circuits should be placed underground where economically feasible, or where natural features viewed by the public would be seriously despoiled by overhead circuits.
11. Consideration should be given to placing distribution, control, and telephone circuits underground.
12. Exterior lighting for switchyards, substations and buildings, adequate for work and for protection of the facilities from sabotage and malicious mischief, should be provided. In some cases exterior lighting may appropriately enhance the aesthetic values of the structured landscape elements.
13. Disconnect switches, fuses, and other similar devices should preferably be located on low ground mounted structures as contrasted with placing them on high substation dead-end structures.
14. Maintenance, warehouse, and storage facilities at the substations should be so located and maintained as to minimize their adverse effect upon the environment.[4]

The Western Systems Coordinating Council in their December 3, 1971, report entitled, "Environmental Guidelines" developed an excellent section on the suggested environmental guidelines for major substations. These are as follows:

MAJOR SUBSTATIONS

1. Location

Substations should be located with consideration for their basic function and for the protection and preservation of scenic, natural, historic, archeological and recreational values and, where possible, should avoid parks and monuments, wildlife refuges, and conspicuous hilltops.

When substations are located near interstate or state primary highways, the exposure of the facility to the public should be considered and adequate screening, landscaping or other esthetic methods used to minimize the visual or physical impact on the landscape.

Requirements for the selection and acquisition of optimal sites for major urban substations do not differ significantly from those applying generally, to substations in other areas. However, the planning and construction of urban substations may be subject to more public involvement and require close cooperation and coordination with local planning and zoning commissions. It is essential that siting, design and construction of an urban station conform with the present and long-range objectives of local community

4. U.S. Department of the Interior, U.S. Department of Agriculture, *Environmental Criteria for Electrical Transmission Systems* (Washington, D.C., U.S. Government Printing Office, Superintendent of Documents, 1970), pp. 31-38.

planning. In order to gain public acceptance, site development must be compatible with existing and future land uses in the area.

2. Land Requirements

The land area of substations should accommodate the ultimate functional and esthetic requirements, including a buffer zone, if necessary.

Cut and fill slopes should be designed to achieve maximum compatibility with the surrounding natural topography.

Multiple level, terraced substations may be used to minimize excavation and provide a facility that will blend effectively with sloping terrain.

In the case of urban areas, land area requirements are usually governed by land availability. In these areas, it is especially desirable to keep stations as small as possible. It may therefore, be necessary to develop unique designs to construct stations of the desired size, including the possible use of architectural enclosures. Plans should include additional area in anticipation of station expansion.

3. Transmission Access

Transmission lines should be planned to be as esthetically pleasing as possible so as to minimize the visual impact and infringement on the landscape.

Transmission line crossovers should be avoided for appearance and reliability purposes.

Consideration should be given to the undergrounding of incoming transmission circuits as well as outgoing feeder circuits for urban substations. In many urban areas, lack of adequate aerial access routes may necessitate the undergrounding of nearly all transmission lines.

4. Transportation Access

Good access by road and railroad is extremely important because of the heavy equipment required for the construction and maintenance of substations. Existing transportation corridors should be utilized wherever possible.

Transportation problems normally associated with urban areas should be given careful consideration in the siting, planning, design, and construction of urban stations. Access routes should be selected to assure that original construction can be accommodated and emergency and repair equipment can be easily moved into and out of the substation area without obstruction.

5. Safeguard Considerations

When designing and locating substations, the following factors should be considered:

(a) Slide areas
(b) Unstable soils
(c) Subsidence
(d) Floods
(e) Earthquakes
(f) Faults

As urban substations must usually be located in areas of high population density, it is especially important that the site selected be free from potential damage due to natural or man-made hazards.

6. Architectural Design

Substations should be designed to be compatible with the area in which they are located and constructed as appropriate from materials which are native to the area or which harmonize with their surroundings.

If distinctive colors are in harmony with the surrounding area they may be used to enhance the appearance of substation structures and identify or express the function of electrical equipment.

Architectural fences, walls or other treatment may be used to complement the surrounding landscape or other features. Architectural fences may complement the architecture of the service or control building within the substation area.

Enclosed stations may be desirable in urban areas to attenuate noise developed by equipment such as high voltage transformers and circuit breakers, and to insure visual acceptability. The current trend is to self-contained equipment or low silhouette design.

Architectural efforts should be directed toward good taste and compatibility and will vary depending on local environment. The objective should be to design the substation so that it is complementary with existing and planned developments.

7. Landscaping and Lighting

Creative landscaping that is compatible with the area should be used to enhance the appearance and public acceptance of the substation installation.

Exterior lighting for substations and buildings should be adequate for work and for protection of the facilities from sabotage and vandalism.

The use of decorative or achitectural lighting in urban substations may also contribute to pleasing facility appearance. In urban areas consideration should be given to the incorporation of small park developments for public use when appropriate and practical.

8. Noise Levels

Potential noise levels should be considered when the locations for major substations are being determined. Noise producing equipment should be located on the site to minimize radiation of sound to the surrounding neighborhood and buffer areas should be provided where necessary.

Noise studies should be made to determine if the substation will cause undesirable noise levels in the surrounding area. Permissible levels can be determined using International Standards Organization (ISO) or other appropriate criteria. Ordinarily the criteria used will be dependent upon the activity being carried out in the adjacent area. Local noise ordinances should also be consulted.

Where it is determined that noise radiated from the substation would be detrimental to the surrounding environment, steps should be taken to minimize and reduce its impact.

The most effective means would be to specify equipment having a lower sound power output. Containment barriers, sound absorption and equipment arrangement should also be utilized to control noise where required.

9. Substation Layout

The use of low profile concepts and simplified structures will enhance the overall appearance of the station. Simplified, functional structures should be considered in preference to conventional lattice types. Bay positions within the substations should be predetermined to eliminate the need for transmission line crossovers. Consideration should be given to placing distribution control and telephone circuits underground. Electrical equipment should be located on low ground mounted structures as contrasted with placing them on high substation structures.

Maintenance, warehouse and storage facilities at the substations should be so located and maintained as to minimize any adverse effect upon the environment. The use of guyed transmission line structures adjacent to substations should be avoided.

10. Construction and Maintenance

Special attention should be given to the following factors:

(a) Clearing and construction should be performed in a manner which will maximize preservation of natural beauty and vegetation, conservation of natural resources, protection of natural habitat for wildlife, minimize scarring of the landscape and prevent silt deposition in water courses.

(b) Clearing and grading of construction areas, such as storage areas, setup sites, should be minimal.

(c) Borrow areas and rock quarries should be located away from the public view, if possible. Upon completion of construction these areas should be restored to such a condition that erosion will be avoided. Borrow pits can be esthetically designed to blend into the natural landscape and in some instances used for fish and wildlife management purposes.

(d) Soil excavated during construction and not removed or used elsewhere should be placed on the cleared area and graded to conform with the adjacent land. Top soil should be saved, replaced and appropriate vegetation planted.

(e) Oil spills, littering and other types of pollution should be avoided while performing work in the vicinity of streams, lakes and reservoirs.

(f) Water taken from streams or bodies of water for construction purposes should be limited to quantities which will cause minimum harm to the environment.

(g) Precaution should be taken to prevent range or forest fires. Special attention should be given to fire prevention planning, training of personnel in fire fighting and fire inspection programs.

(h) Scars, cuts, fills and other unattractive areas should be seeded or reseeded to reduce erosion, and restore natural appearance and provide food and cover for wildlife.

(i) Temporary roads should be rehabilitated to minimize erosion and visual scars. Access roads should be maintained so as to provide proper drainage and erosion control.

(j) Upon completion of use, construction areas, camp sites and storage areas should be cleared and restored to their original and natural condition, and all buildings, equipment and supplies not required for operation removed.

(k) If site factors make it difficult to establish a protective vegetative cover, such as native trees, shrubs, herbs, and grass, other restoration procedures may be advisable, such as the use of gravel, rocks, decomposed granite and concrete.

(l) Chemicals, when used, should be carefully selected to have a minimum effect on indigenous plant life and selective application should be used wherever appropriate to preserve the natural environment in scenic areas. The impact of temporary discolorations of foliage should be considered. Where this factor is critical either mechanical means of vegetative control should be used or the work should be scheduled in the early spring or late fall. Chemicals should be applied in a manner consistent with the protection of the environment, particularly for protection of the health of humans and wildlife.

11. Radio and Television Interference

Prior to construction, ambient radio frequency noise level and received radio and television signal strength measurements should be made immediately adjacent to proposed substation sites.

After construction and energization of equipment and lines, additional ambient radio frequency noise and received radio and television signal strength measurements should be made adjacent to substation sites at the same locations used prior to construction. Further investigation of radio and television interference should be made if complaints are received. Approximately 70% to 85% of all complaints received are caused by customer owned devices and the remaining are usually traced to line hardware or contamination problems and are cleared by line crews.

Any problems located by the repeat measurements should be corrected at once.

12. Existing Installations

Some existing installations require a new look from the standpoint of appearance to minimize the visual impact and physical infringement on the landscape.

As the total of the existing installations far outnumber the anticipated or contemplated new installations for some time to come, the existing installations present the most challenging problem.

In order to improve the appearance of existing substations the following factors should be considered:

(a) When replacement becomes necessary, replace wood pole structures with steel structures, in stations where both types of construction exist.

(b) When the original installation of transmission structures in the vicinity of a substation tends to become disorderly and detracts from the overall appearance of the station, replace and rearrange such structures with more esthetically pleasing structures.

(c) Remove surplus foundations where electrical equipment or steel structures have been removed or relocated.

(d) Improve maintenance yard surfaces and repair and landscape eroded fill and cut slopes about the substation when necessary to preserve the surface or enhance the appearance.

(e) Landscape the entrances at substations and at selected buildings.

(f) Flatten out cut and fill slopes to be more compatible with the natural topography where practicable, round off top of slope.

(g) Clean up approach roads by reshaping roadways, drainage ditches, weed removal and burning, and resurfacing the road as necessary.

(h) Use esthetically pleasing color schemes when repainting existing buildings, structures or equipment.

(i) Clean up and screen storage yards with plantings.

(j) Install attractive lighting.

Special priority should be given to enhancing the appearance and minimizing the environmental impact of existing urban substations. Utilities should evaluate existing stations in terms of the acceptability of present transmission facilities and routes, the effectiveness of noise control treatment, landscaping, and other factors which affect the degree to which the station exists in harmony with the surrounding environment.[5]

The visual impact study conducted by InterDesign, Inc., of Minneapolis, St. Paul for the Northern States Power Company contained a section on substations. The criteria rating definition for substations (the method of formulation of this was mentioned previously) is as follows:

Criteria 1: Siting

Land Use:

++ Industrial-Commercial land uses
0 Open spaces, unused land within industrial-commercial land uses
- -- Residential, open space, unused land and/or agricultural land

Criteria 2: Buffers
(A visual screening of the facility)

Vegetation Buffers:

++ Maximum vegetation buffer
0 Minimum vegetation buffer
-- No vegetation buffer

Topography Buffers:

++ Maximum topographic buffer
0 Minimum topographic buffer
-- No topographic buffer

Structural Buffers:

++ Maximum structural buffer
0 Minumum structural buffer
-- No structural buffer

5. Western Systems Coordinating Council, Environmental Committee, *Environmental Guidelines,* (Los Angeles, California, Western Systems Coordinating Council, 1971), pp. 37-43.

Criteria 3: Topography Destruction

++ No destruction of topography
0 Minimum destruction of topography
-- Maximum destruction of topography

Criteria 4: Vegetation Destruction

++ No destruction of vegetation
0 Minimum destruction of vegetation
-- Maximum destruction of vegetation

Criteria 5: Scale Compatibility
(Size and form of facility in relation to surroundings)

++ In scale with surroundings
0
-- Out scale with surrpundings

Criteria 6: Natural or Historic Features
(Proximity of facility to natural or historic features)

++ No proximity of facility
0 Potential of area destroyed by proximity
-- Features destroyed by proximity

Criteria 7: Viewing
(Exposure of facility to public)

Viewing Level
(Level from which facility is viewed by public)

++ Viewed from below
0 Viewed from eye level
-- Viewed from above

Criteria 8: Arrangement
(Visual order of facility)

Organization:

++ Visually organized
0
-- No visual organization

Criteria 9: Internal Elements
(Functional components of facility)

Clarification
(Visual understandibility of functional components of the facility)

++ Visual clarification of internal elements
0
-- No visual clarification of internal elements.

Criteria 10: Public Exposure

(Viewing from major circulation routes)

Exposure Routes:

++ Located along minor roads

0 Located along major streets

-- Located along Freeways or highways

Criteria 11: Visual By-Products

(Visibility of storage, maintenance and accompanying debris)

Ground Plane Exposure
(Reveals visual by-products)

++ No ground plane exposure

0 Minimal ground plane exposure

-- Maximum ground plane exposure

Criteria 12: Night Visual Continuity

(Lighting levels of facility compatible with surroundings)

++ Night visual continuity with surroundings

0

-- Night visual discontinuity with surroundings

Criteria 13: Creative Use of Color

(Color used to blend, explain or accentuate design)

++ Color used positively
0 Color not used
-- Color used negatively

Criteria 14: Use of Signage
(Visual Communication used to explain function)

++ Coordinated Signage
0 Minimum Signage
-- Unnecessary Signage

Criteria 15: Land Use Continuity

++ Facility Siting does not disrupt existing land use patterns.
0
-- Facility siting disrupts existing land use patterns.

Criteria 16: Day Visual Continuity
(Siting and design are compatible with surrounding character)

++ Day visual continuity
0
-- Day visual discontinuity[6]

6. InterDesign, Inc., *Visual Impact Study,* (St. Paul, Minnesota, for Northern Power Company, 1971), pp. 31-33.

The following chart shows the criteria rating for a number of substations evaluated utilizing these 16 criteria.

Criteria Ratings[7]

++ Desired Condition
o Neutral Condition
-- Undesired Condition

Substations	1		2						3		4	5	6	7		8			9		10	11			12	13	14	15	16
	Land Use Substations	Edge Trans. & Distr.	Vegetation Substations	Topography Substations	Structure Substations	Major Edge Control Service Centers	Minor Edge Control Service Centers	Edge Control Transmission	Topography Destruction Subst. & Serv.	Line Relation To Topography Trans. & Dist.	Vegetation Destruction All Facilities	Scale Compatibility Subst. & Serv.	Natural or Historic All Facilities	Viewing Level Subst. & Serv.	Silhouette Trans. & Distr.	Organization Subst. & Serv.	Number of Poles Viewing Trans. & Distr.	Pole Type Trans. & Distr.	Clarification Substation	Line Attachment Trans. & Distr.	Exposure Routes All Facilities	Ground Plane Exposure Substations	Service Roads Trans. & Distr.	Parking & Storage Service	Night Visual Continuity Subst. & Serv.	Creative Use of Color All Facilities	Use of Signage Subst. & Serv.	Land Use Continuity All Facilities	Day Visual Continuity
1. Campus	--		o	--	--				++		++	++	++	o		--			--		o	o			++	o	o	++	++
2. Cleveland	--		++	--	o				++		++	++	++	o		--			--		o	o			--	o	o	++	--
3. Dayton's Bluff	--		--	--	--				o		o	++	o	--		--			--		--	--			--	o	o	--	--
4. Gleason's Lake	--		++	++	--				o		++	--	++	o		--			--		--	o			--	o	o	--	--
5. Great Northern	++		--	--	--				++		++	++	++	o		--			--		--	--			++	o	o	++	++
6. Inver Grove	--		--	--	--				++		++	--	++	o		--			--		++	--			--	o	o	--	--
7. Main Street	--		--	++	--				--		--	--	--	--		--			--		++	++			--	o	o	--	--
8. Medicine Lake	o		++	o	--				--		--	++	++	o		--			--		++	--			--	o	o	--	--
9. Merriam Park	--		--	--	o				o		++	--	++	++		--			--		o	--			--	o	o	--	--
10. Parker's Lake	--		++	--	--				++		--	--	++	--		--			--		--	--			--	o	o	++	--
11. Red Rock	--		--	--	--				++		--	++	++	--		--			--		--	--			++	o	o	++	--
12. Savage	o		++	--	--				++		--	++	++	o		--			--		--	++			++	o	o	++	++
13. Scott County	--		++	--	--				o		++	--	++	o		--			--		--	--			--	o	o	--	--
14. Southtown	++		--	--	++				++		++	++	++	o		--			--		o	--			--	o	o	++	++
15. St. Louis Park	++		o	++	++				o		++	++	++	++		--			--		--	--			--	o	o	++	--
16. West Coon Rapids	--		--	--	--				++		++	--	++	o		--			--		--	--			--	o	o	++	--

7. Ibid., p. 35.

The following chart shows a satisfaction of criteria for the substations and the criteria previously mentioned.

Satisfaction of Criteria[8]

Percentage of ++
Percentage of o
Percentage of --

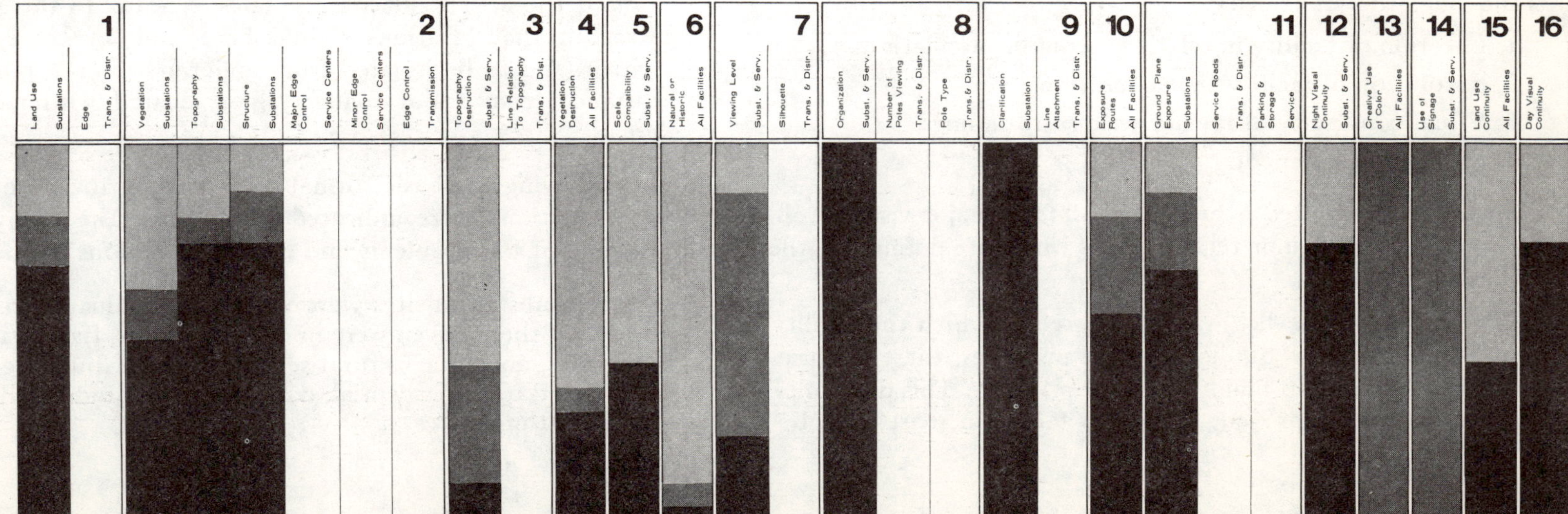

8. Ibid., p. 37.

The following chart shows the high priority criteria for each of these evaluated substations:

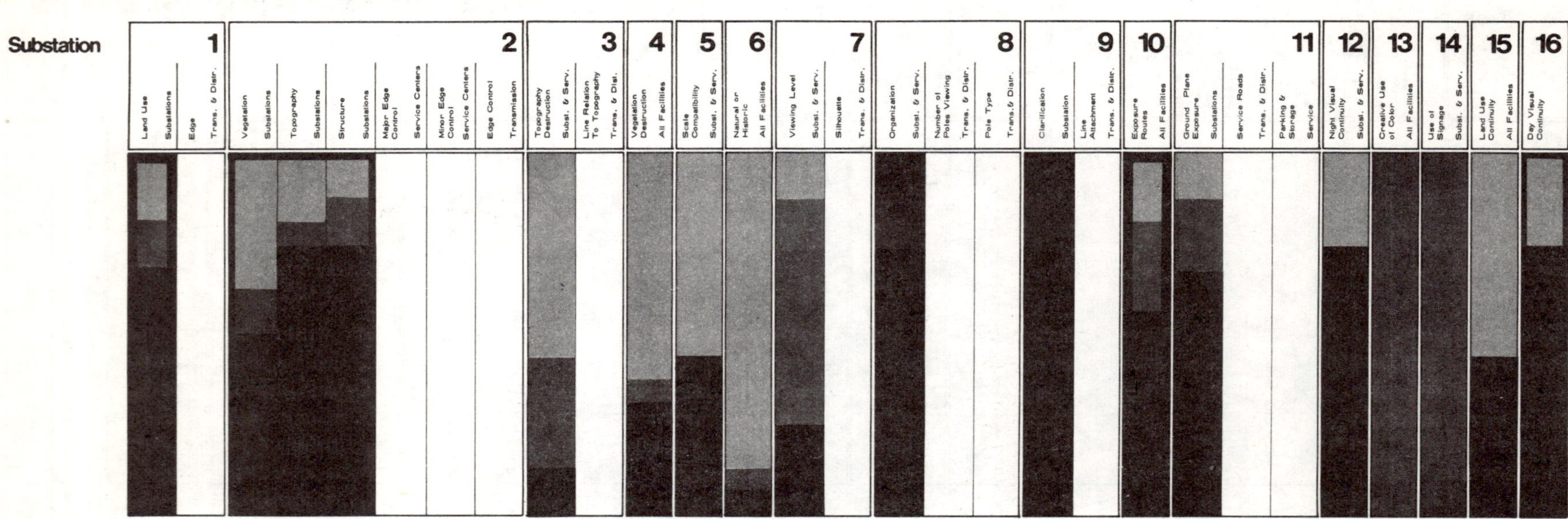

9. Ibid., p. 39.

That same study identified the key problems of substations in the following order of priority.

1. Location of facility in relation to public circulation.
2. Resolution of facility edge conditions.
3. Visual continuity of facility with surroundings.
4. Facility arrangement.

Substations presented a hierarchy of four major visual problems and several minor related issues which are defined on the following pages.

Most crucial on the list was the degree to which the facility was exposed to the public. In most instances, these facilities were located next to major circulation routes. This proximity to public viewing proved a negative factor in most cases because the junction between transmission and distribution often caused the area to become visually chaotic. In this regard, it was felt that locations should be found for these facilities which tend to be isolated from major viewpoints, removed by distance from major roadways, or located within industrial areas.

In following the basic format of this study the problems and recommendations were indicated graphically. The following was the outline of the problems and recommendations of that study:

> Sixteen substation sites were visited and evaluated for this study. Of these, seven were in core city areas, five in suburban areas, and four in rural settings. The surrounding land uses varied from residential, commercial and industrial uses to circulation routes.

The key problems in order of priority:

1. Location of facility in relation to public circulation. (Criteria 1 and 10 with minor criteria 3, 4, 5 and 6.)
2. Resolution of facility edge conditions. (Criteria 2 with minor criteria 7 and 11.)
3. Visual continuity of facility with surroundings. (Criteria 16 with minor criteria 12 and 13.)
4. Facility arrangement. (Criteria 8 with minor criteria 9, 13 and 14.)

Substations

Substations presented a hierarchy of four major visual problems and several minor related issues which are defined on the following pages.

Most crucial on the list was the degree to which the facility was exposed to the public. In most instances, these facilities were located next to major circulation routes. This proximity to public viewing proved a negative factor in most cases because the junction between transmission and distribution often caused the area to become visually chaotic. In this regard, it was felt that locations should be found for these facilities which tend to be isolated from major view points, removed by distance from major roadways, or located within industrial areas.

PROBLEM: Facilities are located adjacent to major public circulation.

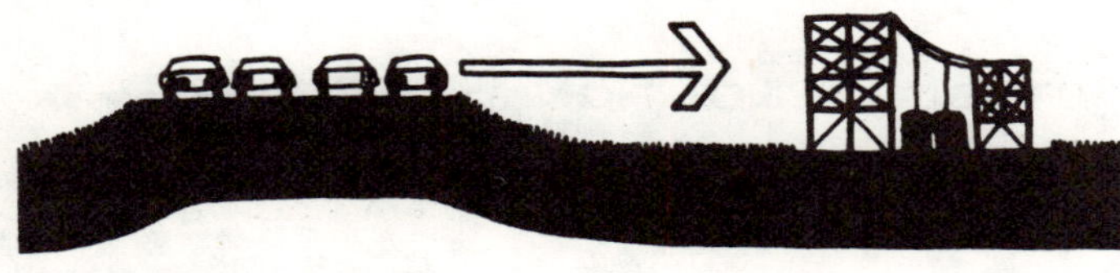

RECOMMENDATION I: Locate facilities away from major public circulation and/or in industrial areas.

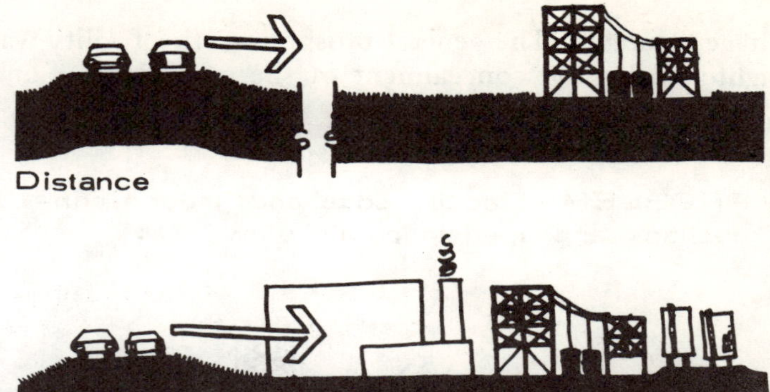

Industrial Area

The second priority problem relates to the first in that it is also concerned with public exposure of the facility. In a high percentage of the sites reviewed, the existing edge of the site was found to allow uncontrolled views into the facility. Often the facility was viewed from eye level or above, which focused the observer's attention on a maze of ground connections and exterior storage areas. In many cases, neither vegetation, topography nor structure was effectively used to screen the variety of elements at eye level.

A more creative organization of the periphery of the facility, using either vegetation, topography or structure as buffer elements would provide a much improved impact. Possibly a more significant improvement would be effected if the land purchased for these facilities would provide a greater peripheral area for screening and blending. This would tend to simplify the visual chaos usually present all around these facilities.

In relation to edge treatment, the Savage substation was a notable exception to this general problem. Although it was difficult to determine what land NSP owned in this instance, the whole station was surrounded by a grove of large trees which effectively screened the ground connections and exposed judiciously some segments of the high voltage structure. The general prospect of this facility was heightened by the concealment of the ground plane and surrounding chaos of wires.

PROBLEM: Facility edge and ground connections are open to public views.

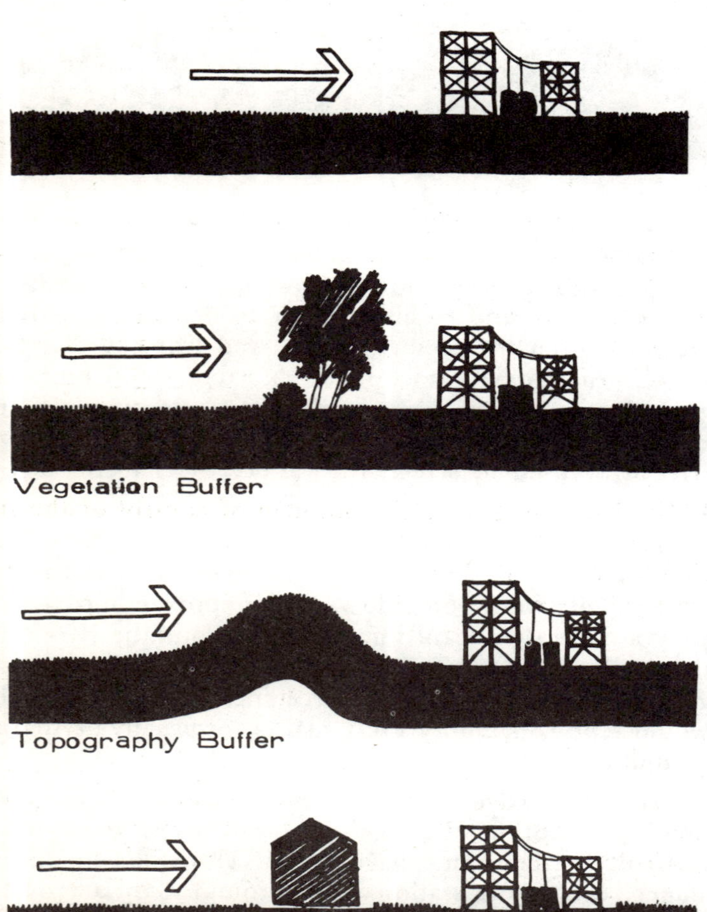

Vegetation Buffer

Topography Buffer

Structure Buffer

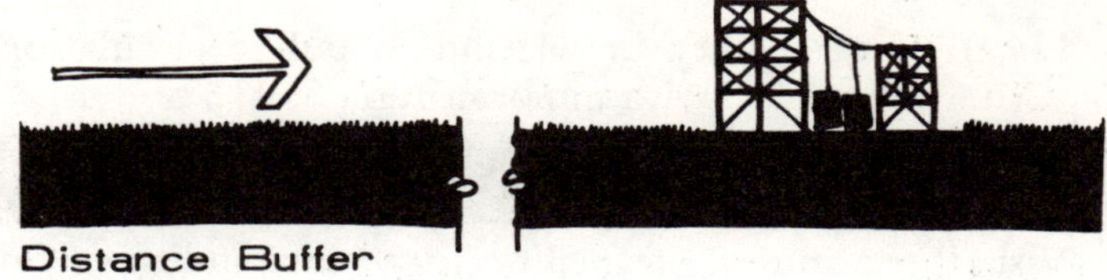

Distance Buffer

PROBLEM: Facility design, scale location and lighting are out of context with the surroundings.

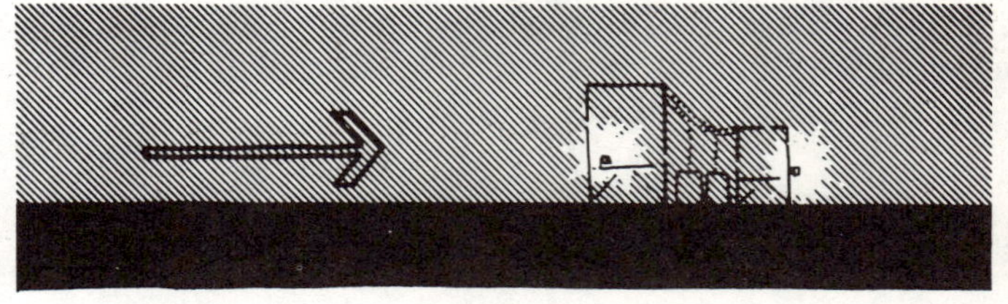

RECOMMENDATION 3: Treat each site uniquely.

Make changes in basic design or layout to

accommodate varying scales available topography, vegetation, structures, or land uses that occur in different locations.

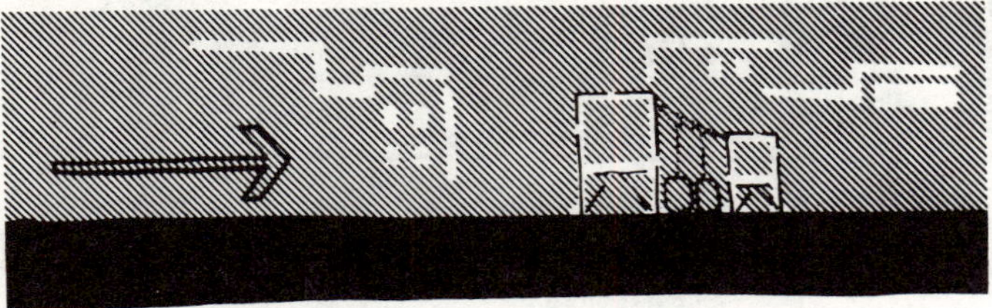

Effect Lighting

Lighting in character of surrounding area

The minimal amount of special treatment for each site was determined to be the third priority problem. The standard design and layout typical of each substation was found to disregard the varying scale of elements within specific areas and provide an objectionable industrial contrast to the related residential land uses when located in these areas. Site locations which attempted to place these facilities within industrial use areas would eliminate a great percentage of the problems. In any case, it was felt that the size and mass of the structures should relate more carefully to the size and mass of structures and other elements in the surrounding areas.

Forms and materials used throughout these facilities were found not to relate to the unique visual form of many surrounding environments. The steel and chain link image makes the facilities stand out negatively when located in residential and commercial areas. The use of brick and plant materials as done at the Cleveland substation begins to tie the development more effectively to its residential-commercial setting. However, the concept could have been extended by continuing the use of brick for screen walls instead of link fence and by adding some major scale plantings to tie the development into the surrounding landscape.

During the evening hours, lighting also distracted from the continuity with the surroundings. The visible light sources provided a glaring distraction in most facilities and were unrelated to the general pattern of lighting in the areas. A more effective use of lighting, concealing the light sources and effectively accenting major elements of the facility, would greatly improve the nighttime image. Patterns of light which tied the development to its surrounding night landscape could be effective in creating a positive element out of these facilities. For example, the Great Northern substation, located in an industrial refining area, could be accentuated with a small scale series of lights similar in pattern to the surrounding lights.

The fourth priority problem determined related to the way in which the facility itself was organized. In a high percentage of the sites reviewed, exterior storage and parking areas were located between the viewer and the facility, thus adding to the visual chaos of the area. Site organization which moved these peripheral developments away from major public viewing would go far toward simplifying the facility and impact.

Related to this issue is the problem of control of the surrounding chaos of development. Often times, the substation structure itself was not that objectionable, but the surrounding incoming and outgoing lines presented a great sense of disorder. Burying the distribution output lines and more carefully ordering the transmission input lines would considerably reduce the negative impact.

Even though the substation structures were not overly complex, they could be given more simple silhouette. The fewer elements a facility has, the less chaotic it will appear and it is recommended that structural elements should be explored in this regard. The use of color might also aid in explaining the various elements of the station and provide a more understandable massing of elements to the layman.

PROBLEM: Facility access, parking, and storage are highly visable surrounding a complicated structural form.

Complicated Elements

RECOMMENDATION 4: Site each facility more carefully as related to the public views of it and its related service access and parking; and take steps to simplify the visual form.

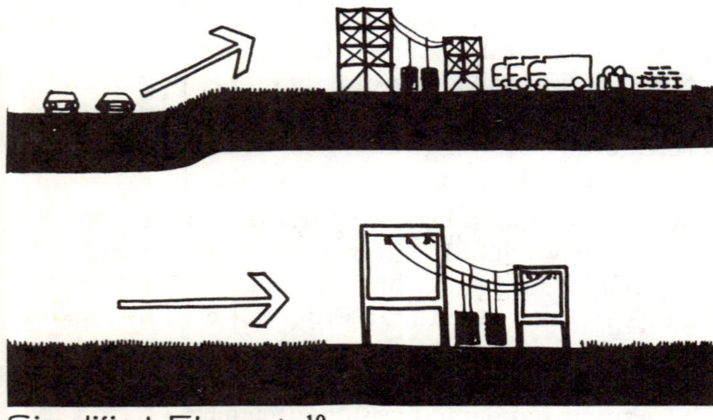

Simplified Elements[10]

10. Ibid., pp. 69-77.

Using the uniform format developed for that visual impact study for Northern States Power, the following figures from that report pertain to substations for transformer facilities.

Northern States Power **VISUAL IMPACT STUDY**[11]	**InterDesign Inc.**
PHASE I Prototype Study **Facility Analysis Criteria and Measures**	**CRITERIA**

Criteria 1. Location of facility in relationship to surrounding uses and public circulation.

a. Substations only

++ Located in an industrial area with minimal adjacent public circulation.

\+

0

\-

-- Located in open space, residential or commercial area with public circulation on all sides.

Comments:

Visualization:

++ --

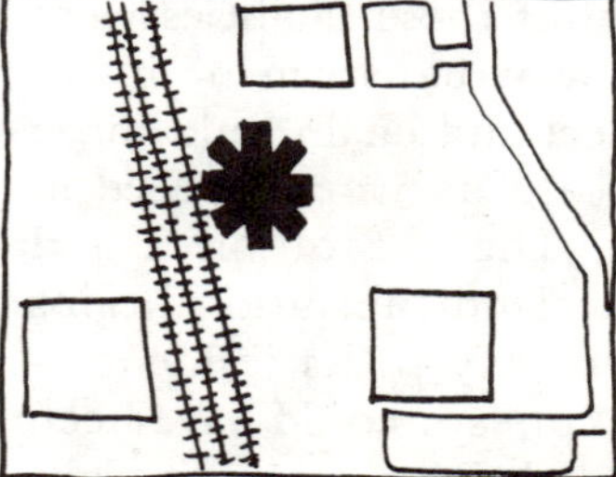

11. Ibid., p. 125.

Northern States Power **InterDesign Inc.**
VISUAL IMPACT STUDY [12]

PHASE I Prototype Study **CRITERIA**
Facility Analysis Criteria and Measures

Criteria 5. Compatability of facility to surrounding structures

a. Substations, service centers

++ Same relative scale (size or mass) or materials.

+

0

-

-- Different scale (size or mass) or materials.

Comments:

Visualization:

++

--

12. Ibid., p. 132.

Northern States Power **InterDesign Inc.**
VISUAL IMPACT STUDY [13]

PHASE I Prototype Study **CRITERIA**
Facility Analysis Criteria and Measures

Criteria 7. Experience from major viewing points to the facility

a. Substations and office, service, storage complexes.

++ Buffer between view points and facility to eliminate view of ground plane and resolve silhouette.

+

0

-

-- No buffer, view from above, with silhouette.

Comments:

Visualization:

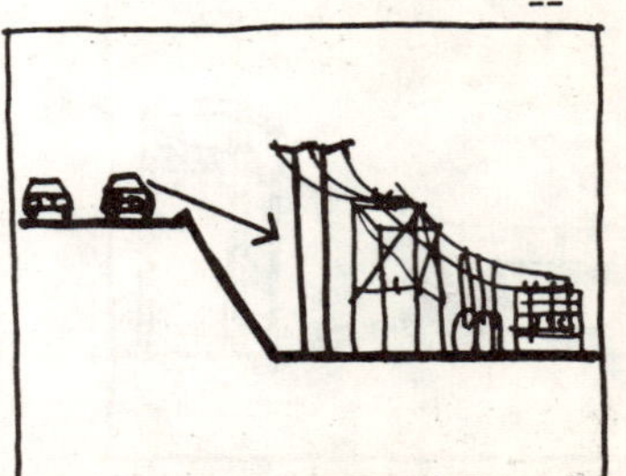

13. Ibid., p. 134.

Northern States Power **InterDesign Inc.**
VISUAL IMPACT STUDY[14]

PHASE I Prototype Study **CRITERIA**
Facility Analysis Criteria and Measures

Criteria 8. Visual arrangement of the general facility

a. Substations

++ Visually organized through the use of a related group of forms for all functional elements and visually organized power lines surrounding the facility.

\+

0

\-

-- No visual organization due to dissimilar forms and intense overlay of functional elements in and around facility.

Comments:

Visualization:

++

--

14. Ibid., p. 136.

Northern States Power **InterDesign Inc.**
VISUAL IMPACT STUDY[15]

PHASE I Prototype Study **CRITERIA**
Facility Analysis Criteria and Measures

Criteria 9. Visual clarification of internal elements.

a. Substations

++ Internal elements clarified and interrelated through the use of color, form or graphics

\+

0

\-

-- No clarification of internal elements.

Comments:

Visualization:

++

--

15. Ibid., p. 139.

Northern States Power **InterDesign Inc.**
VISUAL IMPACT STUDY[16]

PHASE I Prototype Study **CRITERIA**
Facility Analysis Criteria and Measures

Criteria 10. Amount of Public Exposure

a. All facilities

++ Minimum exposure to major automobile routes.

\+

0

\-

-- Maximum exposure to major automobile routes

Comments:

Visualization:

++

--

16. Ibid., p. 141.

Northern States Power **InterDesign Inc.**
VISUAL IMPACT STUDY[17]

PHASE I Prototype Study **CRITERIA**
Facility Analysis Criteria and Measures

Criteria 11. Results of visual by-products (parking and storage)

a. Substations

++ Public edges have buffer; lack of visability of parking, exterior storage and maintenance access; ground connection totally invisible.

\+

0

\-

-- Visability of parking, exterior storage and maintenance access, and connection of facility with ground complicated by chaos of fencing.

Comments:

Visualization:

++

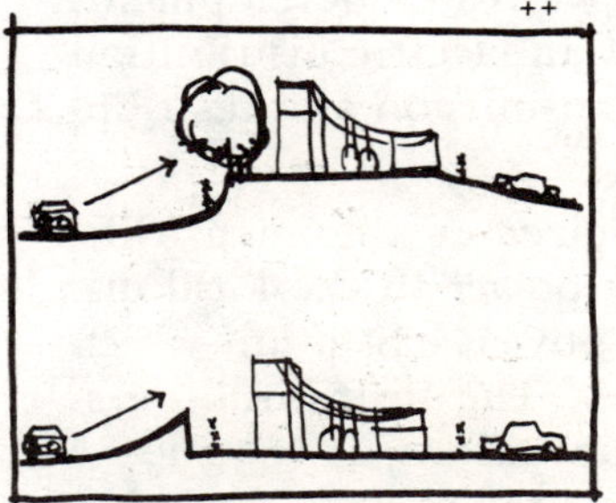

--

17. Ibid., p. 142.

Probably, in the history of the environmental "revolution" the outstanding study done in the United States, if not in the world, dealing with some of the visual environmental problems of the substations or transformer stations is the previously mentioned report developed by the landscape architectural firm of Johnson, Johnson and Roy of Ann Arbor, Michigan for Consumers Power Company of Jackson, Michigan, which was entitled, "Substations, Site Selection and Development." That report dealt in a major way with attempting to solve all of the identified environmental problems created by substations. Therefore, the following section will quote extensively from that report, since it epitomizes probably the highest apex available at the present time of the way that a landscape architectural firm has worked on the environmental problems with a member of the electric utility industry.

The report begins with a discussion of basic requirements for a substation in the following way:

> The expanding development of commercial and residential areas imposes an obligation on the franchised electric utility company to supply the new development with electric power. This is also true in established areas where requirements for power have increased due to installations of new appliances, air conditioners, electric heat, etc.
>
> To fulfill this need, the utility company finds it necessary either to expand its present facilities or to install new facilities to serve an established or new area of development. These facilities are in the form of an electric distribution substation together with its sub-transmission and distribution systems.
>
> The sub-transmission system, energized at a nominal voltage level of 46,000 volts, supplies power to the local distribution substation from a bulk power substation which may be located several miles away. The distribution system, originating at distribution substations, is the means by which electric power is distributed to each consumer throughout the community.
>
> The primary purpose of an electric distribution substation is to reduce the voltage of the bulk power source to a lower voltage for distribution to specific zones of the community. Installed in the basic substation are electrical equipment items such as transformers, voltage regulators, protective devices and switches. It also contains steel structures to terminate the sub-transmission and distribution lines and also to support protective devices, switches and other smaller components which make up the substation.
>
> The function of the transformer is to reduce the voltage of the supply source to a lower voltage for distribution. The size and number of transformers depends on the electrical needs of the community.
>
> The voltage regulator automatically maintains the voltage on the distribution system throughout the community within prescribed limits. It compensates for variations in voltage on the supply source to the substations and for variations in loading on the distribution system which would otherwise affect the voltage available on the system. This ensures each user of electric power of a supply voltage that conforms to the design limitations of his electrical appliances and equipment.
>
> Two types of protective devices are generally installed, one of which is a fuse on the high voltage side of the transformer, for protection of the transformer from possible damage due to overload. When required to operate, it ensures continuity of service to other distribution substations connected to the same bulk power supply since operation of a protective device for the supply source would result in disruption of service to all connected substations. The other protective device installed is an automatic circuit breaker. This device senses troubles or overload conditions on a distribution circuit and operates automatically to interrupt the flow of power. Thus it prevents damage to the circuit and ensures continuity of service to users of electric power connected to other distribution circuits emanating from the same substation by removing from service, the troubled circuit only.
>
> **Location Requirements**
>
> Functionally, there are three basic requirements that influence location of the substation:

1. Geographically the location should be as close as possible to the center of the electric power distribution load (demand) to be served. The future, as well as the present, distribution requirements should be considered. The determinants for current needs can be measured but future demands can only be determined by considering future growth projected by the plans of the community.
2. A factor which sometimes alters the center location is the necessity for the substation to be located near a high voltage transmission line or where right-of-way for a circuit to the high voltage transmission line is available.
3. The service requirements for substation equipment are such that convenient accessibility from a street or highway is important.

Size Requirements

The size and proportion of a site is influenced by the present patterns of equipment layout, safe buffer zones and sufficient area for effective blending and landscaping that harmonizes with the fabric of the community. In addition, the site must be large enough to expand sometimes beyond 100% and have the room to accommodate mobile emergency units.

The minimum area to satisfy these development requirements is 22,500 square feet (approximately 1/2 acre) or 150′ x 150′. This proportion will vary in accordance with land availability and site characteristics. If the cost will permit, the site should more ideally approximate 32,000 square feet, allowing area for adequate space and plant material buffering and for adequate service access drives.

Access and Service Requirements

A service drive minimum width of 14′ is required to provide access for service vehicles, including those delivering a mobile emergency unit to the site. It is desirable, although not absolutely necessary, to allow service vehicles to back around in order to provide a convenient and safe exit.

The location of the access drive will vary with local site conditions. However, care should be given to downgrade its impact without affecting its function.

Fence Enclosures

For reasons of safety and security, substations are enclosed by a barrier or fence which excludes the general public from within the enclosed area.

A sufficient area is enclosed by this fence or barrier so as to permit free movement of personnel within who must operate and maintain substation equipment and also to allow for the entrance of vehicles required to maintain, remove or install additional equipment without the necessity of removing any part of the enclosure.

The enclosure is generally a chain link type fence, 7′ high. As an additional safeguard, two or three strands of barbed wire are installed on top of this fence which increases the overall height to 8′. Drive gates, 14′ in width (two 7′ gates) are installed to permit entrance of vehicles. A walk gate, 3′ wide, is provided at substations located in those sections of the state which experience heavy snowfall. This is to minimize snow removal in front of the gate to gain access into the substation for routine operation and inspection. Assuming normal expansion considerations, a minimum enclosure would be 94′ x 72′.

Setbacks from property lines should conform with the local zoning ordinances with a minimum side yard setback, allowing for an adequate buffering zone, of 18′. Street setbacks will be normally much greater than side yard setbacks. Environmental conditions should be the final determinants in controlling setback. Such elements as adjacent structures, roadways, trees and topography should be considered. Further elaboration of these influencing factors will be covered in Chapter III, Development.

Decorative enclosures, walls, fences and planting, although not required for public safety and protection of equipment, should appropriately be a part of the aesthetic development. This very important subject will be discussed in Chapter III.

Site Drainage and Grading

It is extremely essential that the substation equipment and access drives be well drained. The drainage system should be positive and not dependent upon surface inlets that could be blocked by debris causing even temporary ponding of water affecting access and causing damage to the equipment.

Careful design of the grades will resolve drainage problems as well as conserve natural qualities that provide a harmonious blend with the environment. Among the most important elements worthy of preserving are existing trees that may occur on the site. Extreme care must be taken to retain the existing grade around an existing tree to the breadth of the branches; even the slightest adjustment, either cut or fill, will endanger its life.

Soil Requirements

Soil types should be carefully examined for their drainage characteristics and load bearing capabilities. This would require examination of the immediate environment as well as the site and soil test borings in designated construction areas.

Topsoil for constructing lawns and planting is sometimes scarce and expensive. Therefore, it is important that grading operations include stock piling of existing topsoil that can be used for this purpose. Generous amounts of fertile topsoil should be specified for the installation of plant materials including a minimum of six inches topsoil base for a healthy vigorous lawn, even if sod is specified.

Landscape Requirements

Although one might rationalize that certain plant materials can provide necessary physical barriers and that grass can control erosive particles that deteriorate equipment, the most important contribution of landscaping is improving public image through aesthetics. To the community and its planner, the factor of aesthetics is becoming an essential part of development criteria.

Poles and Guy Wires

Ideally, distribution circuits should be underground but this is not always feasible. When required, transmission poles and guy wires should be placed where they have the least adverse affect on the surrounding area. Background and foreground screening is essential to solving the problem. Overhead wire crossovers should be considered aesthetically and functionally. Clearance regulations are specified in the Michigan Public Service Commission Regulations.

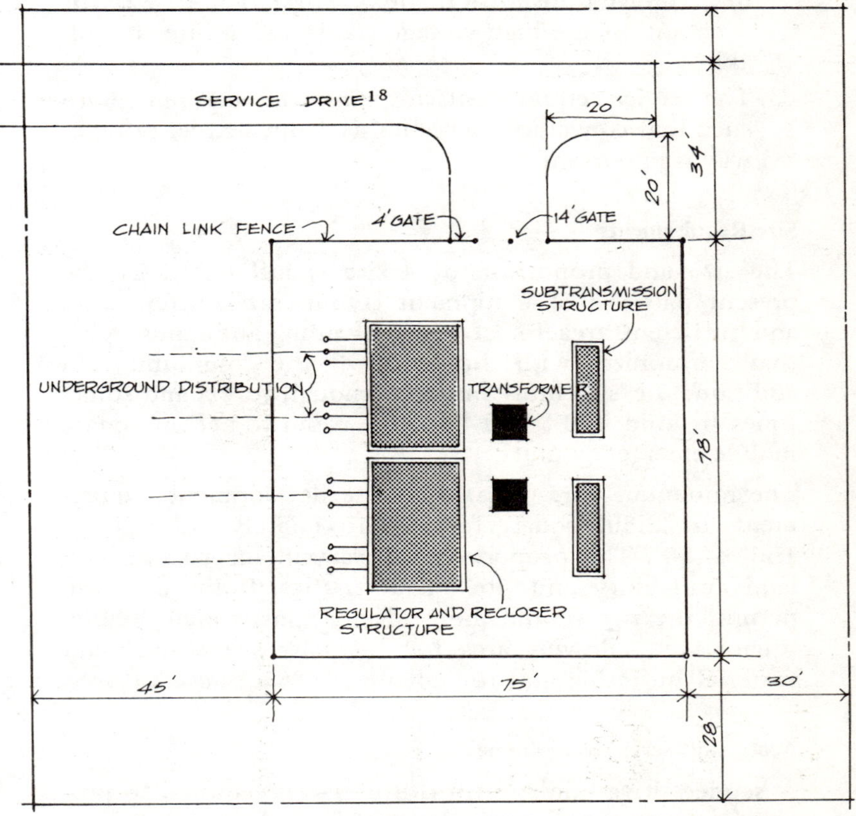

18. William J. Johnson, Carl Johnson and Clarence Roy, *Substation Site Selection and Development*, (Ann Arbor, Michigan, Johnson, Johnson and Roy, for Consumer's Power Company, Jackson, Michigan, 1969), pp. 5-7.

Lighting Requirements

The primary purpose of lighting is to identify the equipment during night time hours for maintenance and operational functions. However, this illumination should be skillfully examined to consider the aesthetics in terms of fixture design and location as well as special effects such as indirect lighting.[19]

The report then goes on to discuss in some detail the difference between the formerly designed high profile substation shown in the following illustration, and the newer low-profile transformer shown in the following illustration:

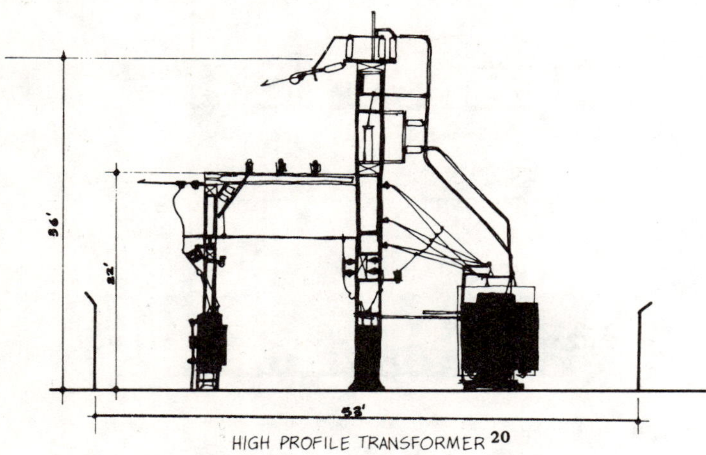

HIGH PROFILE TRANSFORMER[20]

20. Ibid., p. 8.

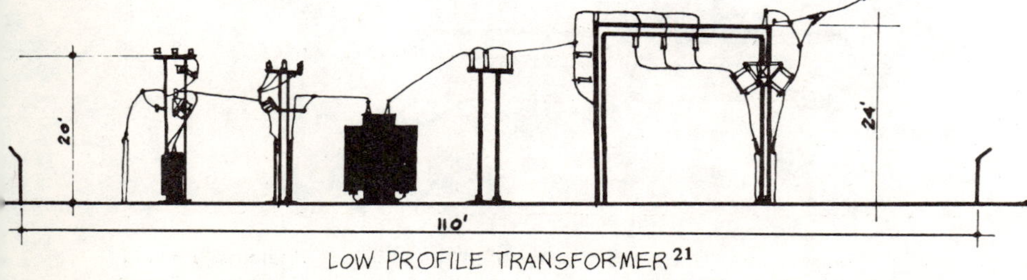

LOW PROFILE TRANSFORMER[21]

21. Ibid., p. 9.

The concept of a Primary Observation Zone (POZ) is then developed with illustrations showing the effect of this concept on the location of landscape elements. The following is a statement from that report.

PRIMARY OBSERVATION ZONE

Definition

Assuming that the public as a whole will not accept the electric substation equipment as an attractive element, efforts to blend the substation into the community become a problem in adequate screening. The primary visual concern is from the primary observation zone, or zones, to be referred to in the text and diagrams of this report as the P.O.Z. Understanding and solving the problem of the P.O.Z. will provide the basis for a valid site development solution and influence site selection.

Low Profile Substation P.O.Z.

The newer, low profile substation also has two basic primary observation zones. The first, the skyline view, is far less dominant because of the lower height (only 20′) and simplified profile of the structural frame.

With the simplification of the structural frame, the eye level view becomes the most dominant zone where most of the equipment occurs. Therefore, the primary concern for screening lies within this area below the top of the fence enclosure.

High Profile Substation P.O.Z.

The high profile substation has two primary observation zones. The most dominant of the two, the skyline view, is a zone which is almost always exposed and is that area which is above the top of the chain link fence that encloses and protects the equipment. Because of intricate patterns of the paraphernalia and the heaviness of the 36 foot high erecticon frame, the impact upon contiguous properties is overpowering and very difficult to screen.

19. Ibid., p. 5.

The second primary observation zone, the eye-level zone, is that area below the top of the fence. This zone is more adaptable to screening and buffering solutions. The fence and more substantive equipment such as transformers contribute to diffusing the paraphernalia. Because of its lower height, and its being absorbed into the background, the area is easier to screen.

Orientation

The position of the sun has a direct influence on the visual impact of the substation, particularly in the skyline zone. When the P.O.Z. faces the sun, the skyline view of the structure is diffused by the sun's rays causing the vertical plane to appear to recede. Conversely, with the sun at your back, the vertical plane is high-lighted causing the vertical structure to appear much closer.

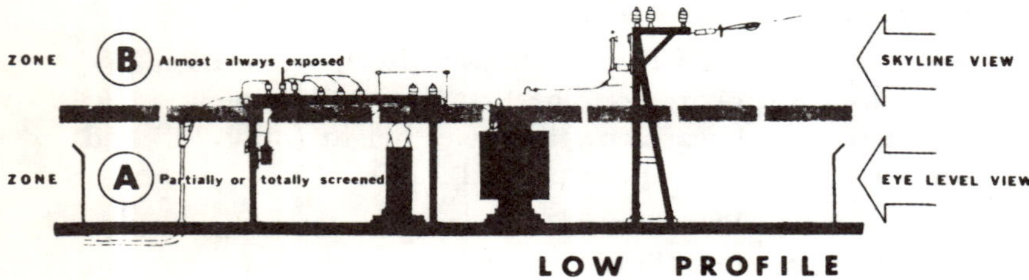

LOW PROFILE

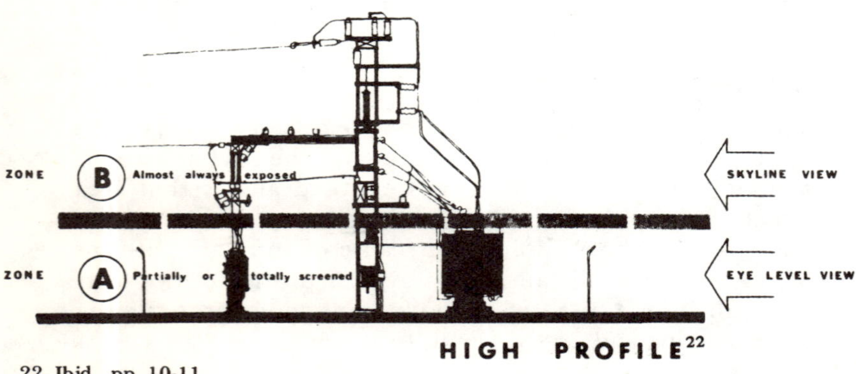

HIGH PROFILE[22]

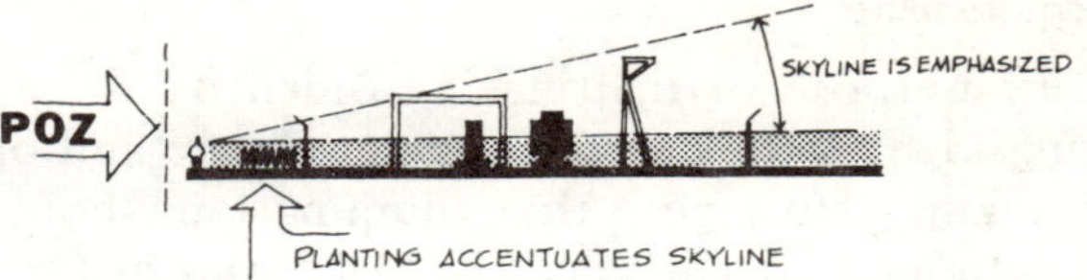

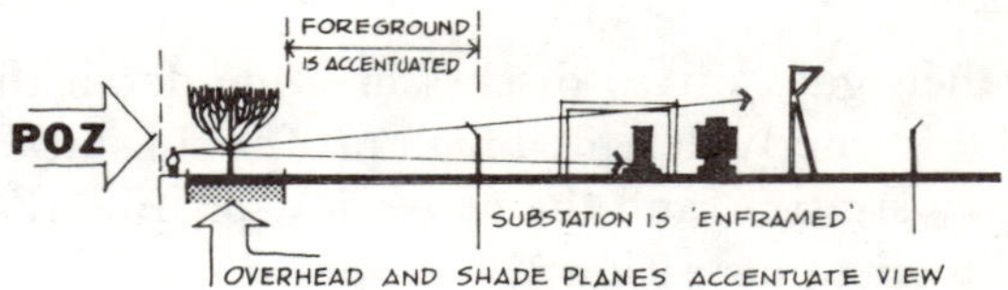

But AS THE PRIMARY OBSERVATION ZONE MOVES AWAY THE SCREENING NEED *lessens*

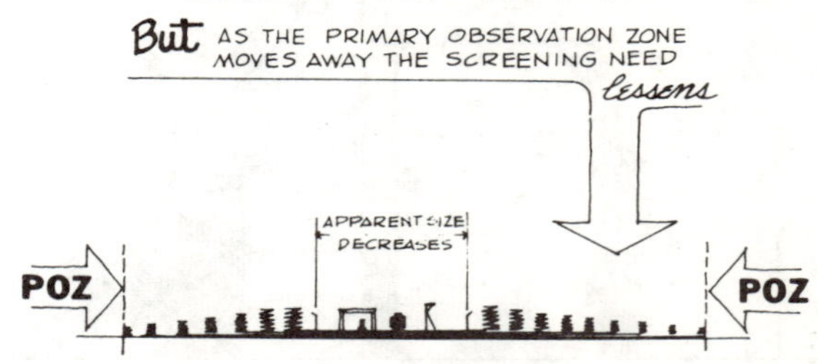

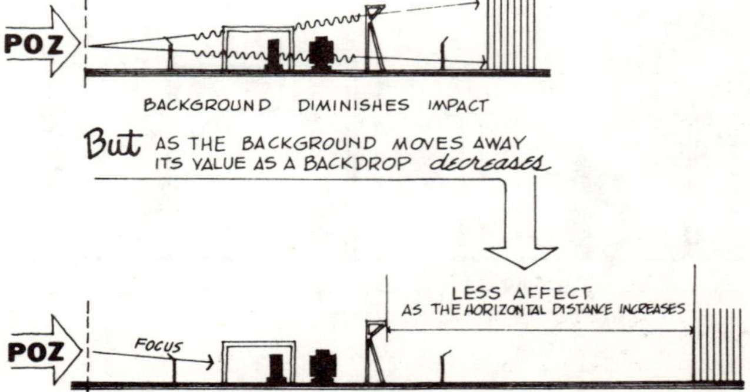

22. Ibid., pp. 10-11.

The basic design principles are then developed and illustrated in the following way:

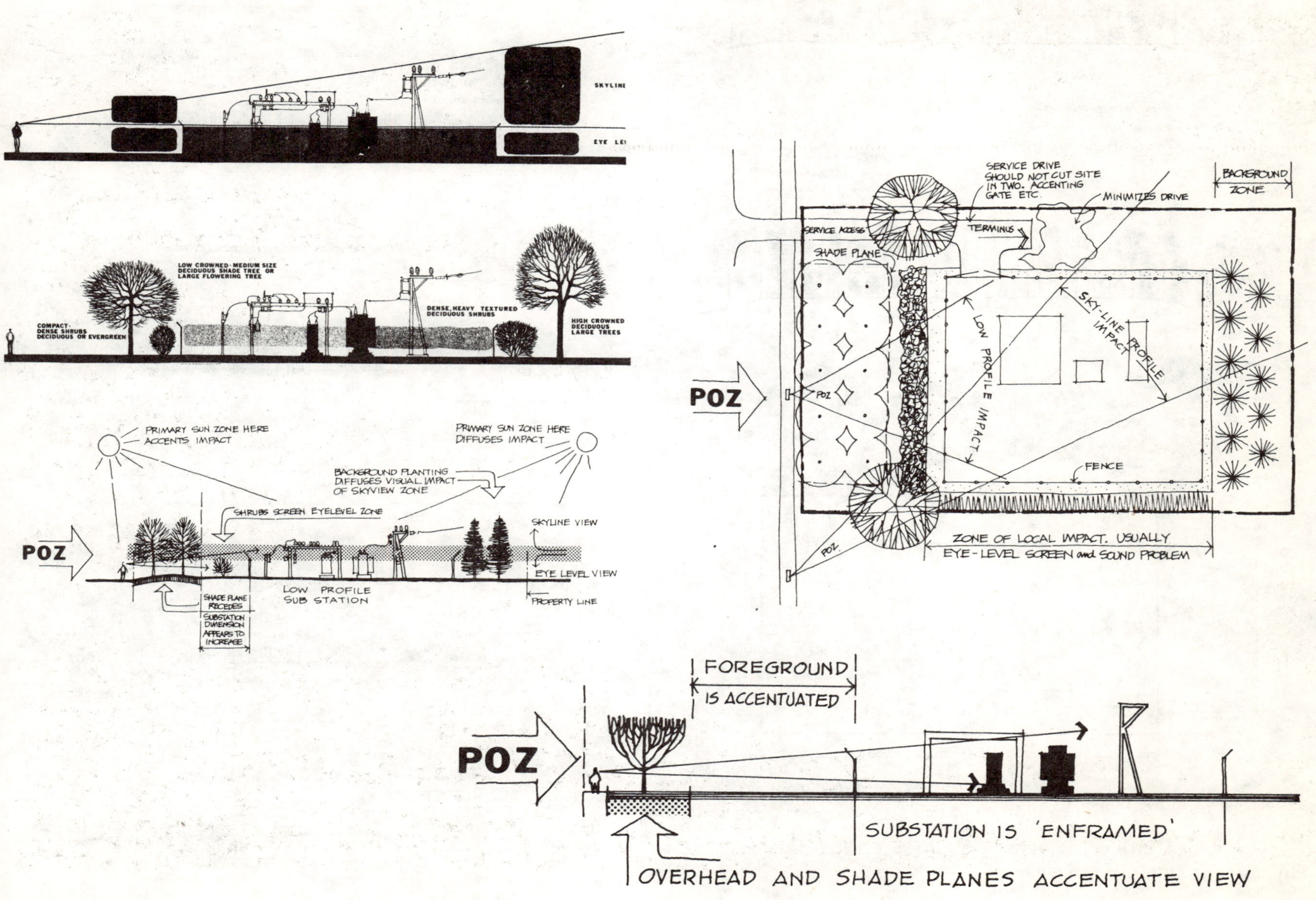

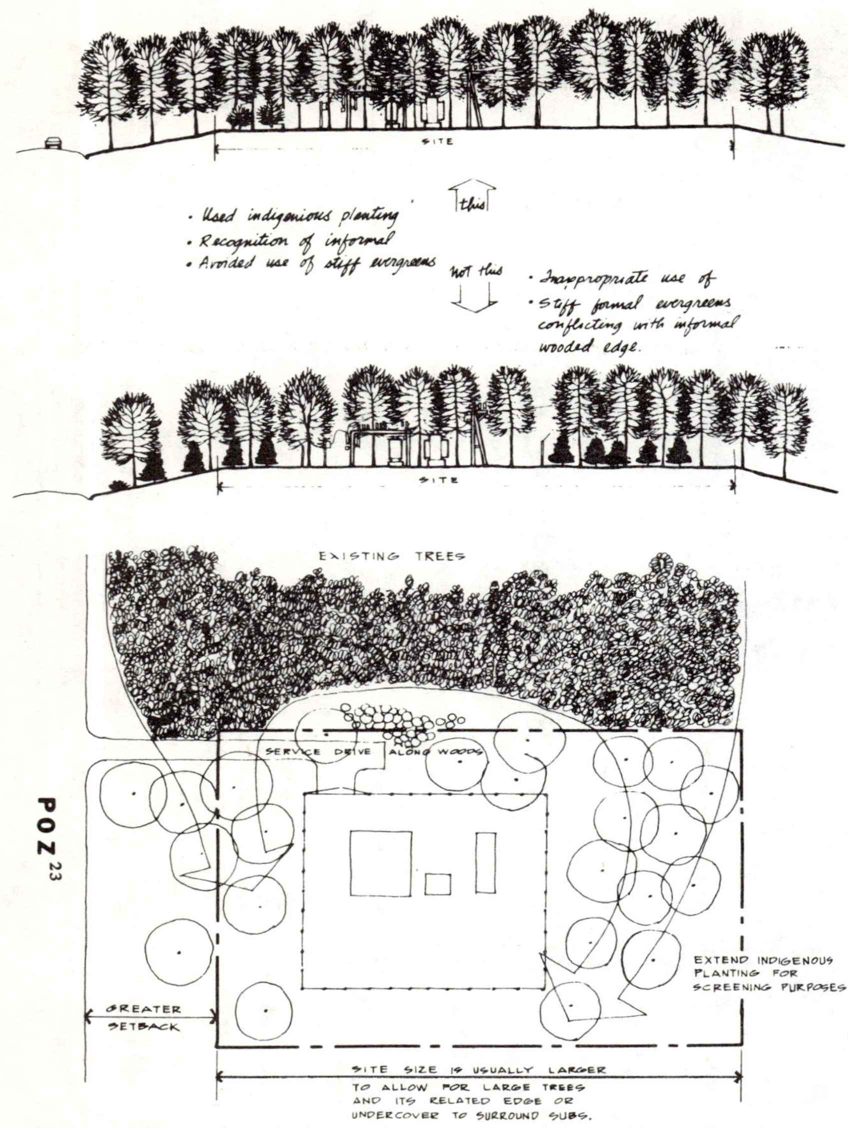

23. Ibid., pp. 14-15, 17, 19.

The design influences of topography are then discussed and illustrated in the following way:

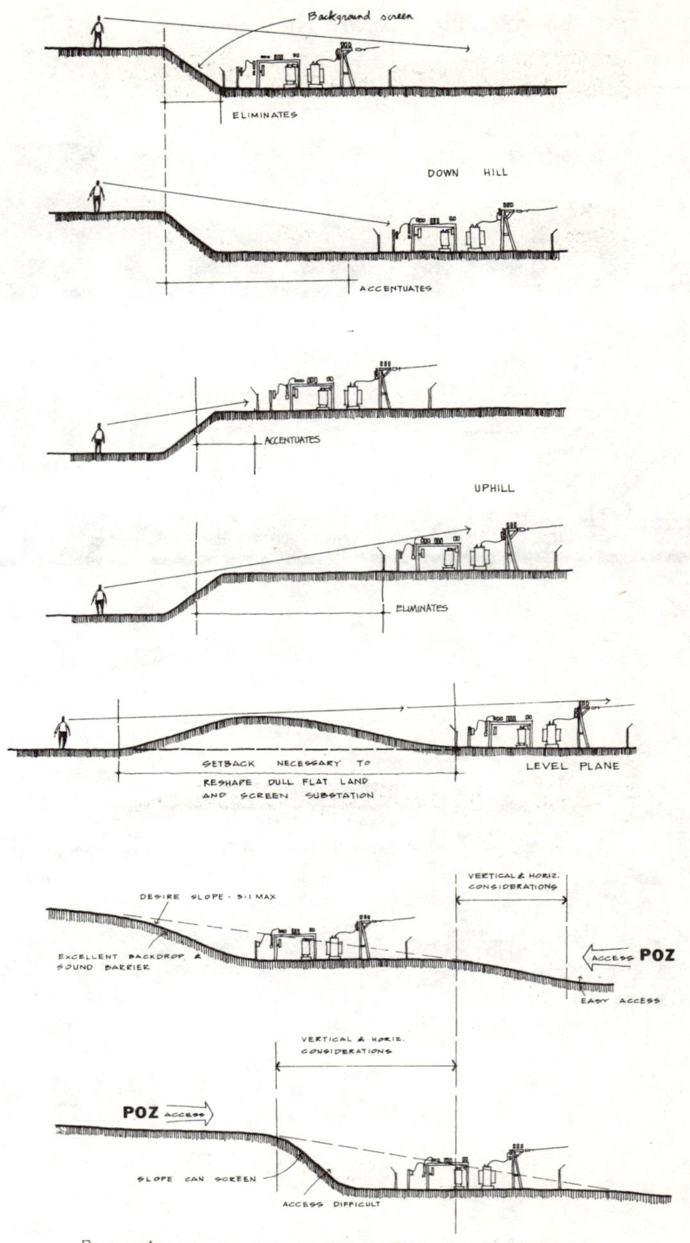

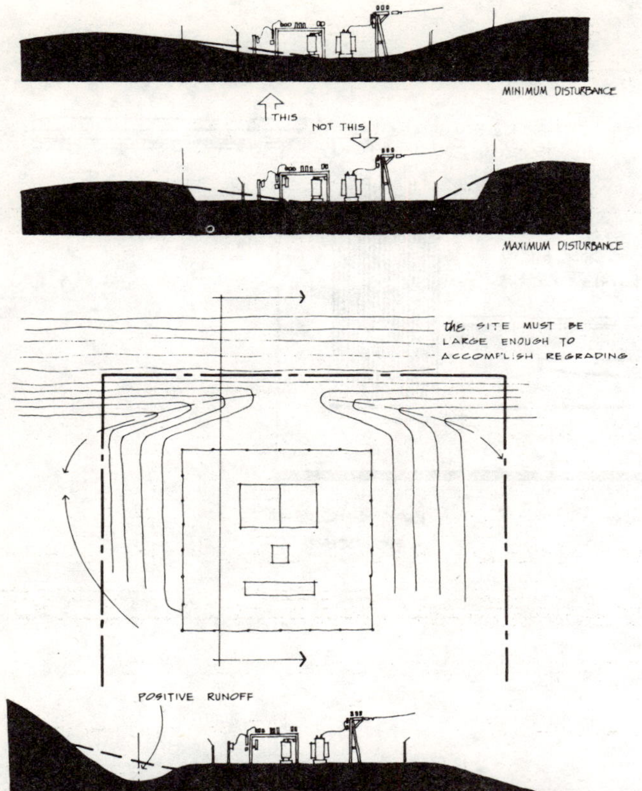

24. Ibid., pp. 20-21.

The discussion is then undertaken attempting to relate the transforming facilities of the substation to an urban, to a suburban, and to a rural situation.

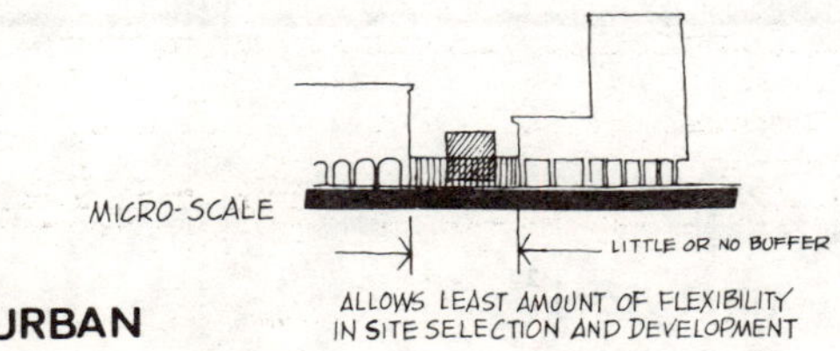

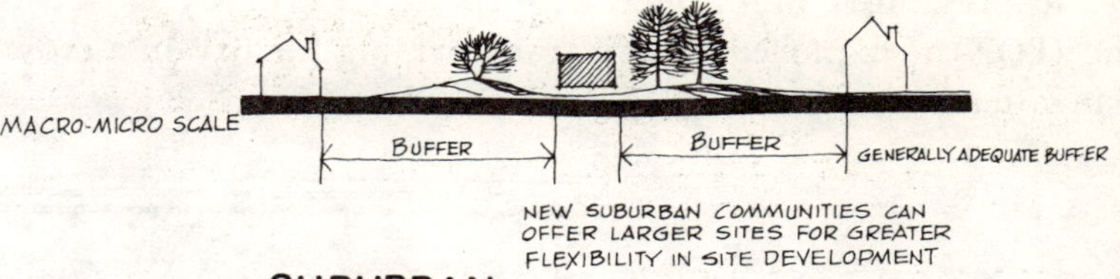

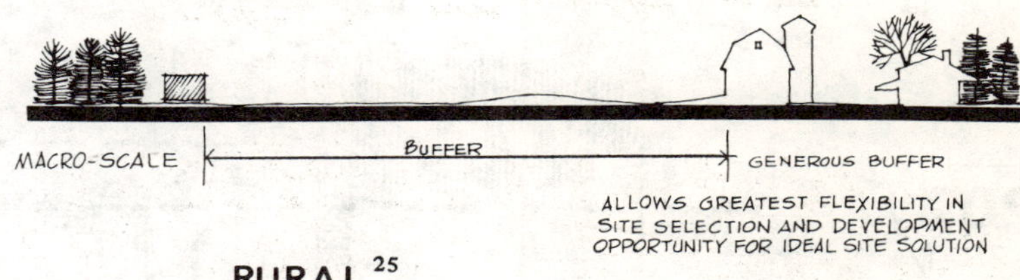

25. Ibid., p. 23.

The visual impact zones are illustrated for urban sites for transforming facilities.

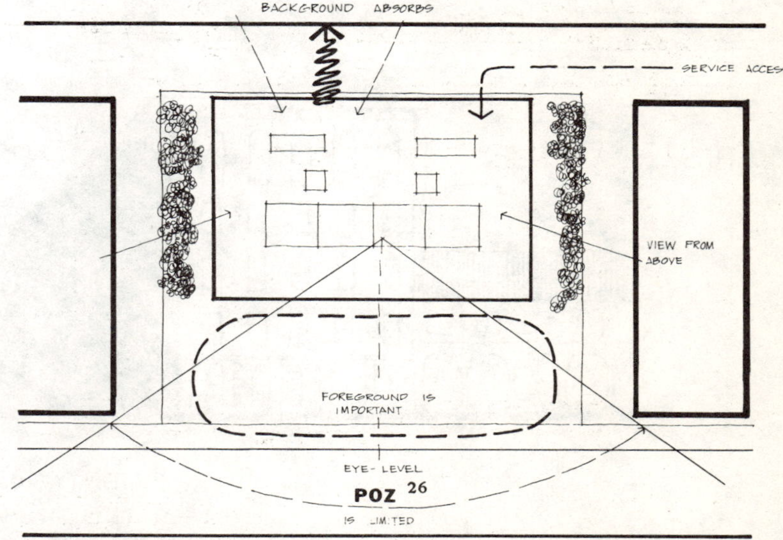

26. Ibid., p. 28.

The previously mentioned concept for a Primary Observation Zone (POZ) is explored with a transforming facility in a crowded urban situation.

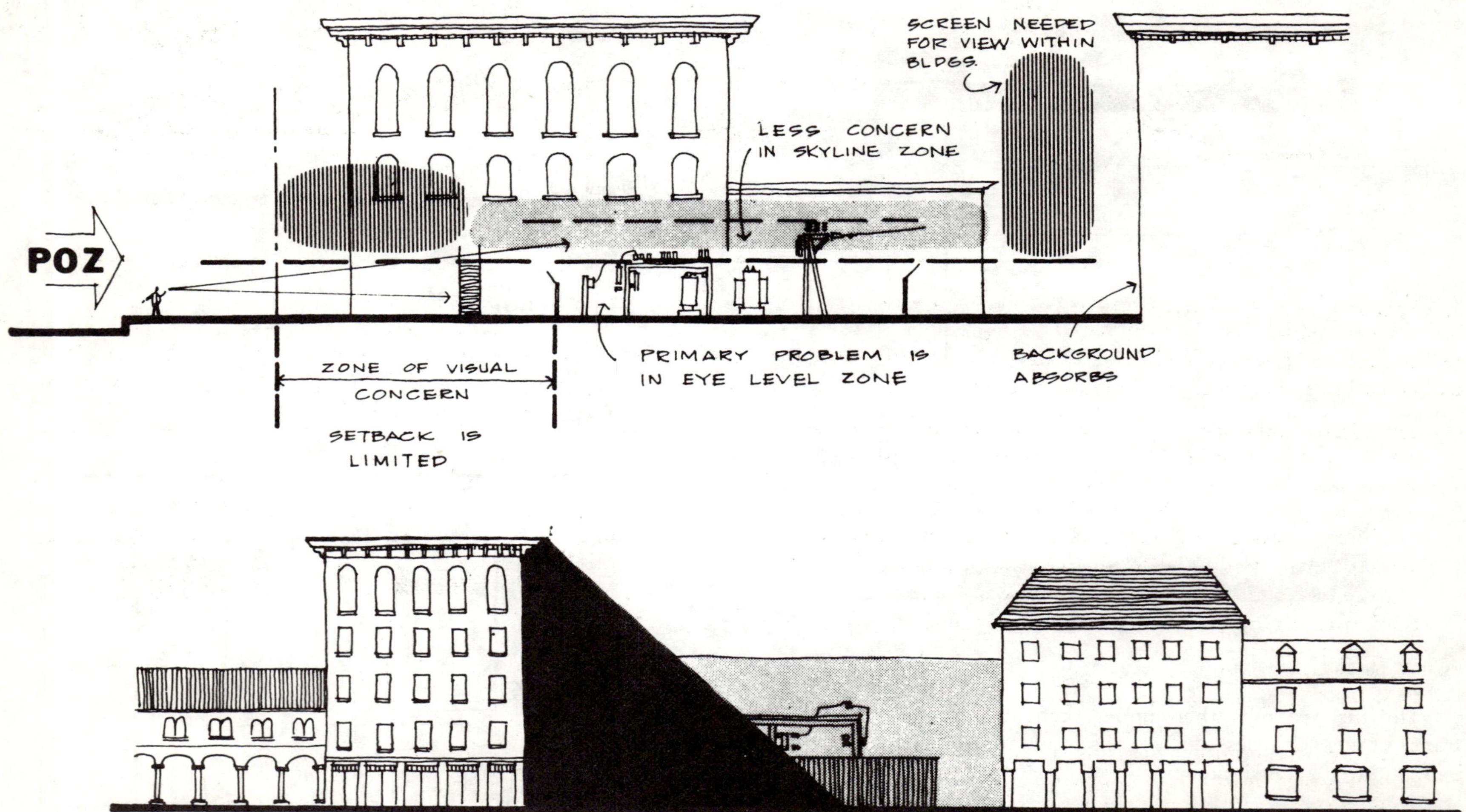

ORIENTATION IS IMPORTANT
SITES IN ALMOST CONSTANT SHADE
ARE MINIMIZED VISUALLY BUT
NEED SPECIAL LANDSCAPE ATTENTION[27]

27. Ibid., p. 29.

The design use of topography is illustrated for use in screening transforming facilities in urban areas.

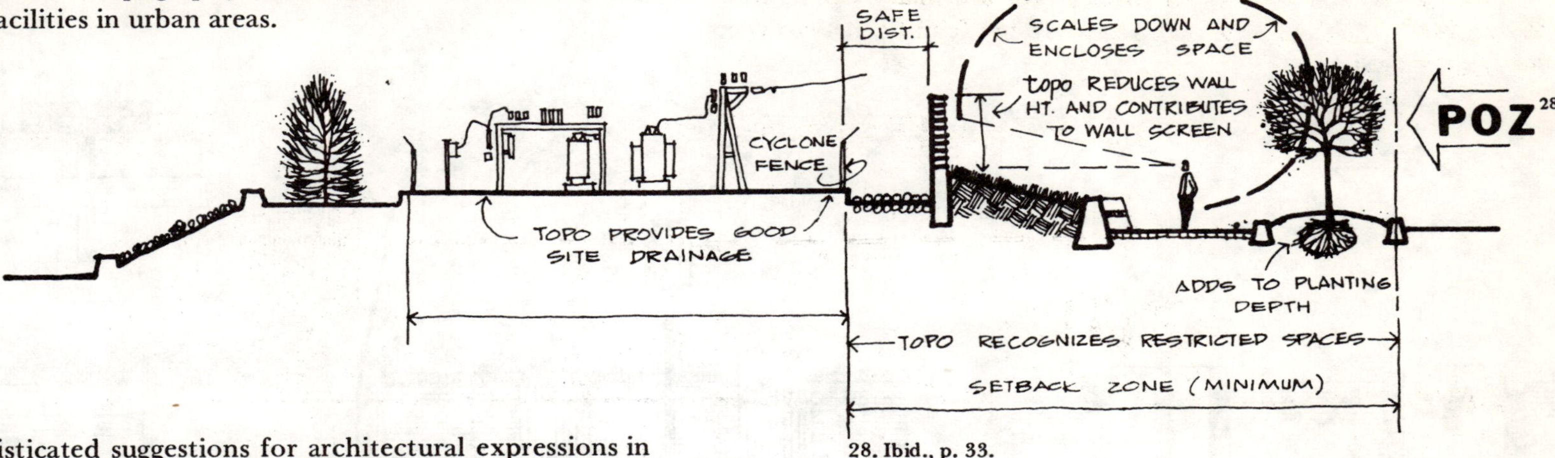

28. Ibid., p. 33.

Very sophisticated suggestions for architectural expressions in order to minimize the environmental disruption of transforming facilities in urban areas are illustrated. The following is a statement by the consultants in regard to architectural expression:

> The rigid framework of buildings and hard surfaces of the urban environment and the restricted dimensions of the substation site are strong influences for architecturally interpreting the design character of the site. This character can be expressed by walls, fences, planters, and paving, and accessories such as benches, planters and fountains.
>
> The material and design of these elements should express the nature of their surroundings either by positive contrast or appropriate blending. For example, within an enclosure of traditional brick buildings, walls, paving and accessories might be constructed with the same or similar brick with a minimum amount of contrasting materials such as metal or concrete. On the other hand the same structural elements of the design might be expressed boldly with strong concrete forms for the purpose of contrasting with the color and texture of the brick. Refined trims and secondary elements might be compatible with the brick to suggest a subtle response to the materials of the environment.
>
> Special accessories such as decorative pools, sculpture and lighting would appropriately relate to the project character and use. It is important however to relate such elements to the design objectives of the community with whom a dialogue should be established in the very early conceptual stages.[29]

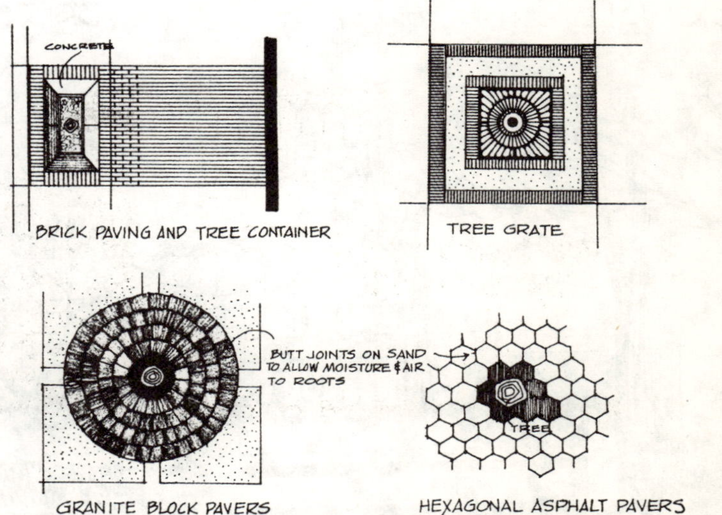

29. Ibid., p. 34.

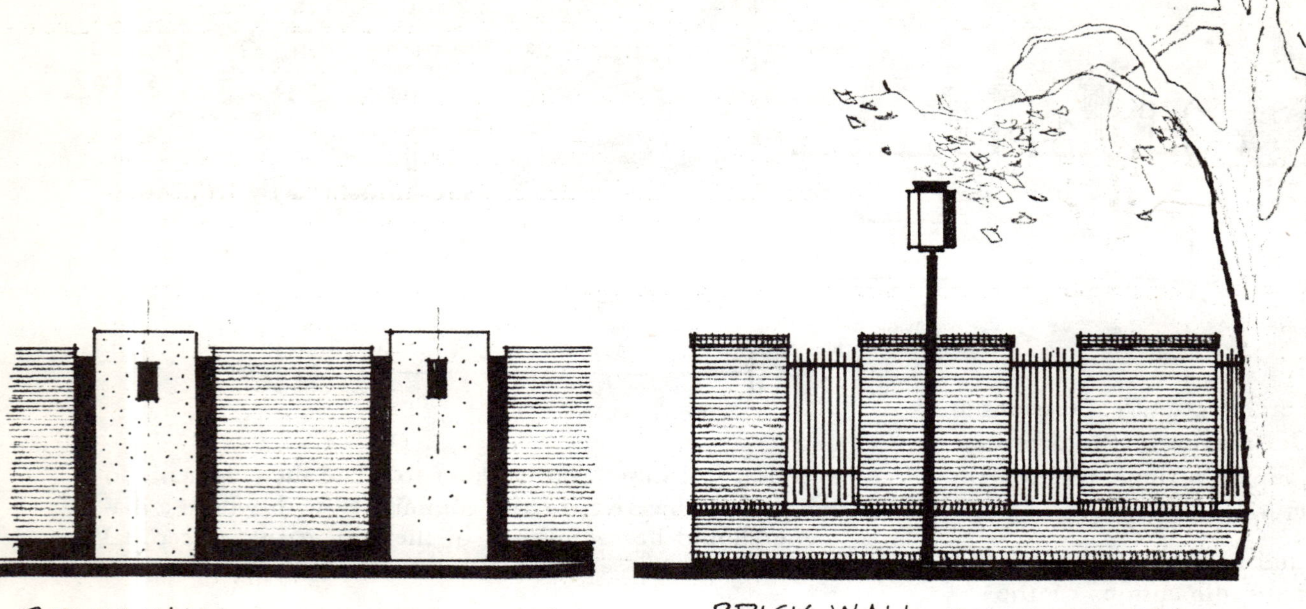

BRICK WALL CONTEMPORARY STYLE

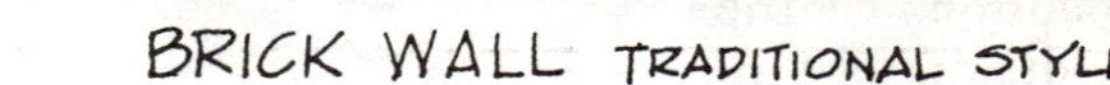

BRICK WALL TRADITIONAL STYLE

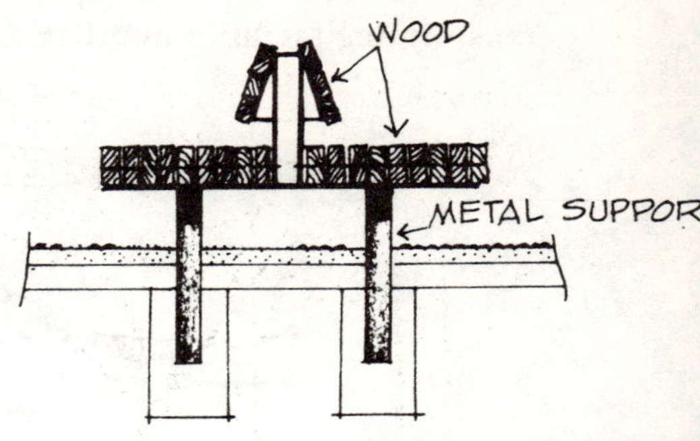

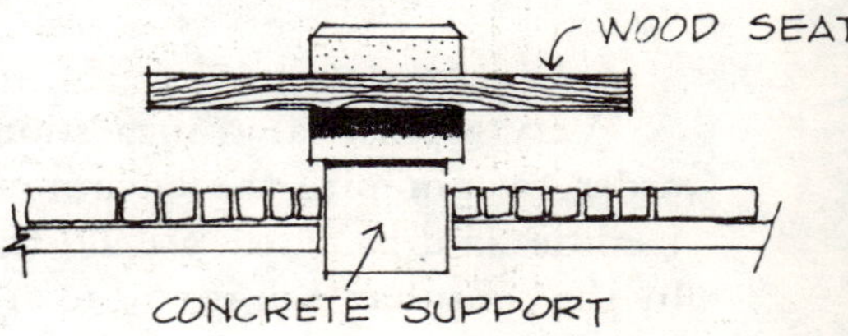

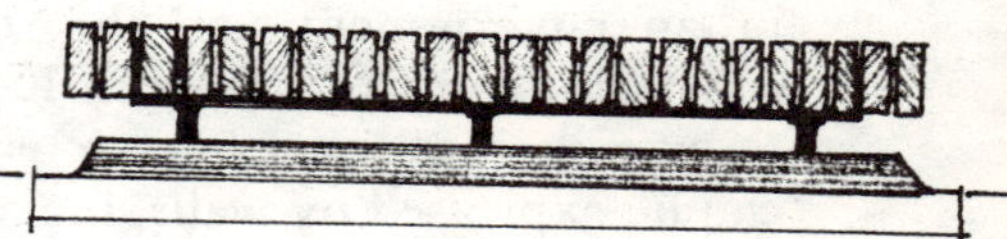

WOOD SLAB ON CONC. BASE

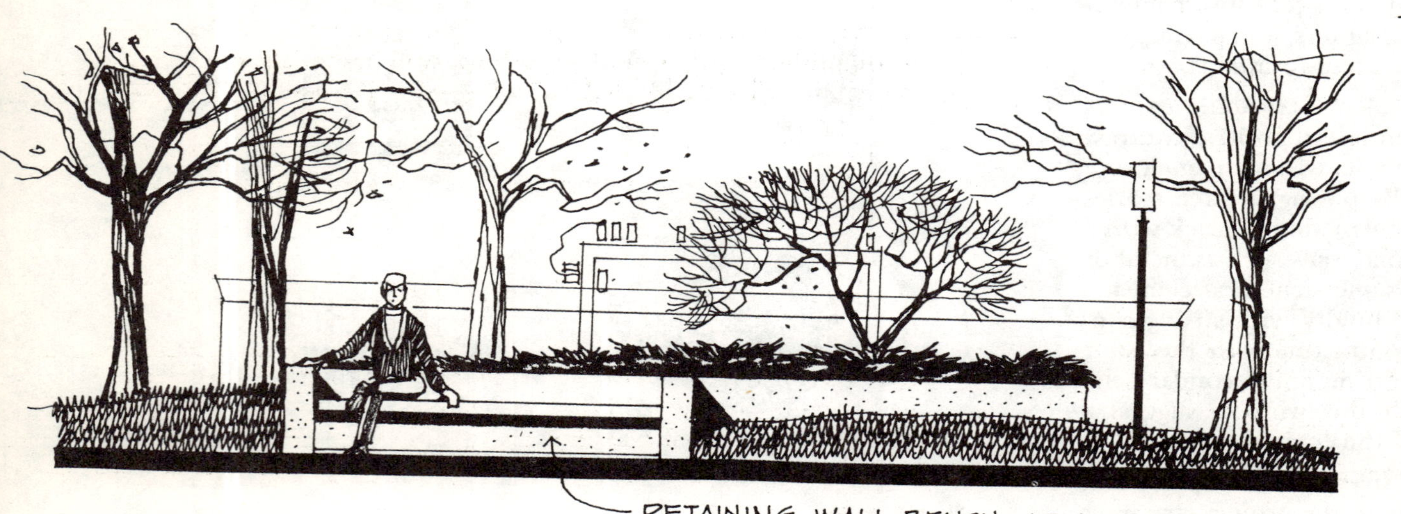

RETAINING WALL BENCH AS A FORE-PLANE

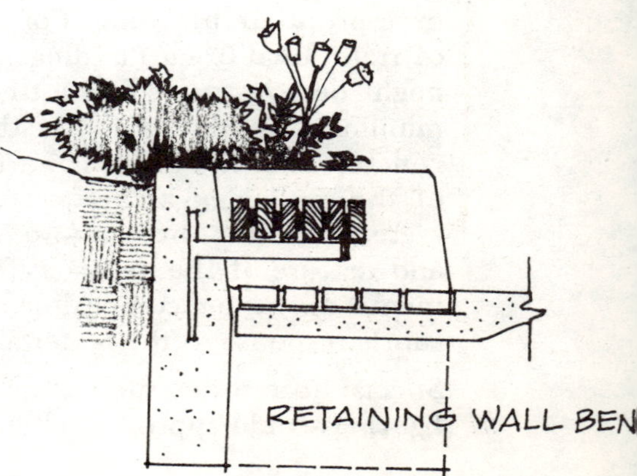

RETAINING WALL BEN

30. Ibid., pp. 34-36.

The Johnson, Johnson and Roy staff then mentioned a special consideration of the urban site in the following words:

SPECIAL CONSIDERATIONS OF THE URBAN SITE

The shape and size of the urban substation site is normally very restrictive. The minimum area will generally be configured so the frontage or width of the site is narrow because land values in the city are influenced by front footage. The site must also be large enough to facilitate future expansion.

Occasionally within the urban situation there will be a need to enclose the switchboard panels and battery control of circuit breakers with a small building approximately ten feet high. The building should be designed within the context of its environment considering such factors as setback, size and materials of adjacent buildings. The enclosure may even fit into decorative walls utilized to enclose the entire structure and apparatus.

In the dense urban environments where poles and wires interfere with a variety of vertical planes and elements, distribution circuits should be installed underground if at all possible.

Abatement of transformer noise will in most situations be accomplished with sound absorbing wall materials since space for planting and earth form buffering is limited.[31]

The complete development of a substation or transformer facilities in an urban site is shown in the illustration on page 252 in a typical way.

The report suggests a checklist of analysis items to be included in the selection of a suburban site for a substation or transforming facility in the following words:

The electrical substation has far more impact on the existing character in the suburban community, comprised primarily of residential use, than in any other zone. Its impact involves and confronts the individual as a property

31. Ibid., p. 36.

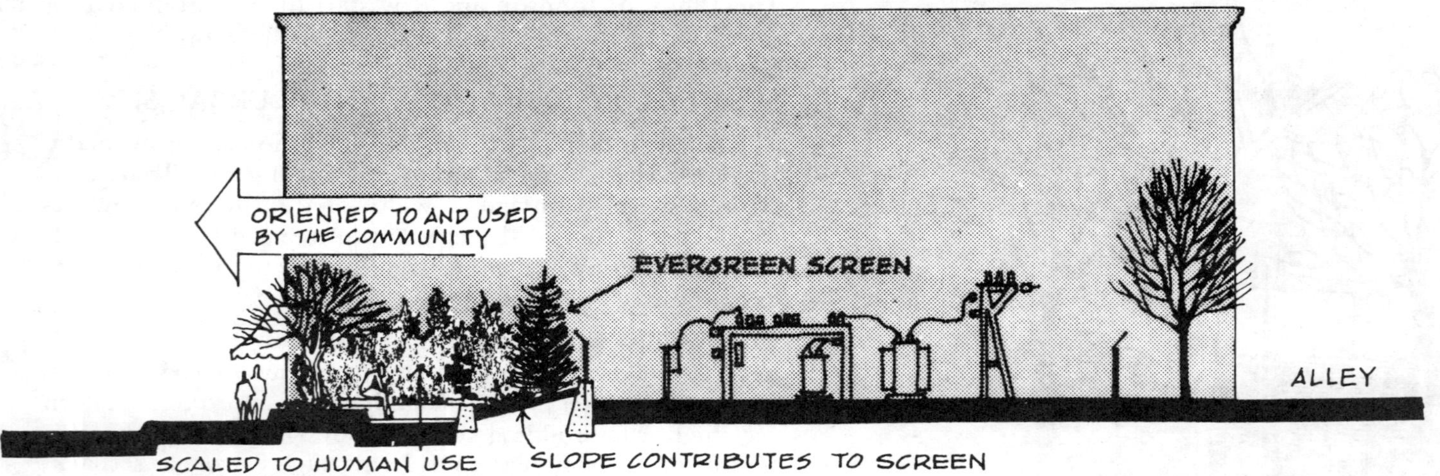

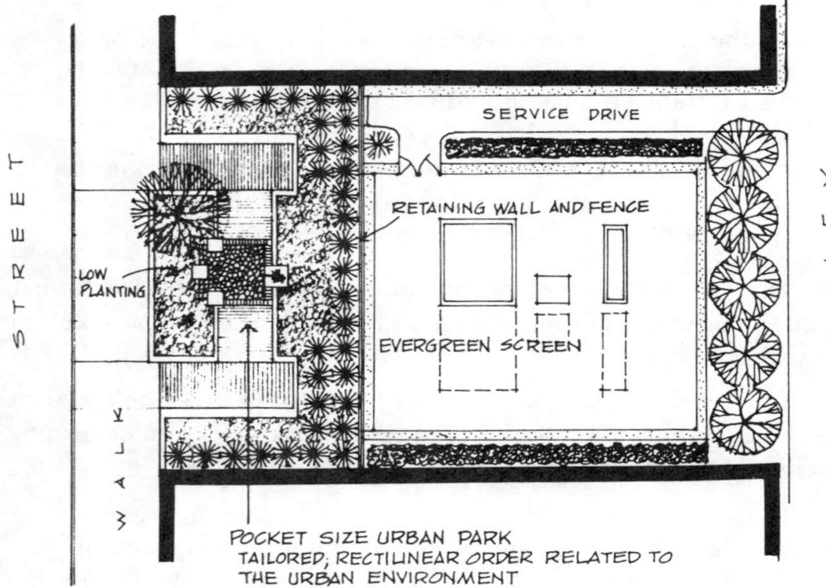

Typical Urban Site Development[32]

32. Ibid., p. 37.

owner whose inherent desire is to protect his investment. Every development recommendation therefore must consider the individual's desires and demands. An analysis of certain community patterns and requirements is necessary to guide site acquisition, including:

1. Use, character, condition and size of the buildings and land within the immediate environment;
2. Vehicular and pedestrian patterns;
3. Required easements for open space and utilities;
4. Zoning and building code requirements;
5. Availability of stormsewers;
6. Future physical land use patterns.

If the site exists in an unstable transitional zone, then thought must be given to potential change in the use of the zone. The site should be developed so that the least amount of adjustment will be needed to adapt to changing community patterns.[33]

The study then deals to a great extent with the development of the suburban site of a transforming facility in the following words:

33. Ibid., p. 40.

The more exposed suburban substation site is generally viewed at close proximity from the front, the sides and rear, but rarely viewed from above. The Primary Observation Zone normally occurs at the street or sidewalk. It is from this point of view that the equally important eye level and skyline views to the substation structure must be screened. The screening materials, whether they are plant materials, ground forms or fences, should be expressed with materials that are compatible to the neighborhood and designed to be viewed from close observation zones.

Although the P.O.Z. of the substation normally occurs along the street, the side property and rear observation zones are of almost equal importance because their impact affects individual property owners. The property owners usually have concern about the negative aesthetic influence of the substation adjacent to or in the vicinity of their property because they feel it affects the value of their home. It is essential that these problems be resolved in a positive way for good community relations.

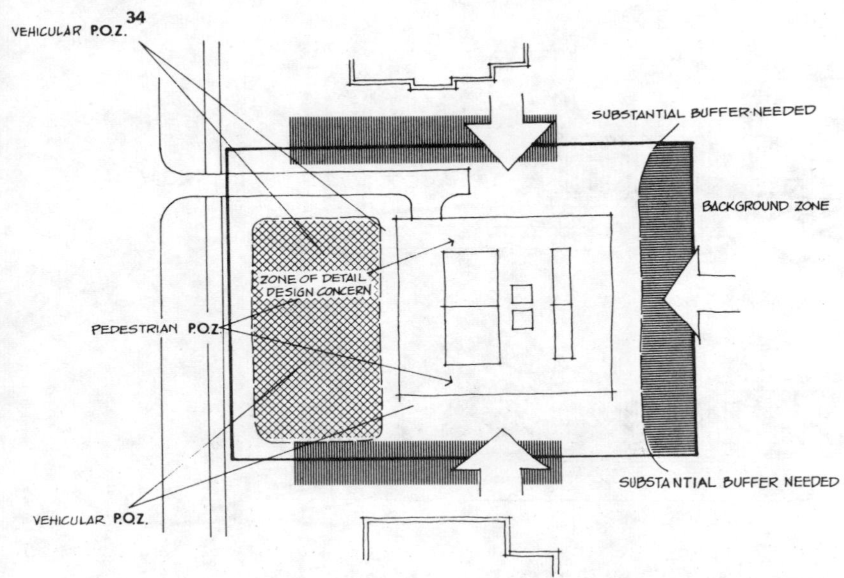

34. Ibid., p. 41.

Consideration is then given to horizontal and vertical screening and visual control. The following illustrations depict the major recommendations. Under this heading suggestions for the treatment of the service drive entrance considerations are then made and illustrated graphically in the following way:

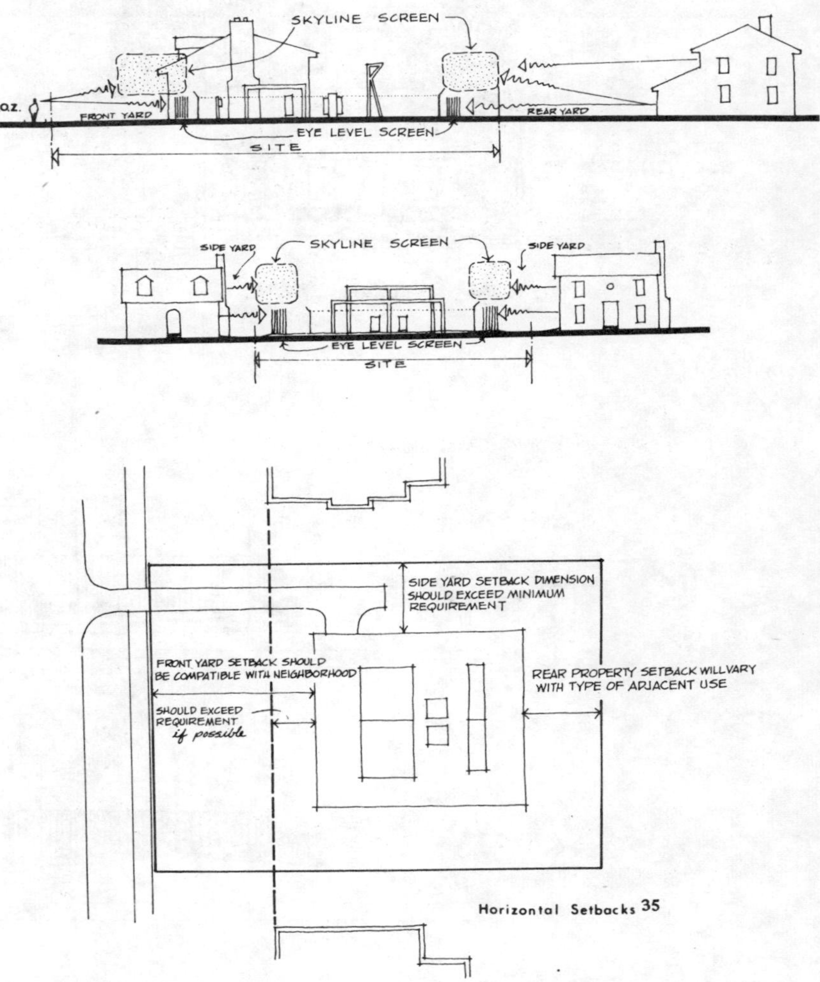

Horizontal Setbacks [35]

35. Ibid., p. 42.

In order to tie the substation site into the surrounding community, the following contextual drawing was made:

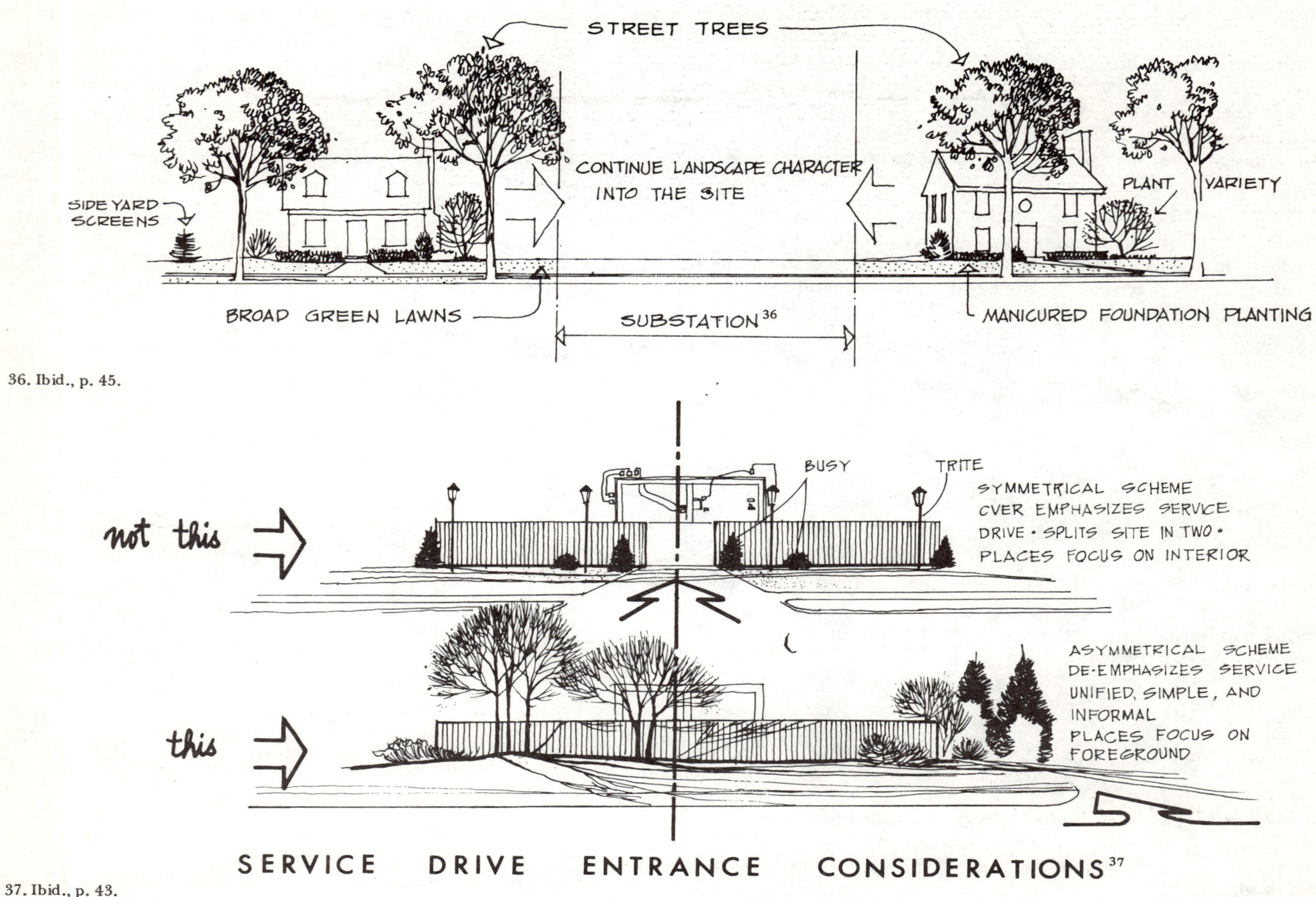

36. Ibid., p. 45.

37. Ibid., p. 43.

The design use of topography in the suburban site and is described and shown in the following illustrations:

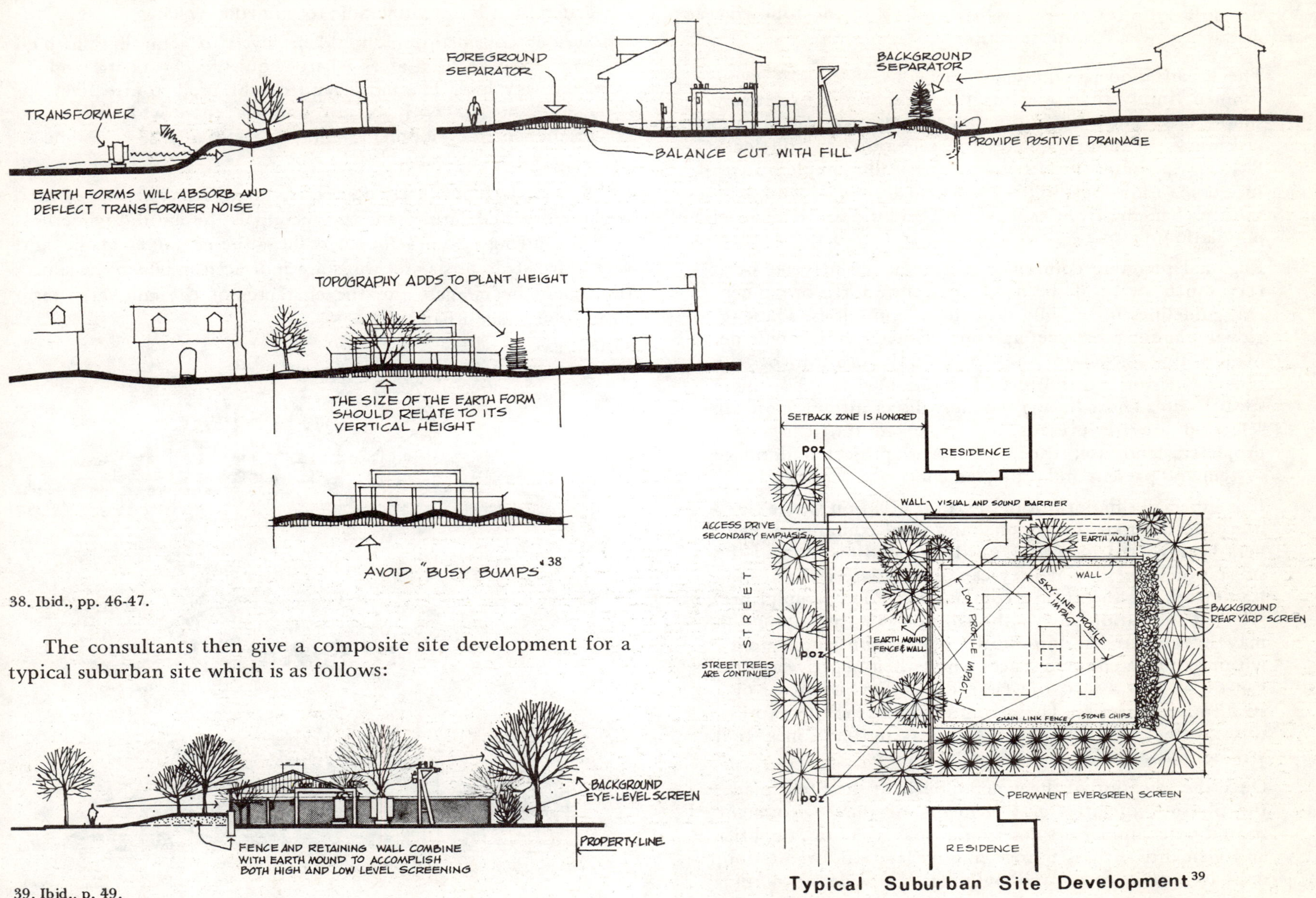

38. Ibid., pp. 46-47.

The consultants then give a composite site development for a typical suburban site which is as follows:

39. Ibid., p. 49.

Typical Suburban Site Development[39]

The selection of the rural site is then discussed at some length by the Johnson, Johnson and Roy staff, and the following admonitions are given for this selection:

> The broad almost unlimited dimensions of the rural environment imply that selecting the rural site is a relatively simple task. The very factor that would seem to make the task easy, flexibility, makes site selection very difficult. Existing community patterns are basically simple but future expansion possibilities are usually vague and constantly changing from extremes of land use to extremes in population densities.
>
> Regional planning commissions have in recent years been very much aware of the need for setting forth broad general guidelines for rapidly growing communities. These regional planning parameters coupled with the growth demands that generate the need and location of the substation can contribute to the rationale for choosing a specific site. Those having the most direct influence on site selection are the present and projected traffic systems, projected land use, the present and projected drainage system and present and projected code requirements.
>
> It is especially important in the rural situation where flexibility occurs that the regional and local community planners be involved in setting up the general location of the substation sites so that they can appropriately fit into the framework of the future community. For example, preselected substation sites, although not exactly pinpointed, may be able to fit into buffer zones contiguous to or within committed open spaces where their negative aspects can be absorbed into the large dimensions of space or a swath of green trees. These zones may be next to expressways, cemeteries, large parks, green buffer zones, and around industrial areas or shopping centers.
>
> Once the selection of the general locale of the site is made, the physical qualities and elements of the environment should be examined to determine final selection, qualities and elements such as topography, vegetation, orientation, drainage, soils, roads and buildings. In wide open rural areas the surrounding environment is a strong influence on the way in which the site is developed and located because generally it is not vulnerable to immediate change.
>
> Serious consideration should be given to acquiring much larger sites in rural areas where acquisition costs are comparatively low. The extra land can be used to effectively merge the site with its surroundings, to provide for unpredictable development, and to eventually serve as the required buffer zone.[40]

Two basic types of rural sites are identified. The first of these is the cultivated farmstead, the second is the natural wilderness. Design drawings are developed to show the character of the cultivated farmstead areas and ways are indicated in which the substation could be merged into the character of the cultivated farmstead type of suburban site.

40. Ibid., p. 52.

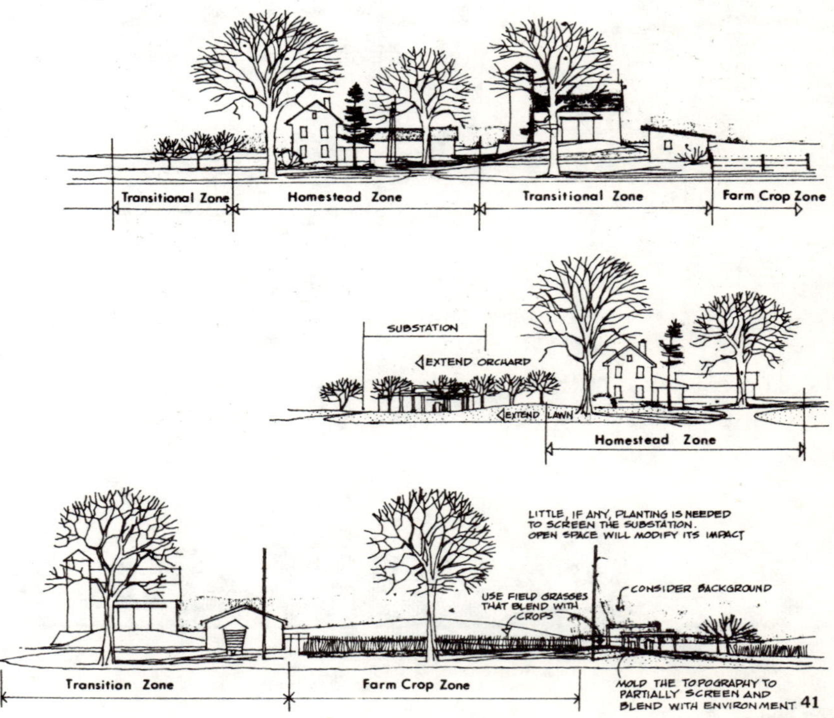

41. Ibid., p. 55.

The typical site development for a substation in a rural situation is then shown in the following illustration:

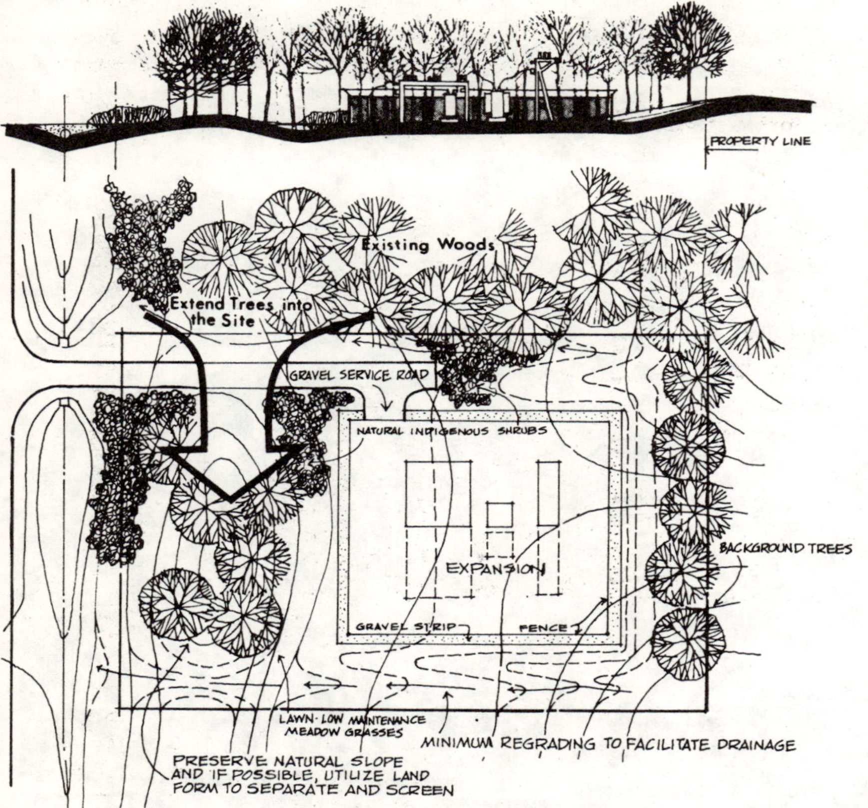

Typical Rural Site Development[42]

The consultants also recommend, or at least suggest, that certain types of existing decorative fencing, such as: brick, redwood, fence ribs, ribbed aluminum siding, decorative metal panels, and erection of structural steel; which have been and continue to be used to screen substation sites in the urban, suburban or rural areas.

As mentioned previously, this is probably the most extensive and definitive study ever conducted and published of some of the methodology of dealing with site selection and development for substations and transforming facilities.

Probably more has been done on the site selection and development of transforming of substation facilities than on any other element in the continuum of accoutrements and facilities associated with the electric power industry.

Since usually in a given neighborhood or locality the electric transforming facilities are your most obvious evidence of the presence and the environmental destruction caused by the electric power industry it is appropriate that this emphasis should be placed in this way. Hopefully, in time to come, a great deal more will be done by more members of the electric utility industry to show the ways that this environmental problem can be rectified, alleviated, or corrected. Hopefully, this section of this publication will prove a guide to those future consultants and members of the electric utility industry in their search for more appropriate solutions to this problem.

The following illustrations show some of the excellent solutions in a variety of locations by a variety of industry representatives to this problem.

42. Ibid., p. 59.

East Pine Substation, Department of Lighting, Seattle, Washington.

East Pine Substation, Department of Lighting, Seattle, Washington.

Camelback Road, Phoenix, Arizona
Before removal of 69 Kv Facilities

Camelback Road, Phoenix, Arizona
After removal of overhead transmission to a less obtrusive corridor

Page Mill Road, Palo Alto, California
Before removal of transmission lines

Page Mill Road, Palo Alto, California
After removal of transmission lines

View of Seattle's Hillcrest division
With overhead distribution lines

View of Seattle's Hillcrest division
After underground distribution had been installed

Power Distribution Facilities

Distribution is the final step in the process of delivering electrical energy from a natural source to the ultimate user. Distribution represents the energy environment problem outside the door of the average citizen. The distribution facilities are the "capillaries" of the overall electrical energy system. As such, environmentally they are the minor annoyances close to the home of the resident.

Much has been written in scattered sources concerning the environmental problems represented in the distribution system. Because of what has been written, it is not essential at this point in this publication to reiterate more than the essentials of the problems and the major directions or approaches to a solution at this particular level. The 1970 Federal Power Survey spoke in terms of distribution substations, which basically were transformer facilities with the following words:

> To fulfill the ever-increasing electric energy requirements of the future, new additional distribution substations will be required. Considerable attention is being given to the location, design, and appearance of these installations to strive for environmental compatibility. The techniques being used include:
>
> 1. Proper site selection based on interdisciplinary analysis of the area.
> 2. Simplified designs that minimize components and the silhouette.
> 3. Reduction in height of structures and equipment.
> 4. Use of equipment colors that harmonize with the surrounding environment.

5. Use of colors to create esthetic effect.
6. Artful landscaping and lighting.
7. Interesting architectural treatment with panels, walls, or enclosures.
8. Use of underground runs for primary feeders to reduce overhead congestion in the immediate vicinity.
9. Site setbacks and appropriate station arrangement and orientation to minimize visual impact.
10. Development of landscaped, public rest spots in setback areas.

In many instances, improvements in the appearance of both new and existing substations can be obtained at nominal cost. The use of extensive architectural treatment, such as complete enclosures, may or may not increase costs of new substations, depending on the siting conditions.

Considerable progress has been made in reducing transformer noise with the result that it is much quieter in the vicinity of new substations than at those installed several years ago.[1]

This, of course, deals primarily with the low-level transformer areas as part of a distribution system.

In the visual impact study conducted by InterDesign, Inc., for the Northern States Power Company in Minnesota, the following statement was made concerning distribution:

Distribution

The most crucial problem for Distribution facilities relates to the viewing points from which the system is apparent to the public. As with Transmission, the location of poles along major public circulation greatly increased the system's impact. Placement on minor vehicle ways, such as alleys, or placement better related to vegetation masses, would reduce this impact. Because of the height of these elements, they can be well hidden when major trees provide a background for the silhouette. The best solution, however, to situations where the system must parallel major circulation would be burial of the lines along these corridors.

Because of the shorter line distance between poles, a major visual impact is created by the overlay of poles, lines and crossarms present in any one view. The reduction of the length of visible segments of this system, then, is a crucial need. Interruption by vegetation, coupled with offset location of long lines, could help to reduce this impact.

The severe pruning of trees, although a detail problem, does create a negative impact. The use of more resilient line materials and a reduction of pruning would reduce this effect.

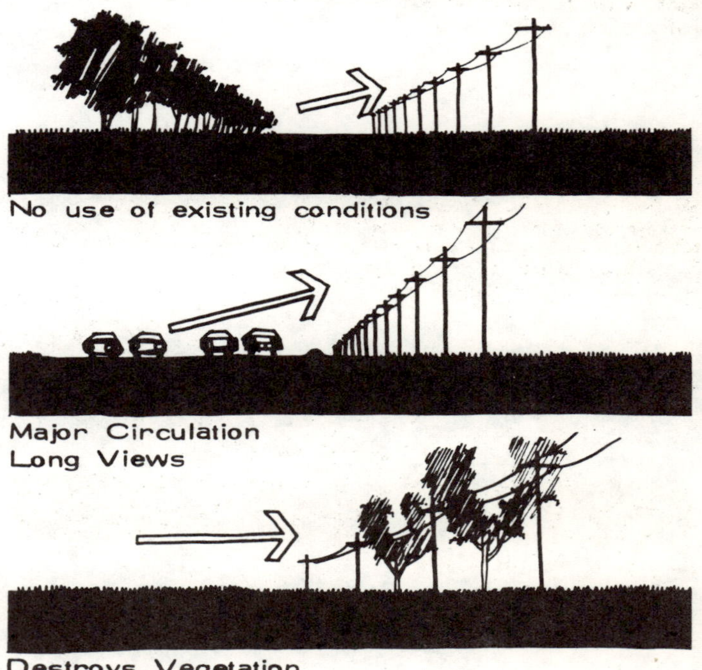

1. Federal Power Commission, *The 1970 National Power Survey: Guidelines for Growth of the Electric Power Industry,* (Washington, D.C., U.S. Government Printing Office, Superintendent of Documents, 1971), p. I-12-5.

RECOMMENDATION 1: Considering public experience of the system from major viewing points; eliminate long views of many poles, eliminate poles along major public circulation, make maximum use of existing vegetation to minimize silhouette of poles and crossarms minimize vegetation destruction, continue use of alleys and lot lines for system instead of streets.

Minimize Silhouette

System buried on major circulation

Lines through foliage
Minimize long views of system

The second priority problem relates to simplicity and design of the poles and line attachments themselves. It was determined that the use of a variety of the pole types with various crossarm connections created a high degree of visual chaos. The selection of one pole type with a unified line attachment device, e.g., a simple insulation connection would minimize the visual chaos, however, the simplest pole-line combination would be a vertical stacking of lines and insulators on a minimum height pole, thereby eliminating any cross pole elements.

The number of poles used on these systems often created an excessive overlay of visual elements and resulted in a crucial visual problem. The use of stronger poles and line materials might well be considered to reduce the number of poles needed and thus remove the resulting impact.

Related to both these issues is the question of underbuilding Distribution lines on Transmission poles as opposed to using separate poles for each system. It was determined desirable to minimize the number of poles wherever possible and, therefore, wise to place both systems on one pole type. When this is done, however, it is recommended to utilize the same method of line attachment for both systems so as to simplify the resulting silhouette.

The ultimate solution to the issue of simplification of the visual aspects of the system is burial, and this would be recommended wherever possible, especially in areas with public visibility and silhouette exposure.

PROBLEM: The great detail complexity of the system in terms of pole numbers, pole types and changing connection details per pole.

RECOMMENDATION 2: Simplify the system arrangement by minimizing poles numbers; minimizing pole types; minimizing multitude of connection details and direction changing details per pole; simplify wood pole; or bury system.

Make poles and lines as transparent as possible.

System underground

The issue of visual continuity of the system within the total surrounding environment proved to be the third priority problem. Here the focus of the criticism related to analyzing special aspects of the total network which contributed to general complexity of the system.

Initially, since the major public exposure to the network occurs along all major circulation ways, a significant problem occurs in the extent of exposure of the network at these points. High priority for burial of lines should be given to the network in relation to these areas.

Another problem of similar nature occurred where the systems crossed public circulation ways. The view of wires at these points created a major negative impact for the system. Burial of lines at intersections of the system with major streets would go far to reducing impact of the total system.

Wherever the line drops occur to several individual units, the chaos of line directions created an excessive negative impact. Burying all line drop connections to individual users so as to eliminate this chaotic point in the system would provide the optimum solution to this problem, and would be an effective use of burial funds in terms of reducing negative visual impact.

Another problem area similar to this relates to the lack of coordination of lighting placed on Distribution poles. A specially designed street light fixture whose attachment relates to the Distribution line attachment would minimize this problem.

PROBLEM: The total visual complexity of the system network dominates and complicates the visual landscape.

RECOMMENDATION 3: Simplify the total complex network.

System underground along public circulation

All street crossings underground

Power drops to individual houses underground.

Coordinate use of poles including lighting.[2]

Of course, much of the current work going on in distribution lines is also concerned with undergrounding. A report of the Working Committee on Utilities to the Vice President and to the President's Council on Recreation and Natural Beauty, dated December 27, 1968, made the following statement concerning undergrounding of distribution lines:

> There is a clear trend in the electric power and communications industries to place new residential distribution lines underground. Some states have promulgated laws and regulations requiring undergrounding of new distribution lines, and both the Citizens Advisory Committee and the Electric Utility Industry Task Force have encouraged the undergrounding of all such lines by 1975. The Citizens Advisory Committee has recommended 'the universal adoption of 1975 as a target date for all new distribution lines serving new residential subdivisions.' Moreover, it has recommended 'the adoption of a policy of undergrounding distribution lines at Federal installations and facilities' and has emphasized that 'the Federal Government, itself should set the precedent for undergrounding.'
>
> The Working Committee on Utilities has considered the 1975 target date established by the Citizens Advisory Committee, but believes that it should be advanced. For the reasons set forth below, we have adopted July 1, 1972, as the target date for undergrounding new distribution lines at Federal installations and facilities, and January 1, 1973, as the target date for undergrounding all new distribution lines serving new residential subdivisions by the other segments of the electric power and communications industries. We believe that these target dates are attainable and that our conclusions are fortified by the substantial progress made by these industries to the present.[3]

That same publication goes on to speak of some of the costs of undergrounding, as used at that particular time with the following statement:

> The capital cost of undergrounding electric distribution facilities was brought to national attention in the National Power Survey of the Federal Power Commission, published in 1964. There, it was noted that public insistence on placing distribution lines underground was increasing and that noteworthy strides were being made toward reducing the cost of placing distribution facilities underground. The National Power Survey included data which disclosed that in new residential areas located in subsoils which were suitable for low-cost trenching, the additional cost of installing underground residential service was, at times, less than fifty percent above the cost of overhead construction.
>
> During the period from 1955 to the present, the cost of underground residential distribution has been greatly reduced. In 1955, the ratio of underground to overhead distribution cost was about 10 to 1, and today this ratio is as low as 1.5 to 1, or even less in areas with especially suitable subsoils. When it is considered that the investment in electric utility plant is in the order of 40 percent for distribution facilities, 25 percent for transmission facilities and 35 percent for generating facilities, the overall effect of this somewhat more expensive underground service on the cost of electricity to residential consumers in these areas would be of a magnitude not likely to affect the annual growth in the use of electric energy. In some locations, however, where the subsoils present problems or where rock is at or just below the surface of the ground, the ratio of underground to overhead residential distribution costs

2. InterDesign, Inc., *Visual Impact Study,* (St. Paul, Minnesota, for Northern States Power Company, 1971), pp. 102-107.

3. Working Committee on Utilities, *Report to the Vice President and to the President's Council on Recreation and Natural Beauty,* (Washington, D.C., Working Committee on Utilities, 1968), p. 55.

cannot be lowered much below 10 to 1. At best, it seems that with existing technology a ratio of 7 to 1 or 6 to 1 can be expected. In these areas, undergrounding would probably not be feasible today.

In those areas where underground residential distribution costs total about 1.5 times as much as overhead costs, the largest contributing cost is for materials such as cables and terminators (20%), switches and transformers (30%) and secondary lines and services (28%). In the high-cost ratio areas, trenching costs are the largest.

One of the principal reasons for the decrease in underground residential distribution costs is the use of synthetic materials for insulation and protection of cable and secondary lines. The favorable early reports on the use of low-cost metallic sodium for conductors of electricity indicate that this technique may cause even a further reduction in undergrounding costs.

In areas having a high incidence of snow, sleet, ice or high winds, or in regions experiencing tornadoes or hurricanes, the operation and maintenance costs for underground residential distribution and other lines may be less than for overhead distribution lines.[4]

The U.S. Forest Service working with Pacific Gas and Electric Company in the Western States has developed an outline of some of the problems and potentials of undergrounding.

- Reduction of vandalism
- Public safety—Ski areas have snowfalls that make overhead lines hazardous due to clearance problems
- Cause less fires

Cost Factors of Undergrounding

- Depends on:
  - Number of transformers
  - Number of cables
  - Size of cables required to supply present and ultimate electric loads
  - Number of substructures
  - Soil conditions
  - Fences, roads, streams to cross
  - Clearing
  - Slope
  - Resurfacing or seeding required
- Can be a cost ratio of 1:1 with overhead but is usually a 3:1 or 2:1 ratio
- A 230 KV underground line in San Mateo, California is on 20:1 ratio

Miscellaneous on Undergrounding

- Line can be laid at rate of 1 to 2 miles per hour
- PG&E provides cable, splice boxes and splices and expects owner to pay for excess cost over equivalent overhead line
- Temporary bypass cable may be laid on ground when repairing outages
- Power line and telephone line can be installed in same operation with plow-in method—12 inches separation is required between telephone and primary voltage power cables

Undergrounding

- Most common type will involve 4, 12, 17, and 21 KV lines with 12 KV and 21 KV Predominant
- The plow-in method is favored for the longer extensions in forest or nonrural areas
- The State of California requires a burial minimum of 24 inches. PG&E uses 36 inch minimum—but can go deeper

Problems of Undergrounding

- Culvert headwalls along highways
- Bedrock
- Locating lines in winter—snow
- Cutting into in emergencies—fires

4. Ibid., pp. 58-59.

- Must install ultimate system initially. Impractical to increase the size of direct burial cables. U.G. lines must be replaced to increase capacity over initial design limits.
- 3000 foot reel length—must splice
- Possibility of electrolysis of aluminum culverts
- Shutdown periods increased—1/2 day to tap in a new line
- When line goes out it is out for a longer period
- Washouts—landslides, changes in road grade

Why Go Underground

- Ordinances and land use permits require it
- Snow loading on lines is eliminated
- A narrower R.O.W. is required
- Lightning damage is reduced
- Wind damage is eliminated
- No danger to aircraft—or kite flying
- Improved appearance[5]

5. U.S. Department of Agriculture, Forest Service, Rocky Mountain Region, Indepartmental memo, pp. 1-2.

Conclusion

In conclusion, it is necessary to reiterate what was said at the beginning of this publication.

> This is not a negative book of fault-finding, but a positive book of hope, of guidance and direction. It is largely the chronicling of one profession dealing with some of the current problems of one industry. It is a sometime perfunctory delineation of the way that the profession of landscape architecture has worked with the electric utility industry in dealing with some of the problems associated with the generation and distribution of the necessary power in an age of increasing concern with the preservation of the environment.

Much has been said in recent years concerning the problems of energy and environment. This book, of course, deals with only a few of these statements from a relatively narrow base of sources. Four other sources need to be mentioned and quoted in an apt conclusion to this publication in order to put all of what has been said previously in perspective.

In testimony in the hearings before the Subcommittee on Communications and Power of the Committee on Interstate and Foreign Commerce of the House of Representatives of the 92nd Congress, First Session, on bills relating to power plant siting and environmental protection, Howard M. Winterson, President of the Atomic Industrial Forum, Inc., made the following statement:

> A major subject of concern to utilities as well as to virtually everyone else in the country, is the effect of an in-

stallation on its environment. A slogan of many conservationists is, 'All Power Pollutes.' This is true, of course, if you consider electricity only at its source. No matter what type of generating plant is built, it has an impact on its immediate environment. At its point of use, however, electricity is perhaps the most positive environmental influence known to men. At its outlet it is a totally clean source of energy that can substitute for individual fires, furnaces, gasoline motors, or other pollution producing devices.[1]

On the other hand, Michael McCloskey, the Executive Director of the Sierra Club wrote in Volume I, No. 3 of Environmental Affairs that:

> Disagreement exists as to the nature of the 'energy crisis.' The industries supplying energy would have us believe that the crisis has arisen from a need to expand supplies and to lessen constraints on growth. Environmentalists, however, submit that the crisis lies instead in the need to halt excessive pressures for increased energy consumption. They contend that present rates of energy growth are unrealistic, environmentally damaging, and artificially induced. Because these rates cannot long continue, they that feel our main task ought to be to bring these rates of growth under control, and that there are reasonable ways of doing this.
>
> Present rates of energy growth are unrealistic for a variety of reasons. These compounding rates of growth cannot be projected very far into the future before they run up against mathematical, physical biological and qualitative limits. Let us examine these rates and some of the limits which they confront.
>
> Today we consume 15 times the energy we did 100 years ago, though our population has only tripled in that time. Over the past decade the average growth rate in the consumption of energy in all its forms has been more than four percent annually, climbing to about five percent annually over the last five years. Growth has been particularly phenomenal in the electrical energy sector, at about seven percent annually in recent years. Projections based on that rate of growth call for a doubling of electric power production about every ten years.
>
> With these growth rates we may soon find that we are reaching absolute limits in physical space for power plants. It has been calculated that even with large 1,000 megawatt power plants, each of which requires an area of only 1,000 feet on a side, in less than 20 doublings (less than 200 years) all the available land space in the United States would be occupied by such plants. In California, where power production is expected to double every eight years, if power were to be supplied by 110 megawatt plants on 80 acre sites, the entire land area would be covered in only 122 years. Similar startling projections could doubtless be made with respect to other forms of energy use, such as the amount of space that will need to be paved over to accommodate our automobile oriented transportation system.
>
> Other physical limits to energy use can be cited. For example, by the end of the century, if growth continues as projected at current rates, one third of our total freshwater run-off might be required for powerplant cooling purposes, if just 'once-through' cooling is used. If 'once-through' cooling is superseded by cooling ponds and towers, then more land will be needed, and the space crunch will come even sooner.
>
> Ultimate limits to growth in energy use also obviously exist in the finite nature of our fuel resources. The fossil fuels now provide by far the greatest part of our energy sources (e.g., almost 96 percent in 1969). Whatever the true situation as to immediate supplies, it is obvious that ultimately these nonrenewable resources will be depleted. Optimistic estimates predict that our fossil fuels as a group will be exhausted within a few hundred years at best, pos-

1. Subcommittee on Communications and Power of the Committee on Interstate and Foreign Commerce, House of Representatives, Ninety-Second Congress, First Session, *Powerplant Siting and Environmental Protection*, Hearings before the Subcommittee on Bills Relating to Powerplant Siting and Environmental Protection from May 4-27, 1971, (Washington, D.C., U.S. Superintendent of Documents, Government Printing Office, Printed for use of the Committee on Interstate and Foreign Commerce, Part 1 (Serial No. 92-31, Part 2 (Serial No. 92-32, Part 3, (Serial No. 92-33, 1972), p. 1059.

sibly much sooner. A recent National Academy of Sciences report, for example, predicts that within approximately 50 years, the great bulk of the world's initial supply of recoverable petroleum liquids and natural gas will be exhausted. Recoverable fuel from the oil shales and tar sands might extend the lifetime of the petroleum group another century. With respect to coal, the report estimates that if used as the principle source of energy at projected demands, it will last no more than two or three centuries.

Though nuclear power is expected to play a major role in future electrical energy production, electrical energy is only a part of the total energy consumed—presently, about one quarter—and the supply of uranium 235 from high grade ores is limited. The NAS report indicates that the production of nuclear power with the present type of reactors and with uranium 235 as the principal energy source can be sustained for only a few decades. Another estimate gives high grade uranium ores a lifetime of under 50 years. Breeder reactors could extend these fuels, but it is not clear what the costs may be, and many operational and environmental problems remain unsolved. Moreover, a practical method of producing electricity from fusion is still only a possibility.

Hydroelectric power has a finite limit in the availability of suitable sites, and is of small importance in the supply picture. While only one-fourth of the potential hydroelectric sites have now been developed in this country, these are the best sites; most of the rest are economically unfeasible.

Another ultimate limit to energy growth is imposed by the problem of dissipating the heat resulting from the production of energy. With energy consumption increasing at an annual rate of five percent, a climatological heat limit—the point at which global climate would be drastically altered—could be reached in less than a century.

By the year 2000, at projected growth rates, the energy produced by man in major urban areas may approximate 30 percent of solar input. Our population centers will turn into giant heat radiators affecting local climates. Increasing attention has also been given to the possible role of two by-products of energy production—carbon dioxide and water vapor—in long range climatological change. Each year fossil fuel combustion adds to the atmosphere an amount of carbon dioxide equal to about 25 percent of the total carbon dioxide in the atmosphere. If the percent growth rate in fuel use continues, there will be an increase of about 170 percent in the carbon dioxide level in the next 150 years. Many scientists fear that the 'greenhouse effect' (the trapping of heat energy which leaves the earth by carbon dioxide in the atmosphere) will have serious repercussions on world climate. For differing reasons, concern is also expressed over adding substantial amounts of water vapor to the atmosphere. Though more research is imperative, these considerations may also place ultimate limits on unrestrained energy consumption.

Long before the ultimate limits are reached the environmental impact of unrestrained energy growth may become unbearable. We do not know where the dividing line between environmental deterioration and irreversible catastrophe may lie, but at the least we can foresee that galloping energy consumption will have a continuing and cumulatively destructive impact upon the environment.

At every stage of energy production and use, unacceptable environmental degradation occurs.[2]

Mr. McCloskey goes on to say in conclusion:

There is no way to foresee precisely the ultimate effects of all of these changes in policy and administration. However, at this time, three points may be made. First, many of these changes are already beginning to occur. Second, the strategy of orchestrating these changes allows many factors to be continually adjusted and corrections to be readily effected. Third, the practical alternatives to such changes are not very attractive. As one alternative, we might defer significant reform until a time when inaction could pro-

2. Michael McCloskey, "The Energy Crisis: Issues and a Proposed Response," in *Environmental Affairs,* Vol. 1, No. 3, (Brighton, Massachusetts, The Environmental Center, Boston College Law School, 1971), pp. 587-589.

duce pressures for revolutionary changes in our basic institutions. As another alternative, we might simply permit the energy system to expand as far as it could and thereby permit the environment to bear enormous and irreversible degradation. Neither is a prudent way to deal with a problem which must be resolved without delay.

One of the most distressing aspects of our dilemma is that we do not know the full extent to which our environment has already been damaged by energy programs. Certainly we should lose no time in pursuing our best option. Our choice need not be between blackouts and governmental orders to turn out the lights. We can still impose restraints on those self-interested and overdeveloped industries which, in their fabrication of artificial crises in energy, are creating genuine crises in the environment.[3]

The publication prepared for the National Rural Electric Cooperative Association by Stanley H. Ruttenburg in August of 1971, entitled, "The Electric Power Crisis, Its Impact on Workers and Consumers," the following statement was made in conclusion:

The welfare of the nation and its people depends on a sufficient supply of electricity. It is a constant companion on which all of us rely for everyday needs at home, work or recreation. However, our continuing adequate supply of electric power is threatened and, in turn, our standard of living and national security are jeopardized.

Faced with such a crisis, a detailed reappraisal of the nation's entire energy situation is an obvious necessity. However, resolving the multitude of complex problems which beset the industry will not be an easy task.

Moreover, ensuring an adequate supply of clean energy at the least possible cost is not solely a problem of industry or government. It is a problem common to all citizens, and, as such, we all must participate in its solution.

Rural and urban residents alike should be vitally concerned about the power crisis and the actions that are taken to correct it. Since the problem is so widespread in scope, the steps required to solve it will involve national action and new legislation.

Society at large must be made aware of the magnitude and imminent nature of the power crisis and of the recommended courses of action which offer the greatest hope of solving the problem in a manner which will best protect the interest of the general public. They should then request their elected officials to take the necessary steps in response to wishes of an informed electorate.[4]

The final quartet of quotes to conclude this publication, comes from Sylvia Crowe in the publication, "Landscape of Power," where she says,

If we accept that a good landscape is as much part of our standard of living as good housing, fast transport and any of the other benefits which we expect power to bring to us, how can we ensure that it will not go by default? The first essential is to be sure that we do accept this point of view. Nothing can be done unless it is recognized that the appearance of our surroundings matters, that without good landscape and townscape our pretensions to a civilized life are futile. Once we have regained our sense of values sufficiently to realize this, the necessary steps to save the landscape are technically relatively simple.

They may be summed up as: co-ordination between users of the land; closer collaboration between scientists and artists, in the widest sense of the two terms; a recognition that civic design has its counterpart in landscape design and a determination to apply the principles of zoning on a national scale and to look at the landscape on a wider basis than that of the local planning authority.

The first would economize the amount of land allocated for each use, and would open up opportunities of relating one structure to another, of pooling the needs for approach roads and fencing and of combining resources for any planting or land formation needed to bring the structures into relation with each other and the landscape, thus enabling an overall pattern of landscape to be re-estab-

3. Ibid., pp. 604-605.
4. Stanley H. Ruttenberg and Associates, Inc., *The Electric Power Crisis: Its Impact on Workers and Consumers,* (Washington, D.C., National Rural Electric Cooperative Association, 1971), p. 58.

lished, within itself all the diverse elements which at present are allowed to tear ragged holes out of the old countryside.[5]

Thanks must be expressed to all of those designers and industrial representatives which have given their permission to allow their projects to be given wider publication and distribution in a publication of this type.

To quote from the Preface of this book:

The scope of this book deals with the physical design, the process, the studies and the guidelines, primarily of the physical accoutrements of the electric utilities. The purpose of this book, as mentioned previously, is a positive one to give credit, guidance, direction and encouragement, to define the condition of the art, and to pull together all of the information available on the subject at the present time. There are certain limitations in an attempt of this type. Some of these are determined by the fact that none of the intricacies of the science and the chemistry of water and air pollution can be dealt with here. The positive contributions of many of the 'hard' scientists are neglected in a study of this type. There is no overview or feedback in this publication regarding the effectiveness of these proposed environmental solutions, since the studies are so recent that it is not possible at this early date to determine their effectiveness. The studies illustrated in this book, however, go far beyond the normal concept of landscape development and illustrate the potential contribution of the landscape architect to some of the current environmental problems. Hopefully, it will be of assistance both to environmental designers and persons in the electric utility industry as well as those with a larger desire to know more of the positive solutions to the problem identified by many responsible persons and groups.

Hopefully this book, which is really a collection and organization of the accomplishments of a number of people and organizations, which need to be given wider distribution and acceptance, is but a link in the chain leading toward a solution. This book has had many precedents and ancestors. Hopefully, it will stimulate much additional work in this area, and will have many offspring. If this publication serves to encourage a larger number of environmental designers to cooperate with an equally larger number of representatives of the electric utility industry to forge new directions and innovative solutions to the problem of energy and environment, it will have served its purpose.

5. Sylvia Crowe, *The Landscape of Power* (London, The Architectural Press, 1958), p. 105.

Appendix A

COUNCIL ON ENVIRONMENTAL QUALITY

Statements on Proposed Federal
Actions Affecting The Environment

(from the Federal Register, Vol. 36, No. 79, April 23, 1971)

Guidelines

1. Purpose. This memorandum provides guidelines to Federal departments, agencies, and establishments for preparing detailed environmental statements on proposals for legislation and other major Federal actions significantly affecting the quality of the human environment as required by section 102(2) (C) of the National Environmental Policy Act (Public Law 91-190) (hereafter "the Act"). Underlying the preparation of such environmental statements is the mandate of both the Act and Executive Order 11514 (35 F.R. 4247) of March 4, 1970, that all Federal agencies, to the fullest extent possible, direct their policies, plans and programs so as to meet national environmental goals. The objective of section 102(2) (C) of the Act and of these guidelines is to build into the agency decision making process an appropriate and careful consideration of the environmental aspects of proposed action and to assist agencies in implementing not only the letter, but the spirit, of the Act. This memorandum also provides guidance on implementation of section 309 of the Clean Air Act, as amended (42 U.S.C. 1857 et seq.).

2. Policy. As early as possible and in all cases prior to agency decision concerning major action or recommendation or a favorable report on legislation that significantly affects the environment, Federal agencies will, in consultation with other appropriate

Federal, State, and local agencies, assess in detail the potential environmental impact in order that adverse effects are avoided, and environmental quality is restored or enhanced, to the fullest extent practicable. In particular, alternative actions that will minimize adverse impact should be explored and both the long- and short-range implications to man, his physical and social surroundings, and to nature, should be evaluated in order to avoid to the fullest extent practicable undesirable consequences for the environment.

3. Agency and OMB procedures. (a) Pursuant to section 2(f) of Executive Order 11514, the heads of Federal agencies have been directed to proceed with measures required by section 102(2) (C) of the Act. Consequently, each agency will establish, in consultation with the Council on Environmental Quality, not later than June 1, 1970 (and, by July 1, 1971, with respect to requirements imposed by revisions in these guidelines, which will apply to draft environmental statements circulated after June 30, 1971), its own formal procedures for (1) identifying those agency actions requiring environmental statements, the appropriate time prior to decision for the consultations required by section 102(2) (C), and the agency review process for which environmental statements are to be available, (2) obtaining information required in their preparation, (3) designating the officials who are to be responsible for the statements, (4) consulting with and taking account of the comments of appropriate Federal, State, and local agencies, including obtaining the comment of the Administrator of the Environmental Protection Agency, whether or not an environmental statement is prepared, when required under section 309 of the Clean Air Act, as amended, and section 8 of these guidelines, and (5) meeting the requirements of section 2(b) of Executive Order 11514 for providing timely public information on Federal plans and programs with environmental impact including procedures responsive to section 10 of these guidelines. These procedures should be consonant with the guidelines contained herein. Each agency should file seven (7) copies of all such procedures with the Council on Environmental Quality, which will provide advice to agencies in the preparation of their procedures and guidance on the application and interpretation of the Council's guidelines. The Environmental Protection Agency will assist in resolving any question relating to section 309 of the Clean Air Act, as amended.

(b) Each Federal agency should consult, with the assistance of the Council on Environmental Quality and the Office of Management and Budget if desired, with other appropriate Federal agencies in the development of the above procedures so as to achieve consistency in dealing with similar activities and to assure effective coordination among agencies in their review of proposed activities.

(c) State and local review of agency procedures, regulations, and policies for the administration of Federal programs of assistance to State and local governments will be conducted pursuant to procedures established by the Office of Management and Budget Circular No. A-85. For agency procedures subject to OMB Circular No. A-85, a 30-day extension in the July 1, 1971, deadline set in section 3(a) is granted.

(d) It is imperative that existing mechanisms for obtaining the views of Federal, State, and local agencies on proposed Federal actions be utilized to the extent practicable in dealing with environmental matters. The Office of Management and Budget will issue instructions, as necessary, to take full advantage of existing mechanisms (relating to procedures for handling legislation, preparation of budgetary materials, new procedures, water resource and other projects, etc.).

4. Federal agencies included. Section 102(2) (C) applies to all agencies of the Federal Government with respect to recommendations or favorable reports on proposals for (i) legislation and (ii) other major Federal actions significantly affecting the quality of the human environment. The phrase "to the fullest extent possible" in section 102(2) (C) is meant to make clear that each agency of the Federal Government shall comply with the requirement unless existing law applicable to the agency's operations expressly prohibits or makes compliance impossible. (Section 105 of the Act provides that "The policies and goals set forth in this Act are supplementary to those set forth in existing authorizations of Federal agencies.")

5. Actions included. The following criteria will be employed by agencies in deciding whether a proposed action requires the preparation of an environmental statement:

(a) "Actions" include but are not limited to:

(i) Recommendations or favorable reports relating to legislation including that for appropriations. The requirement for following the section 102(2) (C) procedure as elaborated in these guidelines applies to both (i) agency recommendations on their own proposals for legislation and (ii) agency reports on legislation initiated elsewhere. (In the latter case only the agency which has primary responsibility for the subject matter involved will prepare an environmental statement.) The Office of Management and Budget will supplement these general guidelines with specific instructions relating to the way in which the section 102(2) (C) procedure fits into its legislative clearance process;

(ii) Projects and continuing activities: directly undertaken by Federal agencies; supported in whole or in part through Federal contracts, grants, subsidies, loans, or other forms of funding assistance; involving a Federal lease, permit, license, certificate or other entitlement for use;

(iii) Policy, regulations, and procedure-making.

(b) The statutory clause "major Federal actions significantly affecting the quality of the human environment" is to be construed by agencies with a view to the overall, cumulative impact of the action proposed (and of further actions comtemplated). Such actions may be localized in their impact, but if there is potential that the environment may be significantly affected, the statement is to be prepared. Proposed actions, the environmental impact of which is likely to be highly controversial, should be covered in all cases. In considering what constitutes major action significantly affecting the environment, agencies should bear in mind that the effect of many Federal decisions about a project or complex of projects can be individually limited but cumulatively considerable. This can occur when one or more agencies over a period of years puts into a project individually minor but collectively major resources, when one decision involving a limited amount of money is a precedent for action in much larger cases or represents a decision in principle about a future major course of action, or when several Government agencies individually make decisions about partial aspects of a major action. The lead agency should prepare an environmental statement if it is reasonable to anticipate a cumulatively significant impact on the environment from Federal action. "Lead agency" refers to the Federal agency which has primary authority for committing the Federal Government to a course of action with significant environmental impact. As necessary, the Council on Environmental Quality will assist in resolving questions of lead agency determination.

(c) Section 101(b) of the Act indicates the broad range of aspects of the environment to be surveyed in any assessment of significant effect. The Act also indicates that adverse significant effects include those that degrade the quality of the environment, curtail the range of beneficial uses of the environment, and serve short-term, to the disadvantage of long-term, environmental goals. Significant effects can also include actions which may have both beneficial and detrimental effects, even if, on balance, the agency believes that the effect will be beneficial. Significant adverse effects on the quality of the human environment include both those that directly affect human beings and those that indirectly affect human beings through adverse effects on the environment.

(d) Because of the Act's legislative history, environmental protective regulatory activities concurred in or taken by the Environmental Protection Agency are not deemed actions which require the preparation of environmental statements under section 102(2) (C) of the Act.

6. Content of environmental statement. (a) The following points are to be covered:

(i) A description of the proposed action including information and technical data adequate to permit a careful assessment of environmental impact by commenting agencies. Where relevant, maps should be provided.

(ii) The probable impact of the proposed action on the environment, including impact on ecological systems such as wildlife, fish, and marine life. Both primary and secondary significant consequences for the environment should be included in the analysis. For example, the implications, if any, of the action for population distribution or concentration should be estimated and an assessment made of the effect of any possible change in population patterns upon the resource base, including land use, water, and public services, of the area in question.

(iii) Any probably adverse environmental effects which cannot be avoided (such as water or air pollution, undesirable land use patterns, damage to life systems, urban congestion, threats to health or other consequences adverse to the environmental goals set out in section 101(b) of the Act).

(iv) Alternatives to the proposed action (section 102(2) (D) of the Act requires the responsible agency to "study, develop, and describe appropriate alternatives to recommended courses of action in any proposal which involves unresolved conflicts concerning alternative uses of available resources"). A rigorous exploration and objective evaluation of alternative actions that might avoid some or all of the adverse environmental effects is essential. Sufficient analysis of such alternatives and their costs and impact on the environment should accompany the proposed action through the agency review process in order not to foreclose prematurely options which might have less detrimental effects.

(v) The relationship between local short-term uses of man's environment and the maintenance and enhancement of long-term productivity. This in essence requires the agency to assess the action for cumulative and long-term effects from the perspective that each generation is trustee of the environment for succeeding generations.

(vi) Any irreversible and irretrievable commitments of resources which would be involved in the proposed action should it be implemented. This requires the agency to identify the extent to which the action curtails the range of beneficial uses of the environment.

(vii) Where appropriate, a discussion of problems and objections raised by other Federal, State, and local agencies and by private organizations and individuals in the review process and the disposition of the issues involved. (The section may be added at the end of the review process in the final text of the environmental statement.)

(b) With respect to water quality aspects of the proposed action which have been previously certified by the appropriate State or interstate organization as being in substantial compliance with applicable water quality standards, the comment of the Environmental Protection Agency should also be requested.

(c) Each environmental statement should be prepared in accordance with the precept in section 102(2) (A) of the Act that all agencies of the Federal Government "utilize a systematic, interdisciplinary approach which will insure the integrated use of the natural and social sciences and the environmental design arts in planning and decision-making which may have an impact on man's environment."

(d) Where an agency follows a practice of declining to favor an alternative until public hearings have been held on a proposed action, a draft environmental statement may be prepared and circulated indicating that two or more alternatives are under consideration.

(e) Appendix 1 prescribes the form of the summary sheet which should accompany each draft and final environmental statement.

7. Federal agencies to be consulted in connection with preparation of environmental statement. A Federal agency considering an action requiring an environmental statement on the basis of (i) a draft environmental statement for which it takes responsibility or (ii) comparable information followed by a hearing subject to the provisions of the Administrative Procedure Act, should consult with, and obtain the comment on the environmental impact of the action of, Federal agencies with jurisdiction by law or special expertise with respect to any environmental impact involved. These Federal agencies include components of (depending on the aspect or aspects of the environment):

Advisory Council on Historic Preservation.
Department of Agriculture.
Department of Commerce.
Department of Defense.
Department of Health, Education, and Welfare.
Department of Housing and Urban Development.
Department of the Interior.
Department of State.
Department of Transportation.
Atomic Energy Commission.
Federal Power Commission.
Environmental Protection Agency.
Office of Economic Opportunity.

For actions specifically affecting the environment of their geographic jurisdictions, the following Federal and Federal-State agencies are also to be consulted:

Tennessee Valley Authority.
Appalachian Regional Commission.
National Capital Planning Commission.
Delaware River Basin Commission.
Susquehanna River Basin Commission.

Agencies seeking comment should determine which one or more of the above listed agencies are appropriate to consult on the basis of the areas of expertise identified in Appendix 2 to these guidelines. It is recommended (i) that the above listed departments and agencies establish contact points, which often are most appropriately regional offices, for providing comments on the environmental statements and (ii) that departments from which comment is solicited coordinate and consolidate the comments of their component entities. The requirement in section 102(2) (C) to obtain comment from Federal agencies having jurisdiction or special expertise is in addition to any specific statutory obligation of any Federal agency to coordinate or consult with any other Federal or State agency. Agencies seeking comment may establish time limits of not less than thirty (30) days for reply, after which it may be presumed, unless the agency consulted requests a specified extension of time, that the agency consulted has no comment to make. Agencies seeking comment should endeavor to comply with requests for extensions of time of up to fifteen (15) days.

8. Interim EPA procedures for implementation of section 309 of the Clean Air Act, as amended. (a) Section 309 of the Clean Air Act, as amended, provides:

SEC. 309. (a) The Administrator shall review and comment in writing on the environmental impact of any matter relating to duties and responsibilities granted pursuant to this Act or other provisions of the authority of the Administrator, contained in any (1) legislation proposed by any Federal department or agency, (2) newly authorized Federal projects for construction and any major Federal agency action (other than a project for construction) to which section 102(2) (C) of Public Law 91-190 applies, and (3) proposed regulations published by any department or agency of the Federal Government. Such written comment shall be made public at the conclusion of any such review.

(b) In the event the Administrator determines that any such legislation, action, or regulation is unsatisfactory from the standpoint of public health or welfare or environmental quality, he shall publish his determination and the matter shall be referred to the Council on Environmental Quality.

(c) Accordingly, wherever an agency action related to air or water quality, noise abatement and control, pesticide regulation, solid waste disposal, radiation criteria and standards, or other provisions of the authority of the Administrator if the Environmental Protection Agency is involved, including his enforcement authority, Federal agencies are required to submit for review and comment by the Administrator in writing: (i) proposals for new Federal construction projects and other major Federal agency actions to which section 102(2) (C) of the National Environmental Policy Act applies and (ii) proposed legislation and regulations, whether or not section 102(2) (C) of the National Environmental Policy Act applies. (Actions requiring review by the Administrator do not include litigation or enforcement proceedings.) The Administrator's comments shall constitute his comments for the purpose of both section 309 of the Clean Air Act and section 102(2) (C) of the National Environmental Policy Act. A period of 45 days shall be allowed for such review. The Administrator's written comment shall be furnished to the responsible Federal department or agency, to the Council on Environmental Quality and summarized in a notice published in the Federal Register. The public may obtain copies of such comment on request from the Environmental Protection Agency.

9. State and local review. Where no public hearing has been held on the proposed action at which the appropriate State and local review has been invited, and where review of the environmental impact of the proposed action by State and local agencies authorized to develop and enforce environmental standards is relevant, such State and local review shall be provided as follows:

(a) For direct Federal development projects and projects assisted under programs listed in Attachment D of the Office of Management and Budget Circular No. A-95, review of draft environmental statements by State and local governments will be through procedures set forth under Part 1 of Circular No. A-95.

(b) Where these procedures are not appropriate and where a proposed action affects matters within their jurisdiction, review of the draft environmental statement on a proposed action by State and local agencies authorized to develop and enforce environmental standards and their comments on the environmental impact of the proposed action may be obtained directly or by distributing the draft environmental statement to the appropriate State, regional and metropolitan clearinghouses unless the Governor of the State involved has designated some other point for obtaining this review.

10. Use of statements in agency review processes; distribution to Council on Environmental Quality; availability to public. (a) Agencies will need to identify at what stage or stages of a series of actions relating to a particular matter the environmental statement procedures of this directive will be applied. It will often be necessary to use the procedures both in the development of a national program and in the review of proposed projects within the national program. However, where a grant-in-aid program does not entail prior approval by Federal agencies of specific projects the view of Federal, State, and local agencies in the legislative process may have to suffice. The principle to be applied is to obtain views of other agencies at the earliest feasible time in the development of program and project proposals. Care should be exercised so as not to duplicate the clearance process, but when actions being considered differ significantly from those that have already been reviewed pursuant to section 192(2) (C) of the Act an environmental statement should be provided.

(b) Ten (10) copies of draft environmental statements (when prepared), ten (10) copies of all comments made thereon (to be forwarded to the Council by the entity making comment at the time comment is forwarded to the responsible agency), and ten (10) copies of the final text of environmental statements (together with all comments received thereon by the responsible agency from Federal, State, and local agencies and from private organizations and individuals) shall be supplied to the Council on Environmental Quality in the Executive Office of the President (This will serve as making environmental statements available to the President). It is important that draft environmental statements be prepared and circulated for comment and furnished to the Council early enough in the agency review process before an action is taken in order to permit meaningful consideration of the environmental issues involved. To the maximum extent practicable no administrative action (i.e., any proposed action to be taken by the agency other than agency proposals for legislation to Congress or agency reports on legislation) subject to section 102(2) (C) is to be taken sooner than ninety (90) days after a draft environmental statement has been circulated for comment, furnished to the Council and, except where advance public disclosure will result in significantly increased costs of procurement to the Government, made available to the public pursuant to these guidelines; neither should such administrative action be taken sooner than thirty (30) days after the final text of an environmental statement (together with comments) has been made available to the Council and the public. If the final text of an environmental statement is filed within ninety (90) days after a draft statement has been circulated for comment, furnished to the Council and made public pursuant to this section of these guidelines, the thirty (30) day period and ninety (90) day period may run concurrently to the extent that they overlap.

(c) With respect to recommendations or reports on proposals for legislation to which section 102(2) (C) applies, the final text of the environmental statement and comments thereon should be available to the Congress and to the public in support of the proposed legislation or report. In cases where the scheduling of congressional hearings on recommendations or reports on proposals for legislation which the Federal agency has forwarded to the Congress does not allow adequate time for the completion of a final text of an environmental statement (together with comments), a draft environmental statement may be furnished to the Congress and made available to the public pending transmittal of the comments as received and the final text.

(d) Where emergency circumstances make it necessary to take an action with significant environmental impact without observing the provisions of these guidelines concerning minimum periods for agency review and advance availability of environmental statements, the Federal agency proposing to take the action should consult with the Council on Environmental Quality about alternative arrangements. Similarly, where there are overriding considerations of expense to the Government or impaired program effectiveness, the responsible agency should consult the Council concerning appropriate modifications of the minimum periods.

(e) In accord with the policy of the National Environmental Policy Act and Executive Order 11514 agencies have a responsibility to develop procedures to insure the fullest practicable provision of timely public information and understanding of Federal plans and programs with environmental impact in order to obtain the views of interested parties. These procedures shall include, whenever appropriate, provision for public hearings, and shall provide the public with relevant information, including information on alternative courses of action. Agencies which hold hearings on proposed administrative actions or legislation should make the draft environmental statement available to the public at least fifteen (15) days prior to the time of the relevant hearings, except where the agency prepares the draft statement on the basis of a hearing subject to the Administrative Procedure Act and preceded by adequate public notice and information to identify the issues and obtain the comments provided for in sections 6-9 of these guidelines.

(f) The agency which prepared the environmental statement is responsible for making the statement and the comments received available and the comments received available to the public pursuant to the provisions of the Freedom of Information Act (5 U.S.C., sec. 552), without regard to the exclusion of interagency memoranda when such memoranda transmit comments of Federal agencies listed in section 7 of these guidelines upon the environmental impact of proposed actions subject to section 102(2) (c).

(g) Agency procedures prepared pursuant to section 3 of these guidelines shall implement these public information requirements and shall include arrangements for availability of environmental statements and comments at the head and appropriate regional offices of the responsible agency and at appropriate State, regional, and metropolitan clearinghouses unless the Governor of the State involved designates some other point for receipt of this information.

11. Application of section 102(2) (c) procedure to existing projects and programs. To the maximum extent practicable the section 102(2) (C) procedure should be applied to further major Federal actions having a significant effect on the environment even though they arise from projects or programs initiated prior to enactment of the Act on January 1, 1970. Where it is not practicable to reassess the basic course of action, it is still important that further incremental major actions be shaped so as to minimize adverse environmental consequences. It is also important in further action that account be taken of environmental consequences not fully evaluated at the outset of the project or program.

12. Supplementary guidelines, evaluation of procedures. (a) The Council on Environmental Quality after examining environmental statements and agency procedures with respect to such statements will issue such supplements to these guidelines as are necessary.

(b) Agencies will continue to assess their experience in the implementation of the section 102(2) (C) provisions of the Act and in conforming with these guidelines and report thereon to the Council on Environmental Quality by December 1, 1971. Such reports should include an identification of the problem areas and suggestions for revision or clarification of these guidelines to achieve effective coordination of views on environmental aspects (and alternatives, where appropriate) of proposed actions without imposing unproductive administrative procedures.

Russell E. Train, Chairman

Appendix B

ENVIRONMENTAL PROTECTION CHECKLIST and GUIDELINES FOR SITE SELECTION

Guidelines

Common to all plants is the requirement of adequate land area for the plant, switchyard, cooling facilities where necessary, and for plant expansion where planned. Coal fired plants require large acreages for fuel storage, ash disposal, and air quality control equipment. A fossil-fueled plant with 3,000 megawatts capacity may require in the range of 1,200 to 1,600 acres if coal fired, 200 to 400 acres if oil fired, and 150 to 300 acres if gas fired. A cooling pond would require an additional 1 to 2 acres per megawatt of capacity.

Siting criteria for nuclear-fueled power plants have been developed by the Atomic Energy Commission on the basis of public safety. Part 100 of Title 10, Code of Federal Regulations, provides criteria for determining an exclusion area surrounding a proposed plant, a low population zone immediately surrounding the exclusion area, and a population center distance representing the distance from the plant to the nearest boundary of a densely populated area containing more than about 25,000 residents. The plant owner must have authority to determine all activities within the exclusion area, including exclusion or removal of personnel and property from the area. A 3,000 megawatt nuclear plant would require a site ranging from 300 to 500 acres in area. For light-water plants a cooling pond would need to be about 50 percent larger than that for a fossil-fueled plant of equal size.

The physical characteristics that must be considered in locating a power plant include the genology, seismology, hydrology, and meteorology of the proposed site. These are factors that are evaluated by the AEC before it approves a proposed nuclear plant site. However, the flexibility of siting may be increased by providing engineering safety features in the plant design to compensate for the natural characteristics of the site. The relationship of the plant site to historical, archaeological, and cultural sites must also be considered.

In developing a land use plan for a proposed site, careful consideration should be given to existing developments in the area and to existing zoning regulations. The location and design of facilities should be such as to minimize the impact on the natural environment. Esthetic treatment can be enhanced by the establishment of buffer zones around the plant. These zones can be used to screen the plant facilities by means of trees, vegetation, and other landscaping.

The esthetics of a power plant can be improved by good architectural design and landscaping treatment. Strategic profiling and positioning of building and structures, use of appropriate materials and colors, and use of screening and blending of landscaping are among the techniques available to moderate visual impact. It is important also that proper consideration be given to the visual effect of transmission lines extending from the plant.

Where practicable, the construction of a thermal power plant should provide beneficial additions to the recreation facilities of the area. In some cases, heated cooling water can be diverted to swimming lagoons or pools. In some areas, sport fishing may be enhanced by the discharge of warm water into a water body. Cooling ponds may provide opportunities for public boating, picnicking, and camping. The exclusion area of nuclear plants may provide camping areas for scouts and similar groups.

Nuclear plants in general can be more attractive than fossil-fueled plants. To take full advantage of their inherently better esthetic qualities, nuclear plants should include imaginative architectural landscaping and suitable recreational facilities or other compatible land uses. In some cases, information centers established at or near plant sites may be desirable. Animated displays and exhibits on atomic energy with emphasis on safety may be featured.

Environmental Protection Checklist

Consistent with the guidelines discussed above, there follows a listing of factors to be considered in site selection relative to land use planning and esthetics of facilities.

1. Land Area and Population—
   (a) Area—Sufficient acreage should be provided for planned facilities, including future expansions. For nuclear plants, the area under control must meet AEC regulations in 10 CFR 100.
   (b) Population—The relationship of nuclear plants to population centers must be in accord with AEC regulations in 10 CFR 100.
2. Physical Characteristics of Site—The suitability of the site should be determined by the following types of studies:
   (a) Hydrologic—To determine ground water gradients and chemical characteristics, interchange between surface and ground water, and soil permeability as a basis of plant design; to determine incidence of inundation of site by floods, tsunamis, or hurricane-driven tides.
   (b) Geologic—To determine the underlying soil and/or rock to establish foundation criteria.
   (c) Seismologic—To analyze the site seismology characteristics to establish seismic design criteria.
   (d) Meteorologic—To determine the susceptibility of a site to winds of hurricane or tornado force and the local diffusion climatological patterns for use in plant design.
3. Relation to Historical, Archeological, or Cultural Areas—Consideration should be given to the proximity and effects on these areas, which are listed in the National Register of districts, sites, structures and objects significant in American history, architecture, archeology, and culture.
4. Land Use Plans—
   (a) Consistency with State and Regional Land Use Plans—The plant site should conform with adopted state and regional land use plans. To the extent possible, utilities should participate in future land use planning.

(b) Consistency with Area Land Use—Site use plans should conform generally with adjacent land uses, now or planned, and with zoning regulations in the area.

(c) Other Uses of Site—Full consideration should be given to recreational and other compatible uses of the plant site. Also, consideration should be given to the need for visitor centers and other facilities to accommodate the visiting public.

5. Esthetics—Architectural Treatment—The architectural design should recognize the engineering requirements so that the plant exterior can reflect the functions of the facility and at the same time convey a pleasing and uncluttered appearance that blends with the surrounding area.

The appearance of the installation can be improved by the provision of a buffer zone around the plant. The judicious positioning of buildings on the site, and the use of screening and blending landscaping.

Water Quality Protection

Guidelines

An important factor in selecting and planning the use of a thermal power plant site is the type of cooling water system to be used. Where adequate water supplies are available, and such use does not degrade the quality of the water below approved water temperature standards, once-through systems are used. Sources of water for such systems include rivers, lakes, reservoirs, estuaries, and the ocean. Where adequate water quantities or flows are not available for once-through use without violating approved water temperature standards, cooling ponds or cooling towers must be provided. In such systems, which may be used for complete cooling, partial cooling, or to supplement once-through systems during certain periods, the water is recirculated and only sufficient makeup water is needed to replace losses by evaporation and blowdown. All cooling water systems should be designed to avoid or minimize adverse heat effects on the receiving body of water.

Under provisions of the Water Quality Act of 1965, the states are charged with establishing water quality standards, including temperature criteria, for interstate and coastal waters, subject to approval by the Secretary of the Interior. Administration of this program is by Interior's Federal Water Pollution Control Administration. The standards are established on the basis of proposed uses of the water. The thermal criteria generally consist of maximum permissible temperatures and maximum permissible changes in temperature, normally to apply outside of an allowable mixing zone. Enforcement of approved standards is the responsibility initially of the states with back-up authority in the federal government. In addition to criteria for interstate and coastal waters, many states have established standards for intrastate waters.

In order to minimize the impact of cooling water use on water bodies and assure compliance with approved water quality standards, a number of pre-construction and post-construction studies are needed. These studies should begin at least two years before plant construction is scheduled in order to properly design the cooling water system and predict the impacts of cooling water use on environmental and biological conditions of the water body. Comparable monitoring studies should continue for at least two years after plant startup in order to verify the effects of plant operation on the water body. Control techniques should be established for operation in accordance with design criteria, and appropriate monitoring should continue during the operation of the plant.

Studies of a water body proposed as a source of cooling water supplies may include measurement of wind direction and velocity, measurement of current patterns in water bodies, measurement of temperatures and other physical and chemical properties of the water, algal studies, macroinvertebrate studies, and fish studies. The biological studies, which should be carried out under the direction of aquatic biologists, should take proper account of seasonal factors and should identify unique or significant forms.

The flow and current patterns in a water body are important in determining the dispersion of the heat load and extent of the mixing zone. Streamflows of many streams are measured by the U.S. Geological Survey. Measurements at additional sites may be made by using the USGS gaging techniques. Currents in water bodies may be measured by current meters. Such current surveys may be particularly important when waste heat is to be discharged into an estuary. Flow patterns in estuaries and other water bodies may also be determined by using dyes or isotope tracers. Post-operational aerial surveys of heat discharges to water bodies using infrared techniques may be useful in predicting the heat dispersion patterns of proposed plants.

Measurements of the properties of the water should follow standard procedures such as those in Standard Methods for the Examination of Water and Wastewater, 12th Edition, American Public Health Association, and might include determination of: temperature, turbidity, biochemical oxygen demand, hydrogen ion concentration, and the content of coliform bacteria, dissolved oxygen, nitrogen, and phosphorus.

Algal studies should determine the population of both free floating (phytoplankton) and attached (periphyton) species. Samples collected with a net equipped with a metering device (such as a Clark-Bumpus net) or a metered pump, should be taken at the bottom, mid-depth, and surface levels near the proposed points of water intake and discharge to determine the quantities present. A midsummer and midwinter 24-hour sampling program should disclose any diurnal vertical migration and the effect of thermal stratification on vertical movement of these organisms.

Macroinvertebrate surveys using Peterson or Ponar dredges should be made to determine the type and numbers of species. The presence of large numbers of such organisms, representing the middle group in the aquatic community, indicates that they have available microscopic organisms for food and, in turn, can provide food for fish. Samples should be collected from selected stations for each type of bottom material found within or immediately adjacent to the expected area of artificial temperature change. Collection of samples from these stations should be made four times per year, preferably coinciding with the midpoint of the seasons. Radioactivity analyses should be made of the captured animals as well as of bottom deposits.

Studies to determine fish populations and seasonal variations in population can be made by the use of gill nets, trap nets, trawls, shoreline seines, and sonar methods. Properly selected sampling stations should be established for such surveys. Use may also be made of a sampling type quantitative creel census carried out on a periodic basis. The fish studies should include a survey of spawning areas.

The prediction of temperature changes in the receiving waters is complicated by many factors affecting heat transfer to the environment. Mathematical models are being developed to predict the dissipation of heat to the atmosphere and the dispersion of heat in a water body after receiving the cooling water discharges. The availability of digital computers facilitates the consideration of the many factors and changing conditions involved.

Hydraulic models are useful in determining the likely mixing and recirculation of plant effluents. Temperature measurements may be made on the models and, after comparison with prototype observations, they may be used to appraise temperature changes in the environment, particularly in the mixing zone. Hydraulic model studies can assist in the assessment of likely water movement, stratification, and mixing processes. Such studies are also useful in designing the cooling water systems, including the planning of intake structures, canals, outlet structures, and dispersion facilities.

Environmental Protection Checklist

Following is a checklist of site selection factors relating to the protection of water quality.

1. Selection of Cooling Water System—
   - (a) One-Through Systems—Used where sufficient cooling water is available to keep within approved temperature criteria and avoid any significant adverse effects on the receiving water body.
   - (b) Cooling Towers—Used to meet water quality standards where water supplies are limited, for complete cooling, partial cooling, or to supplement once-through cooling during certain periods.
   - (c) Cooling Ponds—Used where needed to meet temperature limits of receiving waters and sufficient land area is available at relatively low cost and makeup water supplies can be provided.
   - (d) Reuse of Water—Increased reuse of water can be anticipated, including use of sewage and industrial treatment effluents.

2. Hydrologic Studies—
   - (a) Streamflow—To determine available average and firm flows in streams to be used to supply cooling water.

(b) Current Patterns—To determine dispersion characteristics, particularly for estuarine areas.

3. Physical and Chemical Properties of Water—Using standard techniques, measurements should be made of the following items as appropriate:
   (a) Suspended Solids
   (b) Dissolved Solids
   (c) Alkalinities
   (d) Turbidity
   (e) Total Coliform Bacteria
   (f) Fecal Coliform Bacteria
   (g) Dissolved Oxygen
   (h) Biochemical Oxygen Demand
   (i) Hydrogen Ion Concentration
   (j) Conductivity
   (k) Chlorides
   (l) Organic Nitrogen
   (m) Ammonia Nitrogen
   (n) Nitrite Nitrogen
   (o) Nitrate Nitrogen
   (p) Total Phosphorus
   (q) Soluble Phosphorus
   (r) Temperature
4. Ecological Studies—Such should be made before and after plant startup to determine their seasonal variations and the effects of plant operation.
   (a) Algal Studies—To determine species and quantities present at proposed points of intake and discharge of cooling water supplies.
   (b) Macroinvertebrate Studies—To determine type and numbers of species present.
   (c) Fish Population Studies—To determine species and numbers present and those being caught.
   (d) Unique Systems—Identify any unique or significant ecosystems.
5. Temperature Prediction Studies—These studies are necessary to demonstrate ability of cooling water systems to meet water temperature criteria.
   (a) Mathematical Models—These studies take into account site conditions, proposed plant operating factors, and known physical relationships to predict the changes in temperature and dispersion of waste heat in receiving bodies of water. Digital computers are usually employed in such studies.
6. Design Studies—These design studies of cooling water systems make use of data collected at the site and may utilize both mathematical and hydraulic models.

Air Quality Protection

Air quality protection requires consideration of present and potential air environmental conditions and how these would be affected by planned operation of generating plants. Thus, serious consideration must be given to the air pollution potential in selecting the site for fossil-fueled steam-electric plants. Such plants contribute substantial portions of certain contaminants found in the atmosphere, such as sulfur oxides, nitrogen oxides, and particulate matter. The availability of desirable fuels, such as natural gas and low-sulfur coal for utility use is limited. Supplies of low-sulfur residual fuel oil are increasing on world markets, but they fall far short of the potential demand and their use is limited largely to plants where supplies can be delivered by water transport. Despite the research under way and the apparent initial success of a limited number of processes in prototype installations of removing sulfur oxides from stack gases of coal burning plants, large-scale operation has yet to be demonstrated. Thus, no large-scale process for such purpose is readily available to utilities.

The Air Quality Act of 1967 provided an intergovernmental program for the prevention and control of air pollution on a regional basis. To put this program into operation, the Department of Health, Education and Welfare is required to designate air quality control regions and issue air quality criteria and reports on control techniques. State governments are then expected to establish ambient air quality standards for the air quality control regions and to adopt plans for implementation of the standards. The ambient air quality standards and implementation plans must be submitted to HEW for review and approval. If the states fail to act or their proposed air quality standards are considered inadequate, HEW may establish appropriate standards. Enforcement of the approved standards is primarily the responsibility of the states with some backup authority in HEW. Administration of this program is by HEW's National Air Pollution Control Administration.

Guidelines

The air pollution potential of a particular site depends upon a number of factors, including the characteristics of the fuel, the emission control measures incorporated in the plant design, the stack height, the exit velocity of the flue gases, the topography of the site, and the meteorological conditions in the area. The air quality control standards in effect may influence the choice of plant locations.

Factors related to air quality requiring appraisal in site selection include population distribution, expected growth pattern, existing or expected local industrial pollution sources, terrain over the area of stack gas dispersion, and land-use patterns such as for agriculture, forestry, or recreational purposes. With the present level of technology, a mine-mouth plant in remote, undeveloped, flat terrain might be desirable. However, adoption of national ambient or emission standards would reduce the advantage of such sites.

Meteorology is important to the design of air quality control systems of a plant. Meteorological measurements should include the prevailing wind directions and velocities, ambient temperature ranges, precipitation values, and factors related to temperature inversions.

Studies of a proposed plant site should include an air monitoring program which would begin at least two years prior to plant construction and continue after the plant is in operation. Each air monitoring station should include such instruments as an aerovane wind direction and velocity recorder, recording thermometer, Davis sulfur dioxide analyzer, Gelman particulate sampler, dust fall jar and lead oxide candle, and provision for measuring the coefficient of haze.

Data collected from air sampling stations at operating plants, including meteorological data, can be used to verify the mathematical models used during plant design to predict air quality at specified locations. Also, measurements of the amounts of fuel burned per unit of time, combined with measurements of the carbon dioxide content and temperature of stack gases, give an indication of combustion efficiency.

Determination of expected ground level concentrations at various distances from the plant site with various stack heights is most important in evaluating air quality effects from potential power plants. By dispersing and diluting stack gases before they reach ground level, tall stacks can play an important but incomplete role in air pollution control. Using appropriate formulas and coefficients, such as those from the American Society of Mechanical Engineers' guide for tall stacks, it is possible to estimate the maximum ground level sulfur dioxide concentration which may be expected from a plant where sulfur dioxide emission and stack height are defined. In designing stacks for a particular site, however, use may be made of wind tunnel studies considering the topography of the site and expected meteorological conditions.

Environmental Protection Checklist

The following checklist includes site selection factors relating to air quality protection.

1. Area Population and Industry—
   (a) Population—Determine present population distribution and expected growth patterns.
   (b) Industry—Determine existing and expected industries in the area and likely emissions to the air.
2. Site Conditions—
   (a) Topography—Consider factors of topography affecting dispersion of air emissions.
   (b) Meteorology—Make measurements of prevailing wind directions and velocities, ambient temperature ranges, precipitation values, and factors related to temperature inversions.
   (c) Air Monitoring—Provide for air monitoring before startup and after plant operation, measuring wind direction and velocity, temperature, sulfur dioxide, content, nitrogen oxide content, particulate content, dust fall, and haze.
3. Air Quality Criteria—

(a) Standards—Design and operate facilities to meet applicable air quality emission standards for the plant.

(b) Monitoring—Provide monitoring methods adequate to determine how plant operating emissions conform to any applicable ambient air quality standards.

4. Design Factors—

(a) Stacks—Design the stacks to provide optimum dispersion of stack gases consistent with Federal Aviation Administration regulations concerning stack heights in the area.

(b) Particulate Control—Install facilities necessary to provide optimum control of particulate emissions.

(c) SO_x and NO_x Control—Consider expected composition of any fossil fuel planned for use (coal, oil or gas). Depending upon the included sulfur content, consider best available technology for possible removal of sulfur exceeding proper tolerance level. Pending availability of commercial processes for plant site removal of sulfur, consider use of low sulfur fuel and provision for future plant modification to remove sulfur from normal fuels. Nitrogen oxide control depends on further development of usable processes and equipment.

Control of Radioactive Wastes

For the light-water reactors which are expected to be used predominantly for nuclear power plants for the next 10 to 20 years, neither the reprocessing of fuel nor the disposal of radioactive wastes is conducted at the plant site. However, small controlled quantities of radioactive wastes are released to the atmosphere and the adjacent waterway.

Guidelines

In the operation of a nuclear reactor, radioactive wastes are accumulated as solids, liquids, or gases. Solid wastes are placed in shielded containers and shipped to disposal centers. Liquid wastes are kept in holdup tanks to permit decay of their radioactivity to a lower level and treated by ion exchange and evaporation equipment, but some small amounts of radioactivity are released in the liquid effluent. Gaseous wastes are filtered and kept in holdup tanks to permit decay of their radioactivity to lower levels, after which they are released to the atmosphere in controlled quantities that can be accurately measured. Thus, some radioactivity is released in liquid or gaseous form to the plant site environment.

The Atomic Energy Commission requires, as part of applications for construction and operating permits, detailed analysis of all safety problems and plant design and operating techniques to cope with them. The AEC regulations establish strict limits on the release of radioactive wastes. These regulations are set forth in Part 20 of Title 10, Code of Federal Regulations. Normally, waste disposal is not an important factor in site selection if the reactor facilities can meet the requirements of these regulations. It could be of importance, however, if large installations are planned for locations on the same waterway or in the same airshed.

In order to assure compliance with regulations governing release of radioactive materials and to guard against radiation hazards, a program of radiological monitoring must be established. The levels of background radiation in the environment should be established by measurements taken over a period of at least two years prior to plant operation. These measurements must be used in determining the allowable limits of regulated releases of radioactive wastes. Monitoring of radiological effects on the environment should continue during plant operation to control radioactive waste discharge and to check on the effectiveness of in-plant measurement of releases.

In order to measure possible plant effects, two types of sampling stations should be established. One type, as indicator stations, would be placed where maximum radiation attributable to the plant would be expected. The other type, to measure background radiation, would be placed where radioactive levels of less than one percent of levels at indicator stations would be expected when the plant is releasing a significant fraction of allowable releases. Sufficient samples should be collected and analyzed to assure that significant pathways of radioactive materials to man are evaluated. The samples taken should include air particles, precipitation, external radiation, milk, and well water; also water, sediments, fish, and benthos from the receiving water body.

Environmental Protection Checklist

Following are site selection considerations relating to control of nuclear wastes.

1. Allowable Waste Releases—
   (a) AEC Regulations—Criteria for determining allowable radioactive waste releases are set forth by 10 CFR 20.
   (b) Determination of Proposed Releases—Based on AEC regulations and site conditions, including background radiation, meteorological characteristics, and related factors.
2. Measurements for Design—Through analysis of safety questions and the related engineered safeguards and operating procedures, methods for protection of the environment must be built into the plant. Such analyses and designs will require measurements and studies necessary to predict the effects of project construction and operation on the environment. Such studies may require the use of digital computers.
3. Monitoring Programs—These must be developed to guard against hazards to the environment and to assure compliance with applicable regulations. Stations should be selected to measure radiation where maximum effects of the plant operation are expected and also where background radiation can be determined. Measurements should be made on the following materials sampled in the region of the site to determine the level of radioactivity.
   (a) Airborne Dust
   (b) Precipitation
   (c) External Radiation
   (d) Milk
   (e) Receiving Waters and Sediments
   (f) Receiving Water Benthos
   (g) Receiving Water Fish
   (h) Estuary Oysters
   (i) Well Water

Noise Control

Control of noise from fossil-fueled generating stations has become increasingly important with the use of larger units.

Guidelines

Sound control measures appropriate for the nature of the surroundings should be taken. Plant noise can be disturbing at substantial distances from its source, and the effect of sound levels upon the entire surrounding neighborhood must be considered. Ambient noise levels should be measured as part of the site selection process and used to establish criteria for noise attenuation in the design of the plant. Future development of adjacent properties must be anticipated in order to incorporate proper isolation during the initial design phases. Analysis should be made of the sound generating potential of plant equipment.

Environmental Protection Checklist

The following site selection factors relate to problems of noise control.

1. Site Studies—
   (a) Ambient Levels—Measurements should be made of existing ambient noise levels to establish a baseline to determine the contribution of the proposed plant to noise levels.
   (b) Projected Area Development—Consideration should be given to expected future developments in the area that would contribute to noise levels.
2. Monitoring Studies—Ambient noise levels should be measured after plant startup to determine the plant's contribution to noise levels and to test the effectiveness of attenuation designs.

Appendix C

Appendix material from a Statement of Herbert B. Cohn, Executive Vice President, American Electric Power Service Corp., presented at the hearings before the Subcommittee on Communications and Power of the Committee on Interstate and Foreign Commerce, House of Representatives, Ninety-second Congress, First Session, Bill Relating to Power Plant Siting and Environmental Protection.

Text of Sections 9 and 10 of proposed federal guidelines for power plant siting act drafted jointly by several executive agencies.

Section 9—Evaluative Criteria

In evaluating long-range plans, conducting preliminary site reviews, and evaluating the application for certification of bulk power supply facilities, the certifying agency shall give consideration to the following factors, where applicable:

a. Electric Energy Needs (major emphasis of long-range plan reviews).

   (1) Growth in demand and projection of need.

   (2) Availability and desirability of non-electric alternative sources of energy.

   (3) Availability and desirability of alternative sources of electric power to this facility or to this type of facility.

(4) Promotional activities of the electric entity which may have given rise to the need for this facility.
(5) Socially beneficial uses of the output of this facility, including its use to protect or enhance environmental quality.
(6) Conservation activities which could minimize the need for more power.
(7) Research activities of the electric entity of new technology available to it which might minimize environmental impact.

b. Land Use Impacts (major emphasis of preliminary site reviews).
(1) Area of land required and ultimate use.
(2) Consistency with any State and regional land use plans.
(3) Consistency with existing and projected area land use.
(4) Alternative uses of the site.
(5) Impact on population already in the area; population attracted by construction or operation of the facility itself; impact of availability of power from this facility on growth patterns and population dispersal.
(6) Geologic suitability of the site or route.
(7) Seismologic characteristics.
(8) Construction practices.
(9) Extent of erosion, scouring, wasting of land—both at site and as a result of fossil fuel demands of the facility.
(10) Corridor design and construction precautions for transmission lines.
(11) Scenic impacts.
(12) Effects on natural systems, wildlife, plant life.
(13) Impacts on important historic, architectural, archaeological, and cultural areas and features.
(14) Extent of recreation opportunities and related compatible uses.
(15) Public recreation plan for the project.
(16) Public facilities and accommodation.

c. Water Resources Impacts (Major emphasis during preliminary site reviews and facility certification).
(1) Hydrologic studies of adequacy of water supply and impact of facility on stream flow, estuarine and coastal waters, and lakes and reservoirs.
(2) Hydrologic studies of impact of facilities on ground water.
(3) Cooling system evaluation including consideration of alternatives.
(4) Inventory of effluents including physical, chemical, biological, and radiological characteristics.
(5) Hydrologic studies of effects of effluents on receiving waters, including mixing characteristics of receiving waters, changed evaporation due to temperature differentials, and effect of discharge on bottom.
(6) Relationship to water quality standards.
(7) Effects of changes in quantity and quality on water use by others, including both withdrawal and in situ uses; relationship to projected uses; relationship to water rights.
(8) Effects on plant and animal life, including algae, microinvertebrates, and fish population.
(9) Effects on unique or otherwise significant ecosystems; e.g., wetlands.
(10) Monitoring programs.

d. Air Quality Impacts (major emphasis during preliminary site reviews and facility certification).
(1) Meteorology—wind direction and velocity, ambient temperature ranges, precipitation values, inversion occurrence, other effects on dispersion.
(2) Topography—factors effecting dispersion.
(3) Standards in effect and projected for emissions, design capability to meet standards.
(4) Emissions and controls.
(a) Stack design.
(b) Particulates.
(c) SO_x
(d) NO_x
(5) Relationship to present and projected air quality of the area.

(6) Monitoring program.

e. Solid Wastes Impact (major emphasis during facility certification).
   (1) Solid Waste Inventory.
   (2) Disposal program.
   (3) Relationship of disposal practices to environmental quality standards.
   (4) Capability of disposal sites to accept projected waste loadings.

f. Radiation Impacts (major emphasis during preliminary site review and facility certification).
   (1) Land use controls over development and population.
   (2) Wastes and associated disposal program for solid liquid, and gaseous wastes—criteria set by AEC and EPA.
   (3) Analyses and studies of the adequacy of engineering safeguards and operating procedures—determined by AEC.
   (4) Monitoring—adequacy of devices and sampling techniques.

g. Noise Impacts (major emphasis during facility certification).
   (1) Construction period levels.
   (2) Operational levels.
   (3) Relationship of present and projected noise levels to existing and potential stricter noise standards.
   (4) Monitoring—adequacy of devices and methods.

Section 10—Evaluation of Relative Environmental Effects of Alternative Sites and Routes

To the extent possible, only those sites and routes meeting acceptable standards in relation to the criteria outlined in Section 9 of these Guidelines should receive certification from the appropriate certifying agency. Regularly, however, it will be necessary in order to meet recognized electric power needs that a site or route be chosen from a set of alternatives, all of which will present some adverse environmental effects. In such cases it will be necessary for the certifying agency to establish priorities among the evaluative criteria employed. For example in the case of transmission lines, the priority might be assigned to the land use criteria outlined in Section 9 above; in the case of fossil-fueled power plants the air quality criteria might prevail; and the water quality criteria might prevail for nuclear plants. Assignment of such priorities is not intended to eliminate full consideration of other criteria listed in Section 9, but merely provides guidance for the resolution of difficult cases where a choice among alternatives would otherwise be impossible.

Selected References

American Public Power Association, *Winners A.P.P.A. Awards Program for Utility Design,* 1969, Washington, D.C. 1970, 16 pp.

American Public Power Association, *1971 Winners A.P.P.A. Awards Program for Utility Design,* Washington, D.C. 1971, 12 pp.

"Big Cleanup, The Environmental Crisis 72 (The)," *Newsweek,* June 12, 1972, p. 36-55.

Burgraff, Frank B. Jr., "Power to the People," *Planning News,* New York State Planning Federation, Vol. 35, No. 4, July-August 1971, New York, p. 1 and p. 6.

Citizens Advisory Committee on Environmental Quality, *Report to the President and The Council on Environmental Quality,* Washington, D.C., April 1971, 56 pp.

Crowe, Sylvia, *The Landscape of Power,* The Architectural Press, London, 1958, 115 pp.

Dubos, Rene, *Man and His Environment* in The Fitness of Man's Environment, Smithsonian Annual II, Smithsonian Institution Press, Washington, D.C., 1968.

Eckbo, Dean, Austin and Williams, *Proposed Davenport Power Plant Site Land Use Study,* 144 pp.

Electric Utility Industry Task Force on Environment, *The Electric Utility Industry and the Environment,* a Report to the Citizens Advisory Committee on Recreation and Natural Beauty, New York, N.Y., 1968, 107 pp.

Electric Research Council, *Electric Transmission Structures, A Design Research Program,* EEI Pub. No. 67-61, New York, 1967, 79 pp. (Folder same title issued January 1968).

"Energy Crisis: Are We Running Out?," *Time Magazine,* June 12, 1972, p. 49-55.

"Energy Crisis: Issues and a Proposed Response (The)," Michael McCloskey, Executive Director, The Sierra Club, *Environmental Affairs,* Vol. 1, No. 3, The Environmental Law Center, Boston College Law School, Brighton, Massachusetts 02135, p. 587-603.

Energy Policy Staff, Office of Science and Technology, (The) *Considerations Affecting Steam Power Plant Site Selection,* Superintendent of Documents, G.P.O., Washington, D.C., December 1968, 133 pp.

Energy Policy Staff, Office of Science and Technology (The), *Electric Power and the Environment,* Superintendent of Documents, U.S. Government Printing Office, Washington, D.C., August 1970, 71 pp.

Federal Power Commission, *The 1970 National Power Survey: Guidelines for Growth of the Electric Power Industry,* U.S. Government Printing Office, Washington, D.C., December, 1971, 951 pp.

Halprin, Lawrence and Associates, *The Trojan Nuclear Power Plant,* A Project of the Portland General Electric Co., at Prescott, Oregon, San Francisco, California, 1970, 46 pp.

Howlett, Bruce, Ehniger, Frederick J. with assistance of Corkill, Anthony and Seddon, John for the Hudson River Valley Commission, State of New York, *Power Lines and Scenic Values in the Hudson River Valley,* Hudson River Valley Commission, Tarrytown, N.Y., December, 1968, 23 pp.

InterDesign, Inc., *Visual Impact Study* for Northern States Power Company, St. Paul, Minn., September, 1971, 133 pp.

Johnson, Johnson and Roy, *Substation Site Selection and Development* for Consumers Power Company, Jackson, Michigan, April 1969, 64 pp.

Johnson, Johnson and Roy, *Transmission and Distribution Rights-of-Way Selection and Development,* for Consumers Power Company, Jackson, Michigan, November 1970, 55 pp.

National Academy of Engineering, Committee on Power Plant Siting, *Engineering for Resolution of the Energy-Environment Dilemma: A Summary,* Washington, D.C., 1971, 51 pp.

President's Council on Environmental Quality, *Environmental Quality, The Second Annual Report of the Council on Environmental Quality,* Superintendent of Documents, U.S. Government Printing Office, Washington, D.C., 1971, 360 pp.

Report of the Environmental Pollution Panel, President's Science Advisory Panel, *Restoring the Quality of Our Environment,* U.S. Government Printing Office, Washington, D.C., 1965, 317 pp.

Stanley H. Ruttenberg and Associates, Inc. for National Rural Electric Cooperative Association, *The Electric Power Crisis: Its Impact on Workers and Consumers,* Washington, D.C., August 1971, 62 pp.

Santa Clara County Planning Department, *Greenways,* San Jose, California, July 1961, Reprinted June 1969, folder.

Subcommittee on Communications and Power of the Committee on Interstate and Foreign Commerce, House of Representatives, Ninety-Second Congress, First Session, *Powerplant Siting and Environmental Protection,* Hearings before the Subcommittee on Bills Relating to Powerplant Siting and Environmental Protection from May 4-27, 1971, Printed for the use of the Committee on Interstate and Foreign Commerce, U.S. Superintendent of Documents, Government Printing Office, Washington, D.C. 1972, Part 1 (Serial No. 92-31), Part 2 (Serial No. 92-32) Part 3 (serial No. 92-33).

U.S.D.A. Forest Service, Rocky Mountain Region, *Undergrounding Electric Distribution Lines,* Denver, Colorado, April 1972, 43 pp.

United States Department of the Interior, *Interim Environmental Guidelines for Thermal Power Plant Site Evaluation-Pacific Northwest,* Volume I, Washington, D.C., July 1970, 21 pp.

U.S. Department of the Interior, U.S. Department of Agriculture, *Environmental Criteria for Electric Transmission Systems,* Superintendent of Documents, G.P.O. Washington, D.C., February 1970, 52 pp.

U.S. Department of State, *U.S. National Report on the Human Environment,* prepared for United Nations Conference on Human Environment, June, 1972, Stockholm, Sweden, Superintendent of Documents, U.S. Government Printing Office, Washington, D.C., 1972, 53 pp.

Western Systems Coordinating Council, Environmental Committee, *Environmental Guidelines,* Los Angeles, California, December 3, 1971, 96 pp.

Wirth, Theodore J. and Associates, *A Land Use Plan for North Anna Reservoir,* prepared for Virginia Commission on Outdoor Recreation, Chevy Chase, Maryland, 1971, 61 pp.

Working Committee on Utilities, *Report to the Vice President and to the President's Council on Recreation and Natural Beauty,* Washington, D.C., December 27, 1968, 146 pp.

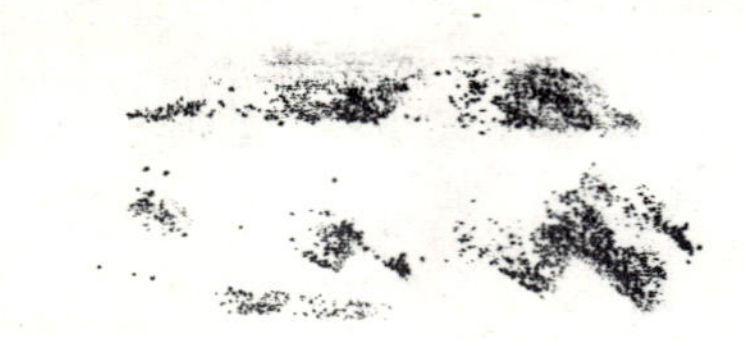